Separation, Extraction and Purification of Natural Products from Plants

Separation, Extraction and Purification of Natural Products from Plants

Editor

Xingchu Gong

Basel • Beijing • Wuhan • Barcelona • Belgrade • Novi Sad • Cluj • Manchester

Editor
Xingchu Gong
Zhejiang University
Hangzhou
China

Editorial Office
MDPI
St. Alban-Anlage 66
4052 Basel, Switzerland

This is a reprint of articles from the Special Issue published online in the open access journal *Separations* (ISSN 2297-8739) (available at: https://www.mdpi.com/journal/separations/special_issues/Preparation_Products).

For citation purposes, cite each article independently as indicated on the article page online and as indicated below:

Lastname, A.A.; Lastname, B.B. Article Title. *Journal Name* **Year**, *Volume Number*, Page Range.

ISBN 978-3-0365-9014-1 (Hbk)
ISBN 978-3-0365-9015-8 (PDF)
doi.org/10.3390/books978-3-0365-9015-8

© 2023 by the authors. Articles in this book are Open Access and distributed under the Creative Commons Attribution (CC BY) license. The book as a whole is distributed by MDPI under the terms and conditions of the Creative Commons Attribution-NonCommercial-NoDerivs (CC BY-NC-ND) license.

Contents

Xinying Chen, Dongyun Guo, Xingchu Gong, Na Wan and Zhenfeng Wu
Optimization of Steam Distillation Process for Volatile Oils from *Forsythia suspensa* and *Lonicera japonica* According to the Concept of Quality by Design
Reprinted from: *Separations* 2023, *10*, 25, doi:10.3390/separations10010025 1

Jing Lan, Gelin Wu, Linlin Wu, Haibin Qu, Ping Gong, Yongjian Xie, et al.
Development of a Quantitative Chromatographic Fingerprint Analysis Method for Sugar Components of Xiaochaihu Capsules Based on Quality by Design Concept
Reprinted from: *Separations* 2023, *10*, 13, doi:10.3390/separations10010013 19

Wenlong Li, Xi Wang, Houliu Chen, Xu Yan and Haibin Qu
In-Line Vis-NIR Spectral Analysis for the Column Chromatographic Processes of the *Ginkgo biloba* L. Leaves. Part II: Batch-to-Batch Consistency Evaluation of the Elution Process
Reprinted from: *Separations* 2022, *9*, 378, doi:10.3390/separations9110378 33

Wenlong Li, Yu Luo, Xi Wang, Xingchu Gong, Wenhua Huang, Guoxiang Wang, et al.
Development and Validation of a Near-Infrared Spectroscopy Method for Multicomponent Quantification during the Second Alcohol Precipitation Process of *Astragali radix*
Reprinted from: *Separations* 2022, *9*, 310, doi:10.3390/separations9100310 45

Tian-Tian Liu, Lin-Jing Gou, Hong Zeng, Gao Zhou, Wan-Rong Dong, Yu Cui, et al.
Inhibitory Effect and Mechanism of Dill Seed Essential Oil on *Neofusicoccum parvum* in Chinese Chestnut
Reprinted from: *Separations* 2022, *9*, 296, doi:10.3390/separations9100296 63

Na Wan, Jing Lan, Zhenfeng Wu, Xinying Chen, Qin Zheng and Xingchu Gong
Optimization of Steam Distillation Process and Chemical Constituents of Volatile Oil from *Angelicae sinensis* Radix
Reprinted from: *Separations* 2022, *9*, 137, doi:10.3390/separations9060137 81

Isabella Bolognino, Antonio Carrieri, Rosa Purgatorio, Marco Catto, Rocco Caliandro, Benedetta Carrozzini, et al.
Enantiomeric Separation and Molecular Modelling of Bioactive 4-Aryl-3,4-dihydropyrimidin-2(1*H*)-one Ester Derivatives on Teicoplanin-Based Chiral Stationary Phase
Reprinted from: *Separations* 2022, *9*, 7, doi:10.3390/separations9010007 97

Qian Cheng, Shuhuan Peng, Fangyi Li, Pengdi Cui, Chunxia Zhao, Xiaohui Yan, et al.
Quality Distinguish of Red Ginseng from Different Origins by HPLC–ELSD/PDA Combined with HPSEC–MALLS–RID, Focus on the Sugar-Markers
Reprinted from: *Separations* 2021, *8*, 198, doi:10.3390/separations8110198 115

Zili Guo, Shuting Xiong, Yuanyuan Xie and Xianrui Liang
The Separation and Purification of Ellagic Acid from *Phyllanthus urinaria* L. by a Combined Mechanochemical- Macroporous Resin Adsorption Method
Reprinted from: *Separations* 2021, *8*, 186, doi:10.3390/separations8100186 127

Xi Wang, Huimin Feng, Halimulati Muhetaer, Zuren Peng, Ping Qiu, Wenlong Li, et al.
Studies on the Separation and Purification of the *Caulis sinomenii* Extract Solution Using Microfiltration and Ultrafiltration
Reprinted from: *Separations* 2021, *8*, 185, doi:10.3390/separations8100185 139

Yanni Tai, Jingjing Pan, Haibin Qu and Xingchu Gong
An Index for Quantitative Evaluation of the Mixing in Ethanol Precipitation of Traditional Chinese Medicine
Reprinted from: *Separations* **2021**, *8*, 181, doi:10.3390/separations8100181 . . . . **151**

Boglárka Páll, Zsuzsa Gyenge, Róbert Kormány and Krisztián Horváth
Determination of Genotoxic Azide Impurity in Cilostazol API by Ion Chromatography with Matrix Elimination
Reprinted from: *Separations* **2021**, *8*, 162, doi:10.3390/separations8100162 . . . . **167**

Guangzheng Xu, Hui Wang, Yingqian Deng, Keyi Xie, Weibo Zhao and Xingchu Gong
Research Progress on Quality Control Methods for Xiaochaihu Preparations
Reprinted from: *Separations* **2021**, *8*, 199, doi:10.3390/separations8110199 . . . . **175**

Article

Optimization of Steam Distillation Process for Volatile Oils from *Forsythia suspensa* and *Lonicera japonica* According to the Concept of Quality by Design

Xinying Chen [1,2], Dongyun Guo [1], Xingchu Gong [2,3,*], Na Wan [1,*] and Zhenfeng Wu [1,4]

1. State Key Laboratory of Innovation Medicine and High Efficiency and Energy Saving Pharmaceutical Equipment, Jiangxi University of Chinese Medicine, Nanchang 330004, China
2. Pharmaceutical Informatics Institute, College of Pharmaceutical Science, Zhejiang University, Hangzhou 310058, China
3. Jinhua Institute of Zhejiang University, Jinhua 321016, China
4. Key Laboratory of Modern Preparation of TCM, Ministry of Education, Jiangxi University of Chinese Medicine, Nanchang 330004, China
* Correspondence: gongxingchu@zju.edu.cn (X.G.); 20152002@jxutcm.edu.cn (N.W.)

Abstract: In this study, the process of steam distillation to collect volatile oils from *Forsythia suspensa* (*F. suspensa*) and *Lonicera japonica* (*L. japonica*) was optimized according to the concept of quality by design. First, the liquid/material ratio, distillation time, and collection temperature were identified as critical process parameters by a review of the literature and single-factor experiments. Then, a Box–Behnken design was used to study the quantitative relationship between the three process parameters, two raw material properties, and the yield of volatile oil. A mathematical model was established with an R^2 value exceeding 0.90. Furthermore, the design space of the volatile oil yield was calculated by a probability-based method. The results of a verification experiment showed that the model was accurate and the design space was reliable. A total of 16 chemical constituents were identified in the volatile oil from mixtures of *F. suspensa* and *L. japonica*. The content of β-pinene was the highest (54.75%), and the composition was similar to that of the volatile oil of *F. suspensa*. The results showed that when *F. suspensa* and *L. japonica* were distilled together, the main contribution to the volatile oil was from *F. suspensa*. The volatile oil yield from the combination of *F. suspensa* and *L. japonica* was not higher than that from *L. japonica*.

Keywords: *Forsythia suspensa*; *Lonicera japonica*; volatile oil; steam distillation; design space

Citation: Chen, X.; Guo, D.; Gong, X.; Wan, N.; Wu, Z. Optimization of Steam Distillation Process for Volatile Oils from *Forsythia suspensa* and *Lonicera japonica* According to the Concept of Quality by Design. *Separations* 2023, 10, 25. https://doi.org/10.3390/separations10010025

Academic Editors: Daniele Naviglio and Carlo Bicchi

Received: 25 September 2022
Revised: 7 November 2022
Accepted: 9 November 2022
Published: 1 January 2023

Copyright: © 2023 by the authors. Licensee MDPI, Basel, Switzerland. This article is an open access article distributed under the terms and conditions of the Creative Commons Attribution (CC BY) license (https:// creativecommons.org/licenses/by/ 4.0/).

1. Introduction

Forsythia suspensa is the dry fruit of *Forsythia suspensa* (Thunb.) Vahl, and *Lonicera japonica* consists of the dry buds or flowers of *Lonicera japonica* Thunb. that bloom early in the spring [1]. Both have the medicinal functions of reducing fevers, detoxification, and dispersing wind-heat and are often used together as a pair in Chinese medicine [2–4]. The 2020 edition of the *Chinese Pharmacopoeia* lists 45 Chinese patent medicines that use *F. suspensa* and *L. japonica* in combination [1]. The volatile oils of *F. suspensa* and *L. japonica* are collected and used in the manufacture of four Chinese patent medicines, including Xiao'er Resuqing Koufuye, Xiao'er Resuqing Keli, Xiao'er Resuqing Tangjiang, and Jinchan Zhiyang Jiaonang [1]. At present, several studies have shown that the volatile oil of *F. suspensa* has antibacterial, antioxidant, antiviral, antipyretic, anti-inflammatory, antitumor, and other pharmacological effects [3,5,6]. The volatile oil of *L. japonica* also has pharmacological effects, such as heat clearing, detoxification, and antibacterial activity [2,4,7].

Steam distillation is a commonly used method to collect volatile oils [8–13], and it has the advantages of simple equipment, easy operation, low cost, and the use of safe solvents [14–16]. However, it is also a time-consuming process with a low volatile oil yield. The steam distillation process is used in the production of the abovementioned four Chinese patent medicines from the volatile oils collected from *F. suspensa* and *L. japonica* [1].

References from the literature concerning the distillation of the volatile oils of *F. suspensa* or *L. japonica* by steam distillation are listed in Table 1. Liu Yan et al. [17] and Wang Yan et al. [18] studied the three parameters of soaking time, distillation time, and material/liquid ratio and found that the influence of each parameter on the volatile oil yield was: material/liquid ratio > distillation time > soaking time. The volatile oil yields from *F. suspensa* and *L. japonica* were approximately 0.93 mL/100 g and 0.34 mL/100 g, respectively. Gu Ke et al. [19] and Li Jianjun et al. [20] studied the collection of volatile oil from *L. japonica* by steam distillation, and the volatile oil yields were about 0.15 g/100 g and 0.17 g/100 g. However, Tong Qiaozhen et al. [21] failed to distillate volatile oil from *L. japonica* by steam distillation. In these published works, the authors suggested that the volatile oil of *L. japonica* has a relatively high solubility in water, which may lead to the loss of *L. japonica* volatile oil in the steam distillation process. When co-distilled with *F. suspensa*, the volatile oil of *F. suspensa* may extract some *L. japonica* volatile oil from the aqueous phase, resulting in the collection of more *L. japonica* volatile oil. However, at present, there are no references in the literature addressing the steam distillation of volatile oil from mixtures of *F. suspensa* and *L. japonica*. Therefore, the authors tried to obtain the volatile oil by the co-distillation of *F. suspensa* and *L. japonica*.

Table 1. Literature on steam distillation.

Medicinal Material	Design of Experiment	Material/Liquid Ratio (g:mL)	Soaking Time (h)	Distillation Time (h)	Optimum Conditions	Volatile Oil Yield	Reference
F. suspensa	Taguchi design	1:8~1:12	1~4	2~6	Material/liquid ratio 1:8, soaking time 2 h, distillation time 6 h.	0.926 mL/100 g	[17]
F. suspensa	Box–Behnken design	1:4~1:10	8~24	8~20	Material/liquid ratio 1:5, soaking time 21 h, distillation time 11 h.	0.342 mL/100 g	[18]
L. japonica	Single-factor design, Box–Behnken design	1:15~1:25	0~5	7~11	Material/liquid ratio 1:20, soaking time 1 h, distillation time 7 h.	0.14998 g/100 g	[19]
L. japonica	Single-factor design, Taguchi design	1:9~1:11	20~28	38~42	Material/liquid ratio 1:10, soaking time 28 h, distillation time 42 h	0.171 g/100 g	[20]

In recent years, the concept of quality by design [22–25] has been used in the optimization of many processes related to Chinese medicines, such as distillation [26], precipitation [27], and column chromatography [28]. In this study, based on the concept of quality by design, the steam distillation process was optimized to distill the volatile oils of *F. suspensa* and *L. japonica*. One of the aims of this work was to find the optimized parameters of steam distillation for collecting a relatively stable amount of volatile oil, which would improve the batch-to-batch consistency of Chinese medicine quality. The critical process parameters were determined through a search of the literature and single-factor experiments. The critical raw material properties of *F. suspensa* were also determined. The quantitative relationship between the process parameters, raw material properties, and volatile oil yield was studied by a Box–Behnken design [29], and a mathematical model was established. The design space of the distillation process of volatile oil from *F. suspensa* and *L. japonica* was calculated by a probability-based method. The operation points inside and outside the design space were selected for verification. Finally, the chemical composition of the volatile oil of *F. suspensa* and the volatile oil of mixtures of *F. suspensa* and *L. japonica* were analyzed and compared.

2. Materials and Methods

2.1. Materials and Reagents

Methanol (chromatographically pure) was obtained from Merck, Germany. Acetonitrile (chromatographically pure) was also obtained from Merck, Germany. Ultrapure water was prepared by an ultrapure water preparation system (Milli-Q, Millipore, Germany). β-Pinene (batch number: C12071236, purity ≥ 98%) was purchased from Shanghai McLean Biochemical Technology Co., Ltd. The medicinal material numbers, origins, and suppliers of *F. suspensa* and *L. japonica* are shown in Table 2.

Table 2. Origin and batch information of medicinal materials.

Medicinal Materials	Number	Origin	Companies
L. japonica	JYH-1	Shandong	Zhejiang Huqing Yutang Materia Medica Co., Ltd.
	JYH-2	Hebei	Hebei Linyitang Pharmaceutical Co., Ltd.
	LQ-1	Shanxi	Haozhou Feimao Pharmaceutical Co., Ltd.
	LQ-2	Shaanxi	Haozhou Chujian Huakai E-Commerce Co., Ltd.
F. suspensa	LQ-3	Shanxi	Nanjing Shangyuantang Pharmaceutical Co., Ltd.
	LQ-4	Shaanxi	Anguo Pharmaceutical Source Trading Co., Ltd.
	LQ-5	Shanxi	Hebei Linyitang Pharmaceutical Co., Ltd.
	LQ-6	Gansu	Sichuan Xunbai Herbal Industry Co., Ltd.

2.2. Steam Distillation

The volatile oil in *F. suspensa* and *L. japonica* was collected by steam distillation. The experimental setup is shown in Figure 1. A low-temperature thermostated bath (THYD-1030 W, Ningbo Tianheng Instrument Factory) was used to lower the temperature of the collection part of the volatile oil extractor [30]. An electronic balance (CN-LQC60002, Kunshan Youkeweiter Electronic Technology Co., Ltd.) was used to weigh 50 g of *F. suspensa* and 50 g of *L. japonica* (Medicinal Material Number: JYH-2). The samples were placed in a 2000 mL flask. A certain amount of water was added, and the flask was shaken to fully wet the medicinal materials. The volatile oil extractor and the condenser pipe were connected. The condensed glycerol and condensed water were fed into the water-cooling jacket and the condenser pipe of the volatile oil extractor, respectively. The volatile oil extractor was filled with water from the top of the condenser until it overflowed into the flask. The electric heater (DZTW 2000 mL, Shaoxing Yuecheng Kechen Instrument and Equipment Co., Ltd.) was turned on, and it started slowly heating the water to boiling. After boiling began, the power of the electric heater was adjusted to 50 W. Starting with the first drop of condensed water dripping into the volatile oil extractor, heating was stopped after heating and refluxing for a certain period of time. After the volatile oil stood for 10 min, its volume was recorded. The volatile oil yield was calculated as shown in Formula (1).

Volatile oil yield (mL/100 g) = volatile oil volume (mL)/medicinal material (100 g)   (1)

2.3. Determination of the Volatile Oil Content in F. suspensa

The thermostat was used to lower the temperature of the collection part of the volatile oil extractor [30]. An electronic balance was used to weigh 100 g of *F. suspensa*, which was then placed in a 2000 mL flask. A certain amount of water was added, and the flask was shaken to fully wet the medicinal materials. The volatile oil extractor and the condenser pipe were connected. The condensed glycerol and condensed water were fed into the water-cooling jacket and the condenser pipe of the volatile oil extractor, respectively. The volatile oil extractor was filled with water from the top of the condenser until it overflowed into the flask. The electric heater was turned on, and it slowly heated the water to boiling. After boiling, the power of the electric heater was adjusted to 50 W. Starting with the first drop of condensed water dripping into the volatile oil extractor, heating was stopped after heating and refluxing for a certain period of time. After the volatile oil stood for 10 min, its volume was recorded. The volatile oil content (mL/100 g) in different batches of *F. suspensa* was calculated as shown in Formula (1).

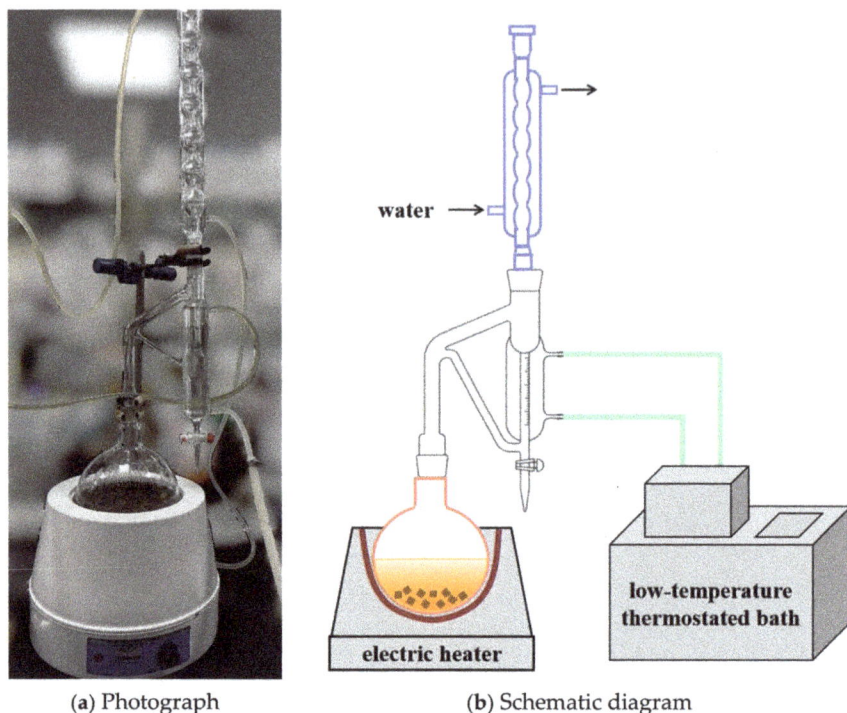

(a) Photograph　　　　　(b) Schematic diagram

Figure 1. Experimental device for steam distillation.

2.4. Determination of β-Pinene Content in F. suspensa

2.4.1. Preparation of Reference Solution and Test Solution

Because the reference substance of β-pinene was a liquid, we precisely measured 83.40 mg of the β-pinene reference substance with a pipette. After that, it was diluted to 10 mL with methanol to obtain the storage reference solution with a concentration of 8340 µg/mL. Then, the storage reference solution was diluted 50, 100, 200, 400, 500, and 800 times with methanol, respectively, to prepare reference solutions with concentrations of 166.8, 83.40, 41.70, 20.85, 16.68, and 10.42 µg/mL.

One hundred microliters of the distilled *F. suspensa* volatile oil was precisely measured, placed in a 10 mL volumetric flask, diluted to volume with methanol, and shaken well to obtain the concentrated stock solution of the test product. A total of 100 µL of the concentrated stock solution of the test product was precisely measured, placed in a 5 mL volumetric flask, diluted to volume with methanol, and shaken well to obtain the test solution.

2.4.2. Chromatographic Conditions

The β-Pinene content in the volatile oil was determined using liquid chromatography (Agilent 1100-DAD, Agilent, USA). The HPLC conditions were modified from the method described in the literature [31]. The details were as follows: Column—Agilent ZORBAX SB-C18, 4.6×250 mm (5-Micron). Mobile phase—0.4% phosphoric acid solution (A)/acetonitrile (B), gradient elution (0 min, 65% B; 0~5 min, 65~70% B; 5~25 min, 70% B; 25~30 min, 70~80% B). Post-run—15 min; injection volume—10 µL; flow rate—1 mL/min; column temperature—25 °C; detection wavelength—202 nm. The chromatograms of the reference solution and test solution are shown in Figure 2.

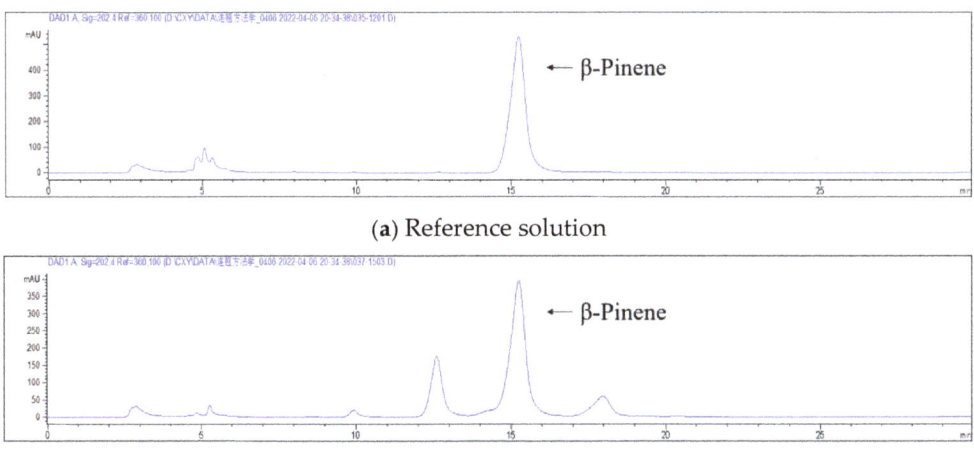

(a) Reference solution

(b) Test solution

Figure 2. HPLC chromatograms of β-pinene.

2.4.3. Methodological Validation

Linear relationship investigation: The β-pinene reference substance was used to investigate the linear relationship. Reference solutions with concentrations of 166.8, 83.40, 41.70, 20.85, 16.68, and 10.42 µg/mL were prepared according to the method described in Section 2.4.1. A 0.22 µm microporous filter membrane was used for filtration, injection, and analysis according to the chromatographic conditions described in Section 2.4.2. Taking the concentration of the reference solution as the abscissa (X, µg/mL) and the peak area as the ordinate (Y, mAU), the linear regression equation and correlation coefficient were obtained.

Precision test: The volatile oil of *F. suspensa* was used to prepare the concentrated stock solution of the test product according to the method outlined in Section 2.4.1. One milliliter of the concentrated stock solution of the test product was accurately absorbed, placed in a 10 mL volumetric flask, diluted to volume with methanol, and shaken well to obtain the test solution. The test solution was filtered with a 0.22 µm microporous membrane. The samples were injected into the column 6 times continuously according to the chromatographic conditions provided in Section 2.4.2, and the RSD value of the peak area was calculated.

Repeatability test: Six replicates of the same batch of *F. suspensa* volatile oil were obtained. The concentrated stock solution of the test product was prepared according to the method in Section 2.4.1. One milliliter of the concentrated stock solution of the test product was accurately absorbed, placed in a 10 mL volumetric flask, diluted to volume with methanol, and shaken well to obtain the test solution. The test solution was filtered with a 0.22 µm microporous membrane. The samples were injected and analyzed according to the chromatographic conditions outlined in Section 2.4.2, and the RSD value of the concentration was calculated.

Stability test: The volatile oil of *F. suspensa* was used to prepare a concentrated stock solution of the test product according to the method described in Section 2.4.1. One milliliter of the concentrated stock solution of the test product was accurately measured, placed in a 10 mL volumetric flask, diluted to volume with methanol, and shaken well to obtain the test solution. The test solution was filtered with a 0.22 µm microporous membrane. The samples were injected and analyzed at 0, 3, 6, 9, 12, and 24 h according to the chromatographic conditions described in Section 2.4.2, and the relative standard deviation (RSD) value of the peak area was calculated.

Sample addition and recovery test: We took 9 aliquots of *F. suspensa* volatile oil from the same batch with a known content of β-pinene and prepared a concentrated stock solution of the test sample according to the method described in Section 2.4.1. One milliliter of the concentrated stock solution of the test sample was accurately absorbed, placed in a 20 mL volumetric flask, diluted to volume with methanol, and shaken well to obtain the test solution (33.2 µg/mL). The reference solution (41.7 µg/mL) was prepared according to the method outlined in Section 2.4.1. The test solution was divided into three groups: low, medium, and high. We precisely measured 1.25 mL of each group and placed it in a 5 mL volumetric flask. The ratios of the added amount of the reference substance to the content of β-pinene in the test solution were controlled at approximately 0.5:1.0, 1.0:1.0, and 1.5:1.0, and 0.5, 1.0, and 1.5 mL of the reference substance was added to the low, medium, and high groups, respectively. Each mixture was diluted to volume with methanol, shaken well, and filtered with a 0.22 µm microporous membrane. The samples were injected and analyzed according to the chromatographic conditions provided in Section 2.4.2, and the β-pinene content and sample recovery rate of each group were calculated.

2.4.4. Determination of Content

An appropriate amount of volatile oil was taken from different batches of *F. suspensa*. The test solutions were prepared according to the method outlined in Section 2.4.1. Then, the test solutions were injected and analyzed according to the chromatographic conditions described in Section 2.4.2. The peak areas were recorded, and the β-pinene contents in different batches of *F. suspensa* were calculated.

2.5. Optimization of Distillation Process Parameters

Based on the literature search and single-factor experiments, it was believed that the critical process parameters affecting the volatile oil yield from *F. suspensa* and *L. japonica* by steam distillation were the amount of water, distillation time, and collection temperature, and the critical properties of the raw materials were the volatile oil content and the β-pinene content of *F. suspensa*.

A Box–Behnken design was adopted, and the water addition (X_1), distillation time (X_2), collection temperature (X_3), volatile oil content (Z_1), and β-pinene content (Z_2) were used as the investigation factors. The volatile oil yield (Y) was used as the evaluation index to optimize the distillation process of volatile oil from *F. suspensa* and *L. japonica*. The factors and levels of the Box–Behnken design are shown in Table 3, and the results are shown in Table 4.

2.6. Analysis of Chemical Constituents of Volatile Oil

A combined total of 20 µL of volatile oil from *F. suspensa* and volatile oil from *F. suspensa* and *L. japonica* was accurately measured; then, the samples were supplemented with 1 mL of diethyl ether for dilution, filtered with a 0.22 µm microporous membrane, and placed into sample bottles. The chemical constituents in the volatile oil of *F. suspensa* and the volatile oil of *F. suspensa* and *L. japonica* were analyzed.

Table 3. The factors and levels of the Box–Behnken design.

Factor	Level		
	Low (−1)	Medium (0)	High (1)
X_1: water addition (mL/g)	8	10	12
X_2: distillation time (h)	3.0	4.5	6.0
X_3: collection temperature (°C)	5	15	25

Table 4. The results of the Box–Behnken design.

No	X_1: Water Addition (mL/g)	X_2: Distillation Time (h)	X_3: Collection Temperature (°C)	Z_1: Volatile Oil Content (mL/100 g)	Z_2: β-Pinene Content (mg/g)	Y: Volatile Oil Yield (mL/100 g)
1	8	3.0	15	1.601	0.857	0.63
2	12	3.0	15	1.599	0.867	0.54
3	8	6.0	15	1.792	1.039	0.92
4	12	6.0	15	1.349	0.812	0.83
5	8	4.5	5	1.969	1.139	0.90
6	12	4.5	5	1.601	0.857	0.80
7	8	4.5	25	1.599	0.867	0.80
8	12	4.5	25	1.792	1.039	0.83
9	10	3.0	5	1.349	0.812	0.74
10	10	6.0	5	1.969	1.139	0.98
11	10	3.0	25	1.601	0.857	0.61
12	10	6.0	25	1.599	0.867	0.71
13	10	4.5	15	1.792	1.039	0.82
14	10	4.5	15	1.349	0.812	0.80
15	10	4.5	15	1.969	1.139	0.89
16	10	4.5	15	1.969	1.139	0.94
17	10	4.5	15	1.969	1.139	0.90

The chemical constituents of the volatile oils were analyzed with a gas chromatography–mass spectrometer [32,33] (GC–MS; Agilent 7890B-7000C, Agilent, USA). The GC–MS conditions were modified from the methods describe in the literature [34,35]. The conditions were as follows: chromatographic column—Agilent HP-5MS, 30 m × 0.25 mm × 0.25 μm; inlet temperature—250 °C; flow—1 mL/min constant flow (He); split ratio—10:1; heating program—40 °C for 4 min, 2 °C/min to 100 °C, hold for 10 min, 10 °C/min to 200 °C, and hold for 5 min; MS detector—detector temperature 250 °C and scan range m/z 30–550.

2.7. Data Processing

2.7.1. Mathematical Model

The results of the Box–Behnken design were analyzed using Design-Expert 12.0.3 software (American Stat-Ease Company). Taking the volatile oil yield (Y) as the evaluation index, quadratic polynomial fitting was performed on the five factors of water addition (X_1), distillation time (X_2), collection temperature (X_3), volatile oil content (Z_1) and β-pinene content (Z_2). The mathematical model is shown as Formula (2).

$$Y = a_0 + \sum_{i=1}^{n} b_i X_i + \sum_{i=1}^{n} b_{ii} X_i^2 + \sum_{i=1}^{n-1}\sum_{j=i+1}^{n} b_{ij} X_i X_j + \sum_{k=1}^{m} d_k Z_k \qquad (2)$$

where X is a process parameter; Z is a raw material property; Y is the evaluation index; superscripts n and m are the number of process parameters and raw material properties, respectively; a_0 is the intercept; and b and d are the partial regression coefficients. The model was simplified using a stepwise regression method with p values of 0.1 for both inclusion and removal from the model.

2.7.2. Design Space

The construction design space was calculated by the probability-based method using MATLAB R2018b (American Math Works Company). Under the conditions of fixed batches of medicinal materials, the effects of the changes in the three process parameters on the volatile oil yield were randomly simulated. The parameter combination with a lower limit of volatile oil yield of 0.60 mL and a probability of reaching the standard of no less than 0.80 was used as the design space. In the calculation, the steps of water addition, distillation time, and collection temperature were set to 0.04, 0.03, and 0.10, respectively. The number of simulations was 1000.

3. Results

3.1. Single-Factor Experiments

The results of the single-factor experiments are shown in Table 5. The results indicated that the effects of crushed particle size and soaking time on the volatile oil yield were not significant. The volatile oil yield decreased with increasing water addition, which may have been due to the increase in water improving the amount of volatile oil dissolved in water, which resulted in a decrease in the volatile oil yield.

Table 5. Single-factor experiments.

Factor	Medicine Number	Particle Size	Soaking Time (h)	Water Addition (mL/g)	Distillation (h)	Volatile Oil Yield (mL/100 g)
Particle size	LQ-5	Not crushed Coarsest flour Coarse flour	0	10	5	1.900 1.820 1.827
Soaking time	LQ-5	Not crushed	0 2 4	10	5	1.900 1.922 1.952
Water addition	LQ-1	Not crushed	0	8 10 12	5	1.485 1.479 1.350

Table 6 shows the results of separately distilling volatile oils from *F. suspensa* and *L. japonica*. The results showed that no volatile oil was distilled from the two different batches of *L. japonica*, which may have been because the *L. japonica* used in this study contained little volatile oil. *F. suspensa* could be distilled to obtain volatile oil. Therefore, in a follow-up study, we will investigate whether the distillation of volatile oil from *L. japonica* can be promoted by the combined steam distillation of *F. suspensa* and *L. japonica*.

Table 6. Separate distillation of volatile oils from *F. suspensa* or *L. japonica*.

Medicine Number	Water Addition (mL/g)	Distillation Time (h)	Collection Time (°C)	Volatile Oil Yield (mL/100 g)
LQ-5	10	5	5	1.864
JYH-1	10	5	5	0
JYH-2	10	5	5	0

The results of the co-distillation of volatile oil from *F. suspensa* and *L. japonica* are shown in Table 7. The effects of water addition and collection temperature on the volatile oil yield were investigated. The results showed that both had a certain influence on the volatile oil yield.

Table 7. Co-distillation of volatile oils from *F. suspensa* and *L. japonica*.

Medicine Number		Water Addition (mL/g)	Distillation Time (h)	Collection Temperature (°C)	Volatile Oil Yield (mL/100 g)
LQ-5	JYH-2	10 12	5 5	Indoor temperature Indoor temperature	0.930 0.851
LQ-1	JYH-2	10 12	5 5	5 5	0.880 0.930

3.2. Methodological Validation

Taking the concentration of the β-pinene reference solution as the abscissa (X, µg/mL) and the corresponding peak area as the ordinate (Y, mAU), linear regression was performed to obtain the linear regression equation: Y = 210.56X + 282.59. The linear range of β-pinene was 10.425–166.8 µg/mL, and the coefficient of determination R^2 in this range was 0.9997, which confirmed a good linear relationship. The results of the precision test showed that the RSD of the peak area of β-pinene was 0.68%, indicating that the precision of the instrument was good. The results of the repeatability test showed that the RSD of the β-pinene concentration was 0.84%, which proved that the method had good repeatability. The results of the stability test showed that the peak-area RSD of β-pinene was 1.84%, indicating that the test solution had good stability within 24 h. The results of the sample addition recovery test are shown in Table S1. The measured recovery rates were between 95 and 102%, the average recovery rate was 98.71%, and the RSD of the three levels was 1.89%, which met the measurement requirements, indicating that the method has good accuracy.

3.3. Characterization of Raw Material Properties of Different Batches of F. suspensa

The process parameters of the steam distillation method were fixed as follows: 100 g raw material, 1000 mL water addition, 5 °C collection temperature, and 6 h distillation time. The contents of volatile oil and β-pinene in different batches of *F. suspensa* were measured as shown in Table 8. The results showed that the raw material properties of different batches of *F. suspensa* were quite different. For example, the volatile oil content of LQ-4 was 1.394 mL/100 g, while the volatile oil content of LQ-5 was 1.969 mL/100 g. The β-pinene content of LQ-6 was 0.746 mg/100 g, while the β-pinene content of LQ-5 reached 1.139 mg/100 g.

Table 8. Characterization of raw material properties of different batches of *F. suspensa*.

Medicine Number	Volatile Oil Content (mL/100 g)	β-Pinene Content (mg/100 g)
LQ-1	1.601	0.857
LQ-2	1.599	0.867
LQ-3	1.792	1.039
LQ-4	1.394	0.812
LQ-5	1.969	1.139
LQ-6	1.500	0.746

3.4. Optimization of Distillation Parameters of Volatile Oil

3.4.1. Data Processing and Model Fitting

The polynomial regression model obtained by modeling the data acquired from the experimental distillation of volatile oil from *F. suspensa* and *L. japonica* is shown in Formula (3).

$$Y = 0.2294 + 0.04977 \times 2 - 0.5199 Z_1 + 1.325 Z_2 - 0.000318 \times X_1 X_3 \qquad (3)$$

The coefficient of determination of the model R^2 was 0.9022, indicating that the model fit well and could explain the changes in the experimental data. The variance analysis of each item in the model is shown in Table 9. The *p* value of the model was less than 0.0001, indicating that the model was extremely significant. In the model, X_2 ($p = 0.0067$), Z_1 ($p = 0.0014$), and Z_2 ($p = 0.0001$) were all extremely significant items, indicating the influence of distillation time and raw material properties (volatile oil content, β-pinene content) on the volatile oil yield. $X_1 X_3$ ($p = 0.0139$) was also a significant item, which indicated that the interaction of the two factors of water addition and collection temperature was significant for volatile oil yield.

Table 9. Regression coefficient and variance analysis of the model.

Factor	Y	
	Coefficient	p Value
Constant	0.2294	
X_2	0.04977	0.0067 **
Z_1	−0.5199	0.0014 **
Z_2	1.325	0.0001 **
X_1X_3	−0.000318	0.0139 *
Model p value	<0.0001	
R^2	0.9022	
R^2_{adj}	0.8695	

* $p < 0.05$, ** $p < 0.01$.

3.4.2. Contour Diagram

The contour maps of the volatile oil yield from *F. suspensa* and *L. japonica* are shown in Figures 3–5. The figures reflect the effect of water addition, distillation time, and collection temperature on the volatile oil yield under the conditions of fixed raw material properties. Under the conditions of a constant collection temperature with increasing distillation time, the volatile oil yield increased gradually. Under the conditions of a constant distillation time with increasing water addition and a decreasing collection temperature, or decreasing water addition and an increasing collection temperature, the volatile oil yield increased. Under the conditions of unchanging water addition with increasing distillation time, the volatile oil yield increased gradually.

3.4.3. Design Space Calculation and Verification

The properties of the raw materials were fixed as LQ-6, and the design space calculated by the probability-based method is shown in Figures 6 and 7. The design space diagram shows that the distillation time had the greatest influence on the volatile oil yield.

A point inside the design space and a point outside the design space were selected for verification. The selection conditions of the verification points in the design space were as follows: the water addition was 10 mL/g, the distillation time was 6 h, and the collection temperature was 10 °C. The probability of reaching the standard was 0.976. Under these conditions, the volatile oil yield predicted by the model was 0.71 mL. According to the above conditions, three parallel experiments were carried out. The obtained volatile oil yields were 0.76 mL, 0.71 mL, and 0.74 mL, respectively. The average volatile oil yield was 0.74 mL, and the relative standard deviation was 2.52%.

The selection conditions of the verification point outside the design space were as follows: the water addition was 10 mL/g, the distillation time was 3 h, and the collection temperature was 10 °C. The probability of reaching the standard was 0.333. Under these conditions, the volatile oil yield predicted by the model was 0.56 mL, and the experimental results were 0.60 mL, 0.60 mL, and 0.58 mL, respectively. The average volatile oil yield was 0.59 mL, and the relative standard deviation was 1.15%. The measured values of the two verification points were relatively close to the model-predicted values, indicating that the mathematical model established according to the Box–Behnken design was accurate. The volatile oil yield at the verification point in the design space was higher than the preset standard, and the volatile oil yield at the verification point outside the design space was lower than the preset standard, indicating that the design space was reliable.

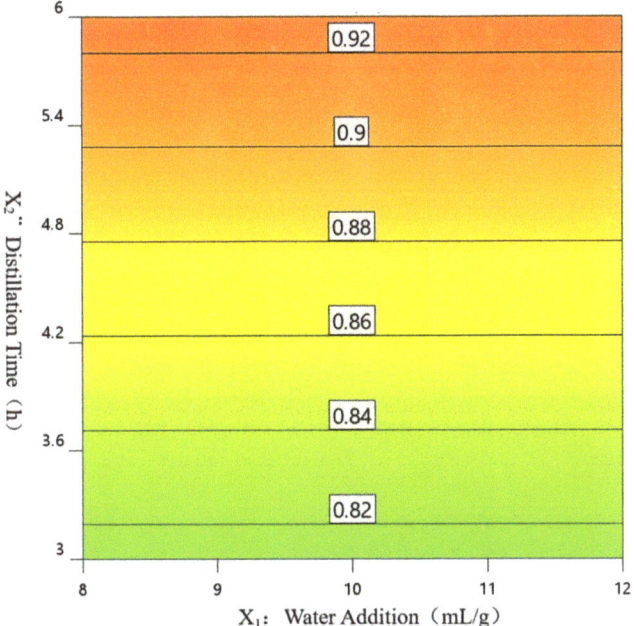

Figure 3. Volatile oil yield at a fixed collection temperature of 10 °C.

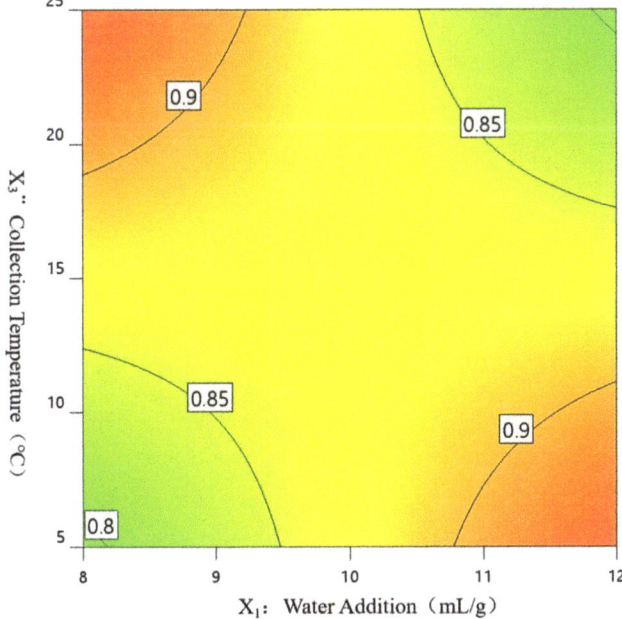

Figure 4. Volatile oil yield when the fixed distillation time was 4.5 h.

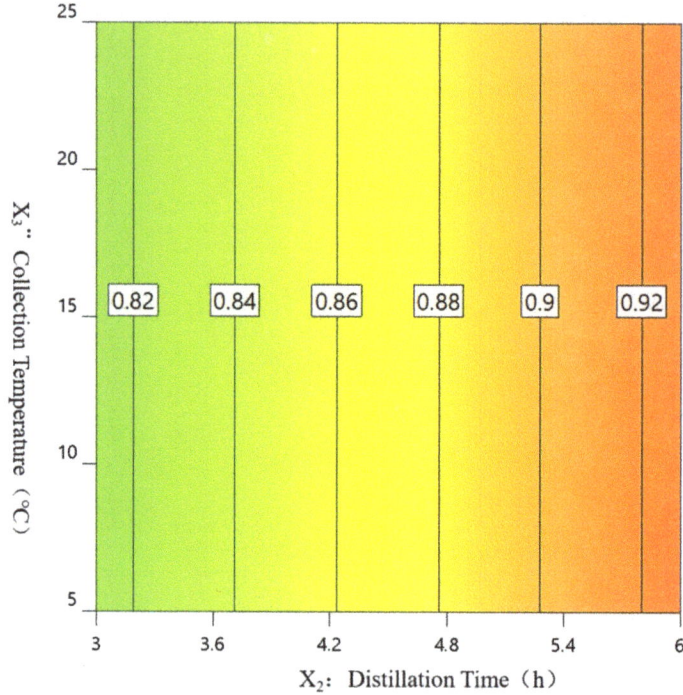

Figure 5. Volatile oil yield when the fixed water addition was 10 mL/g.

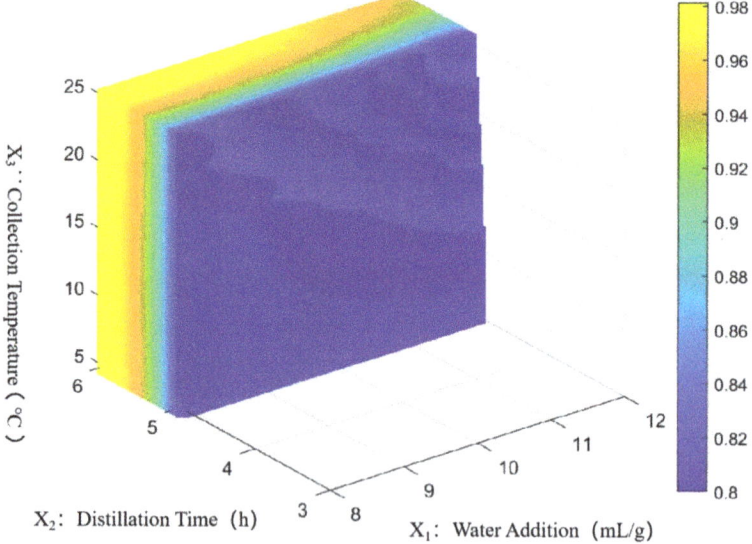

Figure 6. Three-dimensional design space diagram.

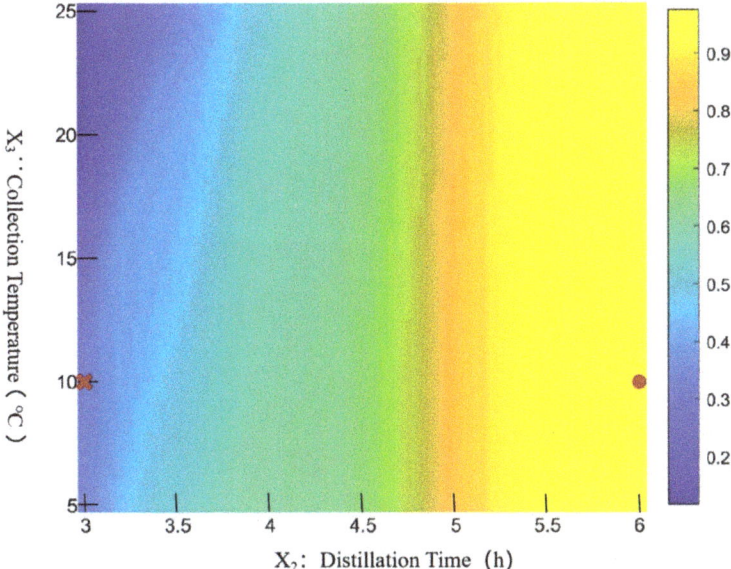

Figure 7. Two-dimensional design space diagram (water addition was fixed at 10 mL/g; "✘" is the verification point outside the design space; "●" is the verification point inside the design space; and the color bar on the right is the probability of meeting the standard).

3.5. Qualitative Analysis of Chemical Constituents of Volatile Oil

The distilled volatile oil of *F. suspensa* and the volatile oil of mixtures of *F. suspensa* and *L. japonica* were analyzed by GC–MS. The total ion chromatogram is shown in Figure 8, and the mass spectrometry results are shown in Tables 10 and 11. Interactions between multiple components can also have an impact on pharmacological activity [36,37]. A total of 16 chemical components were identified in the volatile oil of *F. suspensa*, among which the relative content of β-pinene was the largest (53.71%). Some other components with relatively large contents were α-pinene (21.12%), terpinene-4-ol (5.97%), (+)-limonene (4.36%), and α-thujene (4.25%). A total of 16 chemical components were identified in the volatile oil of *F. suspensa* and *L. japonica*, among which the relative content of β-pinene was also the largest (54.75%). Some other components with relatively large contents were α-pinene (21.38%), terpinene-4-ol (4.98%), (+)-limonene (4.48%), and α-thujene (4.41%).

Tables 10 and 11 show that the chemical composition of the volatile oil of *F. suspensa* was basically the same as that of the mixtures of *F. suspensa* and *L. japonica*. This meant that when *F. suspensa* and *L. japonica* were distilled together, the main contribution of the volatile oil chemical components came from *F. suspensa*, while *L. japonica* basically did not contribute. The combination of the two medicinal materials did not increase the amount of volatile oil distilled from *L. japonica*.

The authors attempted to distill the volatile oil from *L. japonica* but failed to collect any volatile oil. Li Jianjun et al. [20] studied the collection of volatile oil from *L. japonica* by steam distillation, and a total of 79 chemical components were identified in the volatile oil, among which the contents of palmitic acid (46.42%) and linoleic acid (14.32%) were the highest.

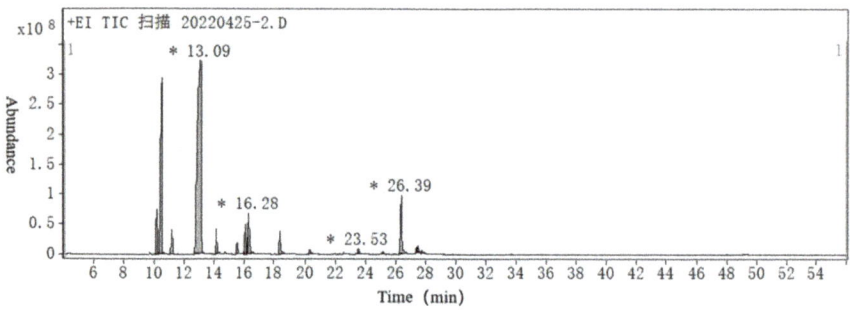

(a) Volatile oil of *F. suspensa*

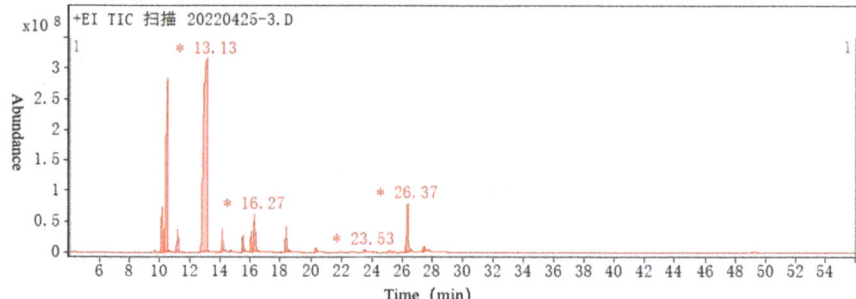

(b) Volatile oil of *F. suspensa* and *L. japonica*

Figure 8. Total ion chromatogram of the GC–MS analysis of volatile oils.

Table 10. The volatile oil of *F. suspensa*.

No	Retention Time (min)	IUPAC Name	Common Name	Chemical Formula	Base Peak (*m/z*)	Relative Content (%)
1	10.14	Bicyclo [3.1.0]hex-2-ene, 2-methyl-5-(1-methylethyl)-	α-Thujene	$C_{10}H_{16}$	93.20	4.25
2	10.51	Bicyclo [3.1.1]hept-2-ene, 2,6,6-trimethyl-	α-Pinene	$C_{10}H_{16}$	93.29	21.12
3	11.15	Bicyclo [2.2.1]heptane, 2,2-dimethyl-3-methylene-	Camphene	$C_{10}H_{16}$	93.20	1.62
4	13.09	Bicyclo [3.1.1]heptane, 6,6-dimethyl-2-methylene-	β-Pinene	$C_{10}H_{16}$	93.29	53.71
5	14.13	1,6-Octadiene, 7-methyl-3-methylene-	β-Myrcene	$C_{10}H_{16}$	93.20	1.66
6	15.49	1,3-Cyclohexadiene, 1-methyl-4-(1-methylethyl)-	α-Terpinene	$C_{10}H_{16}$	121.20	0.82
7	16.04	Benzene, 1-methyl-3-(1-methylethyl)-	m-cymene	$C_{10}H_{14}$	119.20	2.38
8	16.28	Cyclohexene, 1-methyl-4-(1-methylethenyl)-, (4R)-	(+)-Limonene	$C_{10}H_{16}$	93.20	4.36
9	18.35	1,4-Cyclohexadiene, 1-methyl-4-(1-methylethyl)-	γ-Terpinene	$C_{10}H_{16}$	93.20	1.96
10	20.32	Cyclohexene, 1-methyl-4-(1-methylethylidene)-	Terpinolene	$C_{10}H_{16}$	93.20	0.38

Table 10. Cont.

No	Retention Time (min)	IUPAC Name	Common Name	Chemical Formula	Base Peak (m/z)	Relative Content (%)
11	23.53	Bicyclo [3.1.1]heptan-3-ol, 6,6-dimethyl-2-methylene-, (1S,3R,5S)-	(-)-trans-Pinocarveol	$C_{10}H_{16}O$	92.20	0.32
12	25.16	Bicyclo [3.1.1]heptan-3-one, 6,6-dimethyl-2-methylene-	Pinocarvone	$C_{10}H_{14}O$	81.20	0.18
13	26.39	3-Cyclohexen-1-ol, 4-methyl-1-(1-methylethyl)-	Terpinen-4-ol	$C_{10}H_{18}O$	71.20	5.97
14	27.36	3-Cyclohexene-1-methanol, α,α,4-trimethyl-	α-Terpineol	$C_{10}H_{18}O$	59.20	0.40
15	27.47	Bicyclo [3.1.1]hept-2-ene-2-carboxaldehyde, 6,6-dimethyl-	Myrtenal	$C_{10}H_{14}O$	79.20	0.50
16	27.69	Bicyclo [3.1.1]hept-2-ene-2-methanol, 6,6-dimethyl-	Myrtenol	$C_{10}H_{16}O$	79.20	0.39

Table 11. The volatile oil of mixtures of *F. suspensa* and *L. japonica*.

No	Retention Time (min)	IUPAC Name	Common Name	Chemical Formula	Base Peak (m/z)	Relative Content (%)
1	10.13	Bicyclo [3.1.0]hex-2-ene, 2-methyl-5-(1-methylethyl)-	α-Thujene	$C_{10}H_{16}$	93.20	4.48
2	10.50	Bicyclo [3.1.1]hept-2-ene, 2,6,6-trimethyl-	α-Pinene	$C_{10}H_{16}$	93.29	21.38
3	11.15	Bicyclo [2.2.1]heptane, 2,2-dimethyl-3-methylene-	Camphene	$C_{10}H_{16}$	93.20	1.63
4	13.13	Bicyclo [3.1.1]heptane, 6,6-dimethyl-2-methylene-	β-Pinene	$C_{10}H_{16}$	93.29	54.75
5	14.13	1,6-Octadiene, 7-methyl-3-methylene-	β-Myrcene	$C_{10}H_{16}$	93.20	1.56
6	15.49	1,3-Cyclohexadiene, 1-methyl-4-(1-methylethyl)-	α-Terpinene	$C_{10}H_{16}$	121.20	1.23
7	16.03	Benzene, 1-methyl-3-(1-methylethyl)-	m-cymene	$C_{10}H_{14}$	119.20	1.66
8	16.27	Cyclohexene, 1-methyl-4-(1-methylethenyl)-, (4R)-	(+)-Limonene	$C_{10}H_{16}$	93.20	4.41
9	18.35	1,4-Cyclohexadiene, 1-methyl-4-(1-methylethyl)-	γ-Terpinene	$C_{10}H_{16}$	93.20	2.29
10	20.32	Cyclohexene, 1-methyl-4-(1-methylethylidene)-	Terpinolene	$C_{10}H_{16}$	93.20	0.37
11	23.53	Bicyclo [3.1.1]heptan-3-ol, 6,6-dimethyl-2-methylene-, (1S,3R,5S)-	(-)-trans-Pinocarveol	$C_{10}H_{16}O$	92.20	0.28
12	25.17	Bicyclo [3.1.1]heptan-3-one, 6,6-dimethyl-2-methylene-	Pinocarvone	$C_{10}H_{14}O$	81.20	0.14
13	26.37	3-Cyclohexen-1-ol, 4-methyl-1-(1-methylethyl)-	Terpinen-4-ol	$C_{10}H_{18}O$	71.20	4.98
14	27.42	3-Cyclohexene-1-methanol, α,α,4-trimethyl-	α-Terpineol	$C_{10}H_{18}O$	59.20	0.28
15	27.47	Bicyclo [3.1.1]hept-2-ene-2-carboxaldehyde, 6,6-dimethyl-	Myrtenal	$C_{10}H_{14}O$	79.20	0.43
16	27.70	Bicyclo [3.1.1]hept-2-ene-2-methanol, 6,6-dimethyl-	Myrtenol	$C_{10}H_{16}O$	79.20	0.13

4. Conclusions

In this study, the steam distillation process of volatile oil from *F. suspensa* and *L. japonica* was optimized according to the concept of quality by design. First, the liquid/material ratio, distillation time, and collection temperature were identified as critical process parameters by a search of the literature and single-factor experiments. In addition, this study further investigated the effect of different batches of *F. suspensa* on the distillation of volatile oil and determined that the critical raw material properties were the volatile oil content and β-pinene content. HPLC was used to evaluate the different batches. The content of β-pinene in *F. suspensa* was measured, and related methodological verification work was performed. Then, a Box–Behnken design was used to study the quantitative relationship between three process parameters, two raw material properties, and the volatile oil yield. A mathematical model was established with R^2 exceeding 0.90. Furthermore, the design space of the volatile yield was calculated by a probability-based method, and verification experiments were carried out. The measured values of two verification points were relatively close to the model-predicted values, indicating that the mathematical model established according to the Box–Behnken design was accurate. The volatile oil yield at the verification point in the design space was greater than the preset standard, and the volatile oil yield at the verification point outside the design space was lower than the preset standard, indicating that the design space was reliable. In this study, the chemical constituents of the volatile oil of *F. suspensa* and the volatile oil of *F. suspensa* and *L. japonica* were analyzed by GC–MS. A total of 16 chemical constituents were identified in the volatile oil of *F. suspensa*, among which the content of β-pinene was the largest (53.71%). A total of 16 chemical constituents were identified in the volatile oil of *F. suspensa* and *L. japonica*, among which the content of β-pinene was also the largest (54.75%), and the composition was similar to that of the volatile oil of *F. suspensa*. The results showed that when *F. suspensa* and *L. japonica* were distilled together, the main contributor to the volatile oil was *F. suspensa*. The contribution of *L. japonica* was small. The combination of *F. suspensa* and *L. japonica* did not increase the volatile oil yield from *L. japonica*.

Supplementary Materials: The following supporting information can be downloaded at: https://www.mdpi.com/article/10.3390/separations10010025/s1, Table S1: Sample Recovery Test Results.

Author Contributions: Conceptualization, X.G., N.W. and Z.W.; investigation, X.C. and D.G.; data curation, X.C. and X.G.; writing—original draft preparation, X.C., X.G. and N.W.; writing—review and editing, X.G., Z.W. and N.W.; supervision, N.W. and X.G.; funding acquisition, X.G., N.W. and Z.W. All authors have read and agreed to the published version of the manuscript.

Funding: This work was supported by the National Natural Science Foundation of China (82060720); the Open Project of the State Key Laboratory of Innovative Medicine and High-Efficiency and Energy-Saving Pharmaceutical Equipment at Jiangxi University of Chinese Medicine (grant number: GZSYS202001); the Provincial College Students' Innovation and Entrepreneurship Program (S202210412062); the Innovation Team and Talent Cultivation Program of the National Administration of Traditional Chinese Medicine (ZYYCXTD-D-202002); and the Jiangxi Traditional Chinese Medicine Research Program (2018A318).

Institutional Review Board Statement: Not applicable.

Informed Consent Statement: Not applicable.

Data Availability Statement: All data generated or analyzed during this study are included in the published article.

Conflicts of Interest: The authors declare no conflict of interest.

References

1. Chinese Pharmacopoeia Commission. *Pharmacopoeia of the People's Republic of China*; China Medical Science and Technology Publishing House: Beijing, China, 2020.
2. Liu, X.; Li, C.; Xue, J. Research progress on main active components and pharmacological effects of honeysuckle. *J. Xinxiang Med. Univ.* **2021**, *38*, 992–995.

3. Tian, D.; Shi, M.; Wang, Y. Volatile Oil from Forsythia suspense: Chemical Constituents and Pharmacological Effects. *Nat. Prod. Res. Dev.* **2018**, *30*, 1834–1842.
4. Xia, W.; Yu, Y.; Yang, H.; Tan, Z.; Xu, L.; Dong, W.; Lu, H.; Luo, F.; Liang, H. Research Advances on Chemical Constituent and Pharmacology Effects of Honeysuckle. *J. Anhui Agric. Sci.* **2017**, *45*, 126–127+165.
5. Wang, Z.; Xia, Q.; Liu, X.; Liu, W.; Huang, W.; Mei, X.; Luo, J.; Shan, M.; Lin, R.; Zou, D.; et al. Phytochemistry, pharmacology, quality control and future research of Forsythia suspensa (Thunb.) Vahl: A review. *J. Ethnopharmacol.* **2018**, *210*, 318–339. [CrossRef] [PubMed]
6. Shao, S.; Zhang, F.; Yang, Y.; Feng, Z.; Jiang, J.; Zhang, P. Neuroprotective and anti-inflammatory phenylethanoidglycosides from the fruits of Forsythia suspensa. *Bioorg. Chem.* **2021**, *113*, 105025. [CrossRef]
7. Tang, X.; Liu, X.; Zhong, J.; Fang, R. Potential Application of Lonicera japonica Extracts in Animal Production: From the Perspective of Intestinal Health. *Front. Microbiol.* **2021**, *12*, 719877. [CrossRef]
8. Kaya, D.A.; Ghica, M.V.; Dănilă, E.; Öztürk, Ş.; Türkmen, M.; Kaya, M.G.A.; Dinu-Pîrvu, C.-E. Selection of Optimal Operating Conditions for Extraction of Myrtus Communis L. Essential Oil by the Steam Distillation Method. *Molecules* **2020**, *25*, 2399. [CrossRef] [PubMed]
9. Zhang, H.; Huang, T.; Liao, X.; Zhou, Y.; Chen, S.; Chen, J.; Xiong, W. Extraction of Camphor Tree Essential Oil by Steam Distillation and Supercritical CO_2 Extraction. *Molecules* **2022**, *27*, 5385. [CrossRef]
10. Geraci, A.; Stefano, V.D.; Martino, E.D.; Schillaci, D.; Schicchi, R. Essential oil components of orange peels and antimicrobial activity. *Nat. Prod. Res.* **2017**, *31*, 653–659. [CrossRef]
11. Haro-González, J.N.; Castillo-Herrera, G.A.; Martínez-Velázquez, M.; Espinosa-Andrews, H. Clove Essential Oil (Syzygium aromaticum L. Myrtaceae): Extraction, Chemical Composition, Food Applications, and Essential Bioactivity for Human Health. *Molecules* **2021**, *26*, 6387. [CrossRef]
12. Jeliazkova, E.; Zheljazkov, V.D.; Kačániova, M.; Astatkie, T.; Tekwani, B.L. Sequential Elution of Essential Oil Constituents during Steam Distillation of Hops (Humulus lupulus L.) and Influence on Oil Yield and Antimicrobial Activity. *J. Oleo. Sci.* **2018**, *67*, 871–883. [CrossRef]
13. Romanik, G.; Gilgenast, E.; Przyjazny, A.; Kamiński, M. Techniques of preparing plant material for chromatographic separation and analysis. *J. Biochem. Biophys. Methods* **2007**, *70*, 253–261. [CrossRef] [PubMed]
14. Zou, J.; Zhang, X.; Shi, Y.; Guo, D.; Cheng, J.; Cui, C.; Tai, J.; Liang, Y.; Wang, Y.; Wang, M. Kinetic study of extraction of volatile components from turmeric by steam distillation. *China J. Tradit. Chin. Med. Pharm.* **2020**, *35*, 1175–1180.
15. Božović, M.; Navarra, A.; Garzoli, S.; Pepi, F.; Ragno, R. Esential oils extraction: A 24-hour steam distillation systematic methodology. *Nat. Prod. Res.* **2017**, *31*, 2387–2396. [CrossRef] [PubMed]
16. Aziz, Z.A.A.; Ahmad, A.; Setapar, S.H.M.; Karakucuk, A.; Azim, M.M.; Lokhat, D.; Rafatullah, M.; Ganash, M.; Kamal, M.A.; Ashraf, G.M. Essential Oils: Extraction Techniques, Pharmaceutical And Therapeutic Potential - A Review. *Curr. Drug Metab.* **2018**, *19*, 1100–1110. [CrossRef] [PubMed]
17. Liu, Y.; Tian, J.; FU, X.; Fu, C. Technological research on extraction of volatile oil from Lianqiao and inclusion of compunds with β-cyclodextrin. *J. Luzhou Med. Coll.* **2010**, *33*, 382–384.
18. Wang, Y.; Gao, J.; Cui, J.; Wang, F.; Zhang, S.; Yang, Y. Optimization of extraction process of volatile oil from Shanxi Hypericum Perforatum L. and GC-MS analysis of its chemical compositions. *Chem. Bioeng.* **2016**, *33*, 28–32.
19. Gu, K.; Wang, X.; Hu, P.; Li, H.; Wang, H.; Wang, X. Study on the technologies of steam distillation and salting out method for honeysuckle volatile oil extraction. *Asia Pac. Tradit. Med.* **2020**, *16*, 55–58.
20. Li, J.; Ren, M.; Shang, X.; Lian, X.; Wang, H. Extraction of volatile oil from Honeysuckle by distillation and its component analysis. *J. Henan Agric. Sci.* **2017**, *46*, 144–148.
21. Tong, Q.; Zhou, R.; Du, F.; Pei, G.; Peng, F. Study on the extraction technology of volatile oil from honeysuckle. *J. Hunan Univ. Chin. Med.* **2002**, *22*, 24–25.
22. Yu, L.X.; Amidon, G.; Khan, M.A.; Hoag, S.W.; Polli, J.; Raju, G.K.; Woodcock, J. Understanding pharmaceutical quality by design. *AAPS J.* **2014**, *16*, 771–783. [CrossRef] [PubMed]
23. Swain, S.; Parhi, R.; Jena, B.R.; Babu, S.M. Quality by Design: Concept to Applications. *Curr. Drug Discov. Technol.* **2019**, *16*, 240–250. [CrossRef]
24. Kasemiire, A.; Avohou, H.T.; Bleye, C.D.; Sacre, P.-Y.; Dumont, E.; Hubert, P.; Ziemons, E. Design of experiments and design space approaches in the pharmaceutical bioprocess optimization. *Eur. J. Pharm. Biopharm.* **2021**, *166*, 144–154. [CrossRef] [PubMed]
25. Debevec, V.; Srčič, S.; Horvat, M. Scientific, statistical, practical, and regulatory considerations in design space development. *Drug Dev. Ind. Pharm.* **2018**, *44*, 349–364. [CrossRef] [PubMed]
26. Gong, X.; Chen, H.; Pan, J.; Qu, H. Optimization of Panax notoginseng extraction process using a design space approach. *Sep. Purif. Technol.* **2015**, *141*, 197–206. [CrossRef]
27. Tai, Y.; Qu, H.; Gong, X. Design Space Calculation and Continuous Improvement Considering a Noise Parameter: A Case Study of Ethanol Precipitation Process Optimization for Carthami Flos Extract. *Separations* **2021**, *8*, 74. [CrossRef]
28. Chen, T.; Gong, X.; Zhang, Y.; Chen, H.; Qu, H. Optimization of a chromatographic process for the purification of saponins in Panax notoginseng extract using a design space approach. *Sep. Purif. Technol.* **2015**, *154*, 309–319. [CrossRef]
29. Kusuma, H.S.; Mahfud, M. Box-Behnken design for investigation of microwave-assisted extraction of patchouli oil. *Conf. Proc.* **2015**, *1699*, 050014.

30. Wan, N.; Lan, J.; Wu, Z.; Chen, X.; Zheng, Q.; Gong, X. Optimization of Steam Distillation Process and Chemical Constituents of Volatile Oil from Angelicae Sinensis Radix. *Separations* **2022**, *9*, 137. [CrossRef]
31. Chen, Q.; Cao, C.; Wang, L.; He, J. Simultaneous Determination of Seven Components in Blumea balsamifera Oil by HPLC. *J. Chin. Med. Mater.* **2020**, *43*, 2189–2193.
32. Zhao, X.; Zeng, Y.; Zhou, Y.; Li, R.; Yang, M. Gas Chromatography–Mass Spectrometry for Quantitative and Qualitative Analysis of Essential Oil from Curcuma wenyujin Rhizomes. *World J. Tradit. Chin. Med.* **2021**, *7*, 138–145. [CrossRef]
33. Sadgrove, N.J.; Padilla-González, G.F.; Phumthum, M. Fundamental Chemistry of Essential Oils and Volatile Organic Compounds, Methods of Analysis and Authentication. *Plants* **2022**, *11*, 789. [CrossRef] [PubMed]
34. Gong, L.; Jiang, H.; Zhang, H.; Cui, Q.; Rong, R. Analysis of volatile constituents in Forsythia Suspensa by gas chromatography-mass spectrometry. *J. Shandong Univ. Tradit. Chin. Med.* **2015**, *39*, 256–257+276.
35. Dong, M.; Li, Y.; Wang, R.; Ni, Y. Study on volatile oil GC-MS characteristic spectrumin Shanxi Fructus forsythiae. *Chin. J. Hosp. Pharm.* **2011**, *31*, 355–357.
36. Jiang, S.; Wang, M.; Yuan, H.; Xie, Q.; Liu, Y.; Li, B.; Jian, Y.; Liu, C.; Lou, H.; Rahman, A.U.; et al. Medicinal Plant of Bletilla striata: A Review of its Chemical Constituents, Pharmacological Activities, and Quality Control. *World J. Tradit. Chin. Med.* **2020**, *6*, 393–407.
37. Jiang, H.; Wang, X.; Yang, L.; Zhang, J.; Hou, A.; Man, W.; Wang, S.; Yang, B.; Chan, K.; Wang, Q.; et al. The Fruits of Xanthium sibiricum Patr: A Review on Phytochemistry, Pharmacological Activities, and Toxicity. *World J. Tradit. Chin. Med.* **2020**, *6*, 408–422.

Disclaimer/Publisher's Note: The statements, opinions and data contained in all publications are solely those of the individual author(s) and contributor(s) and not of MDPI and/or the editor(s). MDPI and/or the editor(s) disclaim responsibility for any injury to people or property resulting from any ideas, methods, instructions or products referred to in the content.

Article

Development of a Quantitative Chromatographic Fingerprint Analysis Method for Sugar Components of Xiaochaihu Capsules Based on Quality by Design Concept

Jing Lan [1,2], Gelin Wu [3,*], Linlin Wu [1,2], Haibin Qu [1,2], Ping Gong [3], Yongjian Xie [3], Peng Zhou [3] and Xingchu Gong [1,2,4,*]

1. Pharmaceutical Informatics Institute, College of Pharmaceutical Sciences, Zhejiang University, Hangzhou 310058, China
2. Innovation Institute for Artificial Intelligence in Medicine of Zhejiang University, Hangzhou 310018, China
3. Zhejiang Pralife Pharmaceutical Co., Ltd., Taizhou 318000, China
4. Jinhua Institute of Zhejiang University, Jinhua 321299, China
* Correspondence: wugelin1234@163.com (G.W.); gongxingchu@zju.edu.cn (X.G.); Tel.: +86-88133660 (G.W.); +86-88208426 (X.G.)

Abstract: Background: Xiaochaihu capsule is composed of seven traditional Chinese medicines. The pharmacopoeia only focuses on the quantitative detection of baicalin, which cannot fully reflect the quality of the preparation. Some medium polar components were used to establish the fingerprint of Xiaochaihu capsule, but there was no report on the strong polar components. Methods: A high performance liquid chromatography-corona charged aerosol detection technology was used to establish a fingerprint analysis method for Xiaochaihu capsules following an analytical quality by design approach. Definitive screening designed experiments were used to optimize the method parameters. A stepwise regression method was used to build quantitative models. The method operable design region was calculated using the experimental error simulation method. Plackett–Burman designed experiments were carried out to test robustness. Results: The contents of four components were simultaneously determined. There were seven common peaks in the fingerprint. The common peak area accounted for 91.72%. Both fingerprint and quantitative analysis methods were validated as applicable in the methodology study. The quantitative fingerprint analysis method for sugar components can fill the gap in the detection of strong polar components in the existing methods. It provides a new technology for the comprehensive overall evaluation of Xiaochaihu capsule.

Keywords: quality by design; Xiaochaihu capsule; quantitative chromatographic fingerprint; design space; definitive screening design; Plackett–Burman

1. Introduction

In recent years, analytical quality by design (AQbD) has been successfully used in the development of analytical methods for drug discovery [1–7]. AQbD allows analytical methods to be adapted within the method operable design region (MODR). The changes in analytical parameters within MODR do not affect method validity, which meets the method objectives [8–10]. AQbD facilitates the development of robust, effective, and economical analytical methods for the entire product life cycle. This promotes flexibility in the regulatory process of analytical methods [11–13]. AQbD implementation steps include determining the analytical method objective profile, identifying critical method attributes (CMAs) and critical method parameters, establishing mathematical models and MODR, conducting method validation, implementing control strategies, etc. [14–16].

Xiaochaihu capsule is composed of *Bupleurum* root, *Scutellaria* root, *Glycyrrhiza* root and rhizome, *Codonopsis* root, jujube fruit, fresh ginger rhizome, and *Pinellia* rhizome prepared with ginger and aluminum. It is prepared by decoction, percolation, concentration, drying,

mixing, granulation, and capsule filling. It can be used for the treatment of symptoms of exogenous diseases, such as bitter fullness in the chest, loss of appetite, irritability and vomiting, bitter mouth, and dry throat [17–19]. High performance liquid chromatography (HPLC) is used to determine the content of baicalin in Xiaochaihu capsules for quality control in the 2020 edition of the Chinese Pharmacopoeia Vol. 1. The qualitative identification of *Bupleurum* root, liquorice root, and *Glycyrrhiza* root and rhizome is also used as reference herbs [20]. Xiaochaihu capsules are made from seven medicinal materials; therefore, it is obvious that the quantitative determination of baicalin cannot fully reflect Xiaochaihu capsule quality.

Simultaneous quantitative determination of multi-indicator components is more often used in the 2020 edition of the Chinese Pharmacopoeia Vol. 1. It is a widely accepted and feasible method for testing the quality of traditional Chinese medicine (TCM) [21,22]. Fingerprinting is another effective method for the detection of TCM preparations. Similarity and other indicators are used to reflect the overall spectral or peak information of fingerprints [23]. Recently, fingerprint technology has been greatly developed [24–26]. Zhang Xue et al. established a quantitative fingerprint analysis method of the moderately polar components such as glycyrrhizin, baicalin, and chaihu saponin B1 in Xiaochaihu granules [27]. Liu Aoxue et al. also determined the moderately polar components of saponins in Xiaochaihu granules using quantitative analysis of multi-components by a single marker [28]. However, there is no quantitative fingerprint analysis of the sugar components of Xiaochaihu capsules [29].

Because sugar components from Chinese herbs can be easily extracted in the process of decocting with water, they are often the main components of Chinese patent medicines. The chemical composition of Xiaochaihu capsules can be reflected more comprehensively by the detection of sugar components. Stachyose is a functional oligosaccharide [30]. It is naturally found in the Lamiaceae herbs [31]. It can significantly promote the value of beneficial intestinal flora in humans [32]. *Scutellaria* root belongs to the Lamiaceae. *Scutellaria* root is regarded as the minister drug of Xiaochaihu capsules. This may explain the effects of Xiaochaihu capsules in treating loss of appetite, irritability, and vomiting. Ribitol is a reduction product of D-ribose. It is present in the *Bupleurum* root in its free state. It is a characteristic component of *Bupleurum* root. The detection of ribitol can reflect the presence of *Bupleurum* root. This can be used as a supplement to the existing quality control methods of the Pharmacopoeia.

As strongly polar components, sugar components are difficult to analyze with conventional reversed-phase columns. They usually show weak UV absorption. These reasons result in more difficulties in the analysis of sugar components than that of moderately polar components. Therefore, Amino columns and hydrophilic columns are often used to separate sugar components, such as the Prevail Carbohydrate ES (250 mm × 4.6 mm, 5 μm) of Garce Alltech [33], Asahipak NH2P-50 4E column (4.6 × 250 mm, 5 μm) of Shodex [34], and XBridge BEH Amide XP column (3 × 150 mm, 2.5 μm) of Waters [35].

Recently, several researchers have used charged aerosol detection (CAD) to detect sugar components [36–38]. The detection principle of CAD is as follows. The eluent is formed into particles by atomization. Compared to evaporative light-scattering detector, CAD has higher sensitivity, better reproducibility, and a wider linearity range [39]. Thus, CAD is expected to achieve a better separation with a lower detection limit in the quantitative analysis of sugar components of the Xiaochaihu capsule.

In this work, AQbD was used to establish a fingerprint analysis method for Xiaochaihu capsules. Parameters were determined. Definitive screening design (DSD) was used to investigate the relationships between CMAs and method parameters. A stepwise regression method was used to build quantitative models between CMAs and method parameters. The experimental error simulation method was used to calculate and verify the probability-based MODR. After optimizing the analysis conditions, the content determination components were identified, and the quantitative fingerprint was established. Finally, the durability of the analytical method was investigated.

2. Materials and Reagents

Acetonitrile was purchased from Merck (chromatographic purity, Darmstadt, Germany). Ultrapure water was prepared using a Milli-Q purification system (Millipore, Billerica, MA, USA). Ribitol (Lot No. 210916, HPLC > 99%), fructose (Lot No. 210519, HPLC > 99%), sucrose (Lot No. 210620, HPLC > 99%), stachyose (Lot No. 211026, HPLC > 99%), glucose (Lot No. 210917, HPLC > 99%), maltose (Lot No. 2110522, HPLC > 99%), and raffinose (Lot No. 210528, HPLC > 99%) were purchased from Shanghai Ronghe Pharmaceutical Technology Development Co., Ltd. (Shanghai, China).

There were 12 batches of Xiaochaihu capsule samples. The specific source merchant and lot number information are shown in Table S1.

3. Methods

3.1. Sample Preparation

3.1.1. Preparation of the Chemical Reference Solution

The four chemical reference substances were weighed precisely (AB204-N, Mettler Toledo, Zurich, Switzerland) and dissolved in 10 mL 60% acetonitrile solution. Four kinds of single-standard solution were pipetted into the same 50 mL volumetric flask. 60% acetonitrile solution was used for constant volume. The mixed reserve standard solution was diluted 5 times to obtain the mixed standard solution. The mixed standard solution was composed of 0.6512 mg/mL ribitol, 2.189 mg/mL fructose, 1.636 mg/mL sucrose, and 0.6767 mg/mL stachyose.

3.1.2. Preparation of the Sample Solution

The contents of Xiaochaihu capsules were weighed precisely and dissolved in 25 mL 60% acetonitrile solution. After being ultrasonically heated (LMTD15, Lumiere Tech, Beijing, China) and centrifuged (Minispin, Eppendorf, Hamburg, Germany), the supernatant was separated from the solution to obtain the sample solution.

3.2. HPLC Analysis

All HPLC analyses were performed on a Dionex Ultimate 3000 system (Thermo Fisher, Waltham, MA, USA) equipped with an SRD-3600 degasser, an HPG-3400RS pump, a WPS-3000TRS autosampler, a TCC-3000RScolumn thermostat, a photodiode array detector, and a Corona VeoRS CAD. The evaporation temperature was set at 35 °C. Chromatographic separation was carried out on an Asahipak NH2P-50 4E column (4.6 × 250 mm, 5 µm). The column temperature was set at 30 °C. The injection volume was set at 10 µL. The mobile phase consisted of solvent A (water) and solvent B (acetonitrile). The gradient elution program was as follows: 0–10 min, 78–74% B; 10–28 min, 74–50% B; 28–33 min, 50% B. The flow rate was set at 0.6 mL/min.

3.3. Experimental Design

3.3.1. DSD Experiment

Potential critical method parameters were identified with a fishbone diagram, as shown in Figure S1. In Figure 1, an improved AObD process was proposed based on the characteristics of traditional Chinese medicine. Based on the preliminary experiment results, the gradient, column temperature (X_5), and flow rate (X_6) were selected as potential critical method parameters for the experimental design. The other elution conditions are described in Section 3.2.

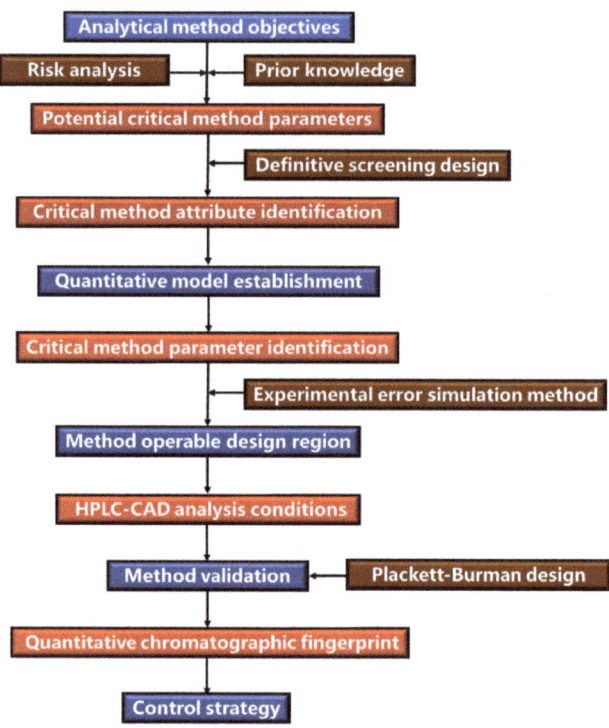

Figure 1. The AQbD process of this work.

As shown in Table 1, the mobile phase gradient was designed as 3 gradients involving 4 parameters (X_1–X_4).

Table 1. HPLC gradient conditions.

t/min	B%
0	X_1
X_2	X_3
X_4	50
$X_4 + 5$	50

A DSD method was employed to determine the relationships between the factors (X_1–X_6) and the response variables. After preliminary experiments, the levels of the factors were defined. The coded and uncoded values of each factor are summarized in Table 2. The CMAs were the peak number (Y_1), percentage of common peak (Y_2), and retention time of the last peak (Y_3). The center point was repeated 3 times. There were 2 additional dummy factors. The total number of experiments was 20. The specific experimental conditions are shown in Table 3.

Table 2. Factors and levels of DSD.

Level	Phase B Content in Mobile Phase at 0 min X_1/%	Closing Time of the First Gradient X_2/min	Phase B Content in Mobile Phase at the Beginning of the Second Gradient X_3/%	Closing Time of the Second Gradient X_4/min	Column Temperature X_5/°C	Flow Rate X_6 /(mL/min)
−1	78.0	8.0	71.0	28.0	26.0	0.60
0	80.0	10.0	73.0	30.0	28.0	0.70
1	82.0	12.0	75.0	32.0	30.0	0.80

Table 3. Experimental conditions and results of DSD.

Run	X_1/%	X_2/min	X_3/%	X_4/min	X_5/°C	X_6/(mL/min)	Y_1	Y_2/%	Y_3/min
1	78.0	8.0	71.0	30.0	30.0	0.60	15	88.86	24.43
2	82.0	10.0	75.0	28.0	30.0	0.60	15	85.26	27.18
3	78.0	10.0	71.0	32.0	26.0	0.80	12	89.63	22.92
4	80.0	10.0	73.0	30.0	28.0	0.70	11	91.21	25.27
5	82.0	8.0	71.0	32.0	28.0	0.80	12	90.45	21.88
6	80.0	12.0	75.0	32.0	30.0	0.80	9	93.88	26.83
7	80.0	8.0	71.0	28.0	26.0	0.60	12	92.43	24.03
8	82.0	12.0	71.0	28.0	26.0	0.70	11	92.16	24.95
9	82.0	12.0	75.0	30.0	26.0	0.80	11	91.93	26.29
10	78.0	8.0	75.0	28.0	26.0	0.80	10	92.51	22.27
11	82.0	12.0	71.0	32.0	30.0	0.60	14	88.22	27.41
12	78.0	12.0	75.0	28.0	28.0	0.60	14	89.58	27.98
13	78.0	8.0	75.0	32.0	30.0	0.70	12	90.51	25.73
14	82.0	8.0	73.0	28.0	30.0	0.80	14	88.44	22.09
15	78.0	12.0	71.0	28.0	30.0	0.80	12	90.4	23.19
16	78.0	12.0	73.0	32.0	26.0	0.60	13	89.47	28.68
17	82.0	8.0	75.0	32.0	26.0	0.60	17	86.24	27.78
18	80.0	10.0	73.0	30.0	28.0	0.70	11	91.59	25.25
19	80.0	10.0	73.0	30.0	28.0	0.70	11	91.32	25.25
20	80.0	10.0	73.0	30.0	28.0	0.70	11	91.63	25.24

3.3.2. Data Processing and Model Validation

The quantitative model between each CMA and the method parameters was developed using Equation (1). The model was simplified with the stepwise backward method ($\alpha = 0.1$) using Minitab software (v19, Minitab, State College, PA, USA).

$$Y = a_0 + \sum_{i=1}^{6} a_i X_i + \sum_{i=1}^{6} a_{ii} X_i^2 + \sum_{i=1}^{5} \sum_{j=i+1}^{6} a_{ij} X_i X_j \quad (1)$$

where a_0 is the constant; a_i, a_{ii} and a_{ij} are the regression coefficients of the primary, secondary and interaction terms, respectively; X refers to a method parameter; and Y is a CMA.

The experimental error simulation method was used to calculate MODR [40]. MATLAB software (R2017b, MathWorks, Natick, MA, USA) was used for program calculations. All parameters were calculated in the form of coded values. The calculation steps of X_1 to X_5 were set at 1. The calculation step of X_6 was set at 0.1. The number of simulations was 500. After obtaining the prediction results with different combinations of method parameters, the corresponding probability values were calculated statistically. The MODR was obtained with the lowest acceptable probability of 0.8.

3.3.3. Plackett–Burman Designed Experiment

In the robustness test, Plackett–Burman designed experiments were carried out to investigate the variation in the response variables with slight changes in analytical conditions to ensure the results had good reproducibility. The parameters and levels were as follows: 78.5 ± 0.5% of phase B content in mobile phase at 0 min (X_1), 8.5 ± 0.5 min of the closing time of the first gradient (X_2), 73.5 ± 0.5% of phase B content in mobile phase at the beginning of the second gradient (X_3), 30.5 ± 0.5 min of the closing time of the second gradient (X_4), 30.0 ± 1.0 °C of column temperature (X_5), 0.60 ± 0.01 mL/min of flow rate (X_6). The experimental design is shown in Table 4.

Table 4. Experimental conditions and results of Plackett–Burman designed experiment.

Run	X_1/%	X_2/min	X_3/%	X_4/min	X_5/°C	X_6/(mL/min)	Y_1	Y_2/%	Y_3/min
1	78.0	9.0	74.0	30.0	31.0	0.59	14	90.05	26.60
2	78.0	8.0	73.0	31.0	31.0	0.61	13	89.78	25.36
3	79.0	8.0	74.0	31.0	29.0	0.61	9	93.69	26.09
4	78.0	9.0	74.0	31.0	29.0	0.61	10	93.16	26.59
5	78.0	9.0	73.0	30.0	29.0	0.61	12	91.10	25.66
6	79.0	9.0	73.0	31.0	31.0	0.59	12	91.34	26.39
7	78.5	8.5	73.5	30.5	30.0	0.60	14	89.26	26.00
8	78.0	8.0	73.0	30.0	29.0	0.59	13	90.37	25.46
9	78.5	8.5	73.5	30.5	30.0	0.60	14	89.85	25.99
10	79.0	8.0	74.0	30.0	29.0	0.59	11	91.85	26.08
11	79.0	8.0	73.0	30.0	31.0	0.61	12	91.86	25.03
12	79.0	9.0	73.0	31.0	29.0	0.59	12	92.46	26.40
13	78.5.0	8.5	73.5	30.5	30.0	0.60	13	90.67	25.94
14	79.0	9.0	74.0	30.0	31.0	0.61	13	90.37	26.14
15	78.0	8.0	74.0	31.0	31.0	0.59	13	91.51	26.21

3.4. LC-Q-TOF-Ms Analysis

A LC-Q-TOF-MS (AB Sciex Triple TOF 5600+, AB Sciex, Framingham, MA, USA) was used to analyze the sugar components in Xiaochaihu capsules. The optimal condition within MODR was selected as the analysis condition. The specific analysis condition is described in Section 3.2. The mass spectrometry conditions are as follows. The electron spray ionization was chosen as ion source. The acquisition mode was set at negative ion. The scan mode was MS. The scan range was set at m/z 100–1500. The drying gas temperature was set at 320 °C. The drying gas flow rate was set at 8 L/min. The nebulizer pressure was set at 35 psi. The sheath gas temperature was set at 350 °C. The sheath gas flow rate was set at 11 L/min. The capillary voltage was set at 3500 V. The nozzle voltage was set at 1000 V. The crushing voltage was set at 175 V. The cone hole voltage was set at 65 V. The octapole radio frequency voltage peak value was set at 750 V.

3.5. Method Validation

Validation of the fingerprinting method was carried out in terms of precision, repeatability, and stability. The validation of the content determination method was carried out in terms of linear examination, precision, repeatability, stability, and recovery. Detailed experimental methods are shown in the supplementary material.

4. Results

4.1. Identification of CMAs

The peak number was chosen as a CMA to fully reflect the chemical composition of samples (Y_1). The percentage of common peak (Y_2) was chosen to ensure the representativeness of the fingerprint. The retention time of the last peak (Y_3) was chosen to optimize the separation time.

The experimental results of DSD are shown in Table 3. The peak number ranged from nine to 17. The percentage of common peak ranged from 85.26 to 93.88%. The retention time of the last peak ranged from 21.884 to 28.681 min. The value of CMAs varied considerably with different method parameters. Thus, method parameters need to be further optimized.

4.2. Influence of Method Parameters

The quantitative mathematical models between each CMA and method parameters were established according to Equation (1). The regression coefficients and analyses of variance (ANOVA) of the models are shown in Table 5. The coefficients of determination (R^2) of the three models were 0.9322, 0.9925, and 0.9999, respectively. The adjusted coefficients of determination (R^2adj) were 0.8712, 0.9795, and 0.9998, respectively. These indicated that the models were all well-fitted and could explain most of the variance.

Contour plots of each response variable can be obtained from the established mathematical models. Some of the contour plots are shown in Figure 2. The *p* value of each parameter was less than 0.1. X_1, X_2, X_5, and X_6 all showed significant effects on the peak number. X_1–X_6 all showed significant effects on the percentage of common peaks. X_1, X_2, X_3, X_4, and X_6 all showed significant effects on the retention time of the last peak.

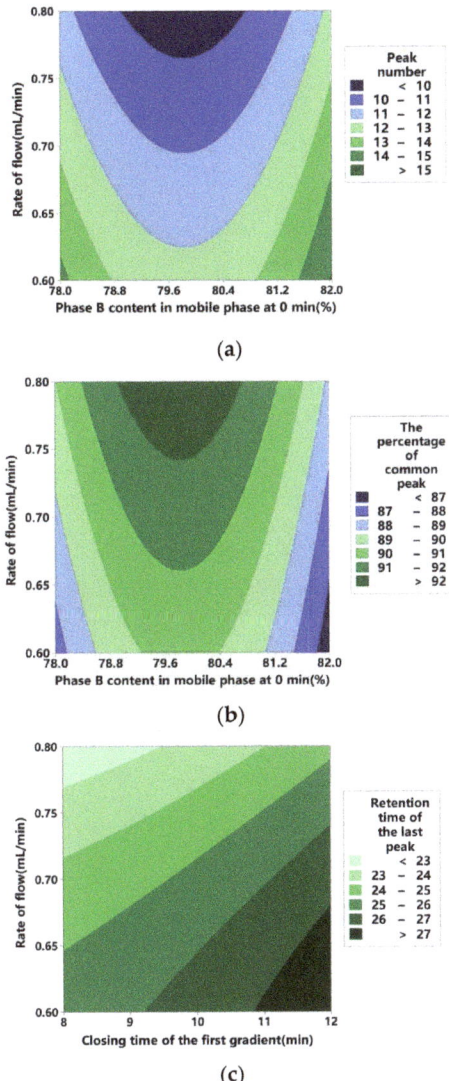

Figure 2. Contour plots of each response variable. In order to better show the relationship between the parameters and the response values, the other parameters are fixed (**a**) Peak number. Closing time of the first gradient was 10 min; closing time of the second gradient was 30 min; column temperature was 28 °C; (**b**) The percentage of common peak. Closing time of the first gradient was 10 min; column temperature was 28 °C; (**c**) Retention time of the last peak. Phase B content in mobile phase at 0 min was 80%; phase B content in mobile phase at the beginning of the second gradient was 73%; closing time of the second gradient was 30 min; column temperature was 28 °C.

Table 5. Regression coefficients and ANOVA for each model.

Item	Y_1 Coefficient	p Value	Y_2/% Coefficient	p Value	Y_3/min Coefficient	p Value
Constants	10.942	0.000	91.463	0.000	25.254	0.000
X_1	0.429	0.036	−0.590	0.000	0.170	0.000
X_2	−0.571	0.009	0.443	0.001	1.223	0.000
X_3	-	-	−0.160	0.092	1.088	0.000
X_4	-	-	−0.170	0.077	0.682	0.000
X_5	0.357	0.073	−0.629	0.000	-	-
X_6	−1.429	0.000	1.227	0.000	−1.573	0.000
X_1^2	2.337	0.000	−3.323	0.000	−0.208	0.000
X_2^2	-	-	2.788	0.000	0.236	0.000
X_3^2	-	-	-	-	0.223	0.000
X_4^2	−1.288	0.020	0.543	0.044	−0.460	0.000
X_5^2	0.962	0.068	−1.690	0.000	0.554	0.000
X_6^2	-	-	-	-	−0.374	0.000
$X_1 X_2$	−1.125	0.001	1.106	0.000	−0.323	0.000
$X_1 X_4$	-	-	−0.401	0.008	−0.062	0.002

4.3. MODR and Validation

A larger peak number indicates more information in the fingerprint. Thus, the lower limit was set at 14. The percentage of the common peak should be larger. Therefore, the lower limit was set at 0.87. The upper limit of the retention time of the last peak was set at 27 min in order to shorten the analysis time.

To be able to better demonstrate the MODR, three of the parameters were fixed. The calculated MODR is shown in Figure 3.

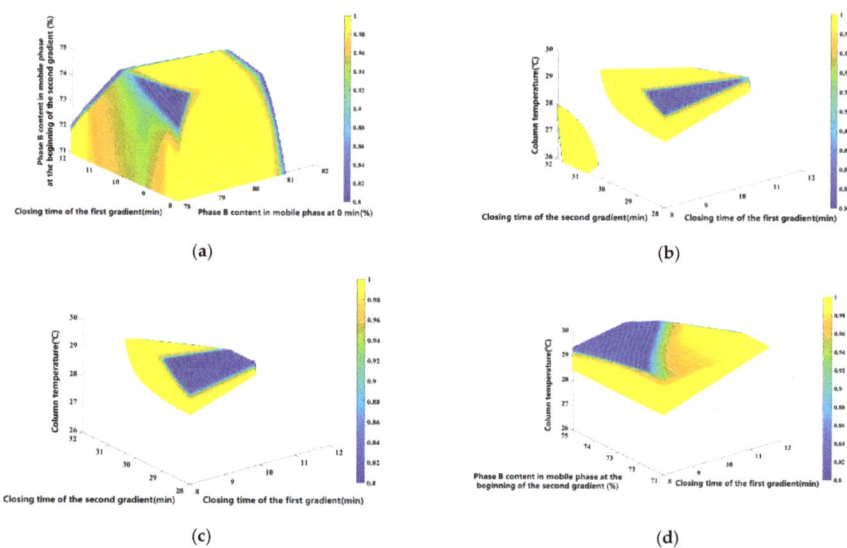

Figure 3. The MODR calculated using the experimental error simulation method. (**a**) Closing time of the second gradient was 28 min; column temperature was 30 °C; rate of flow was 0.6 mL/min; (**b**) Phase B content in mobile phase at 0 min was 81%; phase B content in mobile phase at the beginning of the second gradient was 74%; rate of flow was 0.6 mL/min; (**c**) Phase B content in mobile phase at 0 min was 81%; phase B content in mobile phase at the beginning of the second gradient was 75%; rate of flow was 0.6 mL/min; (**d**) Phase B content in mobile phase at 0 min was 81%; column temperature was 28 °C; rate of flow was 0.6 mL/min.

Within the MODR, three optimal combinations of method parameters were selected for validation experiments. The conditions and results of the validation method are shown in Table 6. Among them, the column temperature was set at 30 °C and the flow rate was set at 0.60 mL/min. The measured values were closer to the predicted values. Most of the indicators met the range requirements of the MODR, indicating that the established MODR is reliable.

Table 6. Validation of experimental conditions and results.

Methods	Gradient 1		Gradient 2		Peak Number		Percentage of Common Peak/%		Retention Time of the Last Peak/min	
	Phase B Content in Mobile Phase at the Beginning/%	Closing Time/min	Phase B Content in Mobile Phase at the Beginning/%	Closing Time/min	Predicted Value	Measured Value	Predicted Value	Measured Value	Predicted Value	Measured Value
A	78.0	10.0	74.0	28.0	14	14	88.26	91.72	26.025	26.683
B	78.5	8.5	73.5	30.5	14	14	90.19	91.50	26.236	25.504
C	78.0	9.0	74.0	30.0	15	14	89.43	88.65	26.515	26.445

4.4. Plackett–Burman Designed Experiment Result

The results of the Plackett–Burman designed experiment are shown in Table 4. Most groups of experiments showed that the peak number was greater than or equal to 12, the percentage of the common peak was greater than 89%, and the retention time of the last peak was less than 27 min. The CMAs obtained can still meet the analytical requirements when the analytical parameters are varied within MODR. In other words, the established analytical method has good robustness.

4.5. LC-Q-TOF-MS Analysis

The total ion chromatogram obtained using LC-Q-TOF-MS is shown in Figure S2. Based on the accurate relative molecular masses and chemical reference substances, seven compounds were inferred. Their numbers and inferred results are shown in Table S2. Peaks 2–9 were inferred to be ribitol, fructose, glucose, sucrose, maltose, raffinose, and stachyose, respectively.

4.6. Method Validation

4.6.1. Fingerprint Method Validation

Different batches of Xiaochaihu capsules numbered S1–S10 were studied. Seven common peaks were identified under the conditions. The peak of fructose (No. 3) was designated as the reference peak for its relatively large peak area and good separation from neighboring peaks. The results were expressed as the relative standard deviation (RSD) of the relative retention time and relative peak areas of each common peak with respect to the reference peaks. As shown in Tables S4 and S5, in the precision, repeatability, and stability tests of injection, the RSD values of relative retention time and relative retention peak area of each peak were less than 4%. The results met the requirements of chromatography fingerprinting, indicating that the test solution was stable within 24 h.

4.6.2. Application of Fingerprinting

Ten batches of Xiaochaihu capsules were prepared into sample solution according to Section 3.1. They were analyzed under the conditions of Section 3.2 to establish fingerprints. As shown in Figure 4, the original data of ten batches (S1–S10) of Xiaochaihu capsules were imported into the similarity evaluation system software (v2012.130723, Chinese Pharmacopoeia Commission, Beijing, China). The reference fingerprint was generated using the average method. The similarity results between the reference fingerprint and the sample fingerprints are shown in Table S3. The values of similarity were all above 0.90. They indicated that the sugar components of Xiaochaihu capsules from each batch had good-quality consistency.

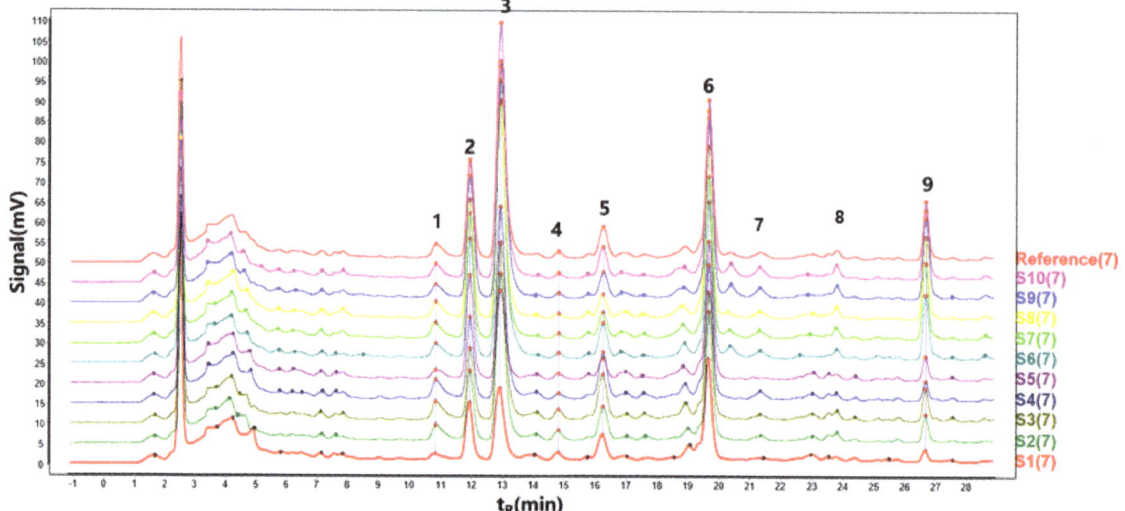

Figure 4. Reference fingerprint of Xiaochaihu capsule. Peaks 2, 3 and 5–9 were inferred to be ribitol, fructose, glucose, sucrose, maltose, raffinose, and stachyose, respectively.

4.6.3. Content Determination Method Validation

According to the retention times of the chemical reference substances and mass spectrometry, peaks 2, 3, 5, 6, 7, 8, and 9 were identified as ribitol, fructose, glucose, sucrose, maltose, raffinose, and stachyose, respectively. According to the principle of establishing fingerprint, fructose and sucrose are the main components of four components in Xiaochaihu capsules. Their detection can reflect the chemical composition of the Xiaochaihu capsule. This follows the principle of systematicity. Ribitol is a characteristic component of *Bupleurum* root. The detection of ribitol reflects the medicinal material of *Bupleurum* root. It is in line with the principle of characterization. Stachyose has the effect of regulating intestinal flora. This can be used to explain the effect of Xiaochaihu capsules in the treatment of loss of appetite, irritability, vomiting, etc. The detection of the effective ingredient is also in line with the principle of systematicity. In summary, ribitol, fructose, sucrose and stachyose were selected as the components for content determination.

The regression equations, linear range, limit of detection (LOD), and limit of quantitation (LOQ) of content determination components are shown in Table S6. The linear fit results were all greater than 0.999. The results of the injection precision experiments are shown in Tables S7 and S8. The results of the reproducibility experiments are shown in Table S9. The results of the solution stability experiments are shown in Table S10. The RSD values of precision, reproducibility, and stability were less than 3%. The results were in accordance with the requirements of the Chinese Pharmacopoeia. The results of the recovery experiments are shown in Table 7. The average recovery value of each component met the requirements, and the RSD values were less than 4%. These proved that the optimized method is accurate and reliable and can be used for the determination of sugar components in Xiaochaihu capsules.

The control strategy of the analysis method can be realized in the following two ways. First, the system suitability needs to be paid attention to before tests, including system precision, signal–noise ratio, tailing factor, and other parameters. Second, the parallel sample and reference substances can be used to observe whether the retention time is offset. When the chromatographic analyzer works abnormally, it is important to act in time.

Table 7. Results of recovery experiments.

Concentration Level	Ribitol	Fructose	Sucrose	Stachyose
Low level recovery rate (%)	102.4	105.4	106.6	103.5
	103.4	105.0	106.4	99.64
	105.3	103.8	105.4	99.71
Medium level recovery rate (%)	101.1	99.95	101.5	95.20
	102.9	103.3	103.9	102.1
	104.8	102.8	103.1	101.2
High level recovery rate (%)	97.39	97.59	100.6	98.58
	97.78	96.87	98.86	98.26
	97.14	94.50	99.28	98.82
Average recovery rate (%)	101.4	101.0	102.9	99.66
RSD (%)	3.142	3.896	2.880	2.414

4.6.4. Applications of Content Determination

Twelve batches of Xiaochaihu capsules were prepared into sample solution according to Section 3.1. They were analyzed under the conditions of Section 3.2. The content determination results of the quantitative component are shown in Table 8. In each batch of Xiaochaihu capsules, the content of ribitol ranged from 0.8985 to 2.281%, fructose ranged from 1.815 to 9.018%, and sucrose ranged from 2.054 to 5.320%. The content of stachyose was lower than 2.430%. Among them, the contents of fructose and sucrose were higher.

Table 8. Quantitative component content determination results for 12 batches of Xiaochaihu capsules.

Sample Number	Ribitol (%)	Fructose (%)	Sucrose (%)	Stachyose (%)
S1	2.004	2.012	4.666	0.5578
S2	1.205	4.819	3.504	0.6880
S3	1.206	5.021	3.413	0.8859
S4	1.411	5.737	3.750	0.4003
S5	1.051	2.263	3.877	0.5058
S6	2.200	9.018	4.549	1.815
S7	2.281	7.209	4.838	2.255
S8	2.209	6.478	5.109	2.411
S9	2.236	7.244	5.316	2.430
S10	2.185	7.935	5.320	2.310
S11	0.8985	1.815	2.735	0 *
S12	1.042	2.303	2.054	0.412

* 0 means the result is below LOQ.

5. Conclusions

In this study, a HPLC-CAD analytical method for quantitative fingerprinting of the sugar components of Xiaochaihu capsules was established based on AQbD. First, the peak number, the percentage of common peak, and the retention time of the last peak were chosen as CMAs. According to the results of definitive screening designed experiments, the critical parameters affecting the peak number were phase B content in mobile phase at 0 min, closing time of the first gradient, column temperature, and flow rate. The critical parameters affecting the percentage of common peak were phase B content in mobile phase at 0 min, closing time of the first gradient, phase B content in mobile phase at the beginning of the second gradient, closing time of the second gradient, column temperature, and flow rate. The critical parameters affecting the retention time of the last peak were phase B content in mobile phase at 0 min, closing time of the first gradient, phase B content in mobile phase at the beginning of the second gradient, closing time of the second gradient, and flow rate. Then, quantitative mathematical models between each CMA and each method parameter were established using multiple regression analysis. The R^2 of all three models exceeded 0.93, which could explain most of the variation. The MODR was calculated using the experimental error simulation method. Three experimental conditions within it were selected and successfully validated, indicating that the established MODR was reliable. Considering various factors, ribitol, fructose, sucrose, and stachyose were identified as the content determination components. The HPLC conditions for quantitative fingerprint analysis were as follows: The mobile phase consisted of solvent A (water) and solvent B (acetonitrile). The gradient elution program was as follows: 0–10 min, 78–74% B; 10–28 min, 74–50% B; 28–33 min, 50% B. The flow rate was set at 0.6 mL/min. The injection volume was set at 10 µL. The column temperature was set at 30 °C. The evaporation temperature was set at 35 °C. The peak of fructose was chosen as a reference peak to establish the fingerprint with seven common peaks. The method validation results showed that the performance of the fingerprint and content determination methods were good. Twelve batches of Xiaochaihu capsule samples were determined using the developed analysis method. The results showed that the content of fructose and sucrose were higher. In the established analytical method, the influence of the analytical parameter variation on the method's performance has been investigated. Most groups of experiments showed that the CMAs obtained can still meet the analytical requirements when the analytical parameters are varied within MODR. The proposed method is expected to be robust in the quality control of Xiaochaihu capsules.

Supplementary Materials: The following supporting information can be downloaded at: https://www.mdpi.com/article/10.3390/separations10010013/s1, Table S1: Source merchant and lot number information of Xiaochaihu capsules; Table S2: LC-Q-TOF-MS analysis of some sugar components of Xiaochaihu capsules; Table S3: Fingerprint similarity evaluation results of 10 batches of Xiaochaihu capsule sample solution; Table S4: The relative retention time of injection precision, method repeatability, and sample stability; Table S5: The relative peak areas of each common peak of injection precision, method repeatability, and sample stability; Table S6: The linear equation, coefficient of determination, and analytical range of each component; Table S7: Injection precision of the peak area; Table S8: Injection precision of retention time; Table S9: Method repeatability of content determination; Table S10: Sample stability of content determination; Figure S1: Fishbone diagram of potential critical method parameters; Figure S2 The total ion chromatogram of LC-Q-TOF-MS.

Author Contributions: Conceptualization, X.G. and G.W.; methodology, J.L.; software, J.L.; validation, J.L. and L.W.; formal analysis, J.L.; investigation, P.G., P.Z. and Y.X.; resources, Y.X., P.G. and P.Z.; data curation, J.L.; writing—original draft preparation, J.L.; writing—review and editing, J.L., L.W., X.G. and H.Q.; visualization, J.L. and P.Z.; supervision, X.G. and H.Q.; project administration, G.W. and X.G.; funding acquisition, H.Q. and X.G. All authors have read and agreed to the published version of the manuscript.

Funding: This research was funded by the Innovation Team and Talent Cultivation Program of the National Administration of Traditional Chinese Medicine (ZYYCXTD-D-202002) and the Fundamental Research Funds for the Central Universities (226-2022-00226). The APC was funded by Xingchu, Gong.

Institutional Review Board Statement: Not applicable.

Informed Consent Statement: Not applicable.

Data Availability Statement: Not applicable.

Acknowledgments: The authors are grateful for the support of Yanni Tai and Feng Ding.

Conflicts of Interest: The authors declare no conflict of interest. The funders had no role in the design of the study; in the collection, analyses, or interpretation of data; in the writing of the manuscript; or in the decision to publish the results.

References

1. Xingchu, G.; Ying, Z.; Huali, H.; Teng, C.; Jianyang, P.; Xiaoyu, W.; Haibin, Q. Development of an Analytical Method by Defining a Design Space: A Case Study of Saponin Determination for Panax Notoginseng Extracts. *Anal. Methods* **2016**, *8*, 2282–2289. [CrossRef]
2. Kyoungmin, L.; Wokchul, Y.; Jin Hyun, J. Analytical Method Development for 19 Alkyl Halides as Potential Genotoxic Impurities by Analytical Quality by Design. *Molecules* **2022**, *27*, 4437. [CrossRef]

3. Dispas, A.; Avohou, H.T.; Lebrun, P.; Hubert, P.; Hubert, C. 'Quality by Design' Approach for the Analysis of Impurities in Pharmaceutical Drug Products and Drug Substances. *TrAC, Trends Anal. Chem.* **2018**, *101*, 24–33. [CrossRef]
4. Taevernier, L.; Wynendaele, E.; D Hondt, M.; De Spiegeleer, B. Analytical Quality-by-Design Approach for Sample Treatment of BSA-containing Solutions. *J. Pharm. Anal.* **2015**, *5*, 27–32. [CrossRef]
5. Jayagopal, B.; Murugesh, S. QbD-mediated RP-UPLC Method Development Invoking an FMEA-based Risk Assessment to Estimate Nintedanib Degradation Products and Their Pathways. *Arabian J. Chem.* **2020**, *13*, 7087–7103. [CrossRef]
6. Shangxin, G.; Jing, L.; Bo, L.; Baixiu, Z.; Xingchu, G.; Xiaohui, F. Continuous Flow Synthesis of N-Doped Carbon Quantum Dots for Total Phenol Content Detection. *Chemosensors* **2022**, *10*, 334. [CrossRef]
7. Jingyuan, S.; Wen, C.; Haibin, Q.; Jianyang, P.; Xingchu, G. A Novel Quality by Design Approach for Developing an HPLC Method to Analyze Herbal Extracts: A Case Study of Sugar Content Analysis. *PLoS ONE* **2018**, *13*, e0198515. [CrossRef]
8. Musters, J.; Van Den Bos, L.; Kellenbach, E. Applying QbD Principles to Develop a Generic UHPLC Method Which Facilitates Continual Improvement and Innovation Throughout the Product Lifecycle for a Commercial API. *Org. Process Res. Dev.* **2013**, *17*, 87–96. [CrossRef]
9. Vogt, F.G.; Kord, A.S. Development of Quality-by-Design Analytical Methods. *J. Pharm. Sci.* **2011**, *100*, 797–812. [CrossRef]
10. Rozet, E.; Lebrun, P.; Debrus, B.; Boulanger, B.; Hubert, P. Design Spaces for Analytical Methods. *TrAC Trends Anal. Chem.* **2013**, *42*, 157–167. [CrossRef]
11. Raman, N.; Mallu, U.R.; Bapatu, H.R. Analytical Quality by Design Approach to Test Method Development and Validation in Drug Substance Manufacturing. *J. Chem.* **2015**, *2015*, 435129. [CrossRef]
12. Peraman, R.; Bhadraya, K.; Reddy, Y.P. Analytical Quality by Design: A Tool for Regulatory Flexibility and Robust Analytics. *Int. J. Anal. Chem.* **2015**, *2015*, 868727. [CrossRef]
13. Sangshetti, J.N.; Deshpande, M.; Zaheer, Z.; Shinde, D.B.; Arote, R. Quality by Design Approach: Regulatory Need. *Arabian J. Chem.* **2017**, *10*, S3412–S3425. [CrossRef]
14. Bastogne, T.; Caputo, F.; Prina-Mello, A.; Borgos, S.; Barberi-Heyob, M. A State of the Art in Analytical Quality-by-Design and Perspectives in Characterization of Nano-Enabled Medicinal Products. *J. Pharm. Biomed. Anal.* **2022**, *219*, 114911. [CrossRef]
15. Volta e Sousa, L.; Goncalves, R.; Menezes, J.C.; Ramos, A. Analytical Method Lifecycle Management in Pharmaceutical Industry: A Review. *Aaps Pharmscitech* **2021**, *22*, 128. [CrossRef]
16. Rathore, A.S. Roadmap for Implementation of Quality by Design (QbD) for Biotechnology Products. *Trends Biotechnol.* **2009**, *27*, 546–553. [CrossRef]
17. Shukun, Z.; Naiqiang, C.; Yuzhen, Z.; Jiangong, H.; Junhong, L.; Dihua, L.; Lihua, C. Modified Xiaochaihu Decoction Promotes Collagen Degradation and Inhibits Pancreatic Fibrosis in Chronic Pancreatitis Rats. *Chin. J. Integr. Med.* **2020**, *26*, 599–603. [CrossRef]
18. Kaibin, S.; Xinyu, Z.; Jing, L.; Rong, S. Network Pharmacological Analysis and Mechanism Prediction of Xiaochaihu Decoction in Treatment of COVID-19 with Syndrome of Pathogenic Heat Lingering in Lung and Obstructive Cardinalate. *Zhongcaoyao* **2020**, *51*, 1750–1760. [CrossRef]
19. Shanyong, W.; Ninghua, J.; Yanbo, S.; Xibin, Z.; Zhili, G.; Jintao, L.; Yanyan, H. Effects of Xiaochaihu Decoction Combined with Irbesartan on Intestinal Flora and Lipid Metabolism in Patients with Hypertension. *Zhonghua Zhongyiyao Xuekan* **2022**, *40*, 169–172. [CrossRef]
20. Commission, C.P. *Pharmacopoeia of the People's Republic of China: Part I*; China Medical Science Press: Beijing, China, 2020.
21. Yujiao, H.; Fen, X.; Shijun, Z. Difference of Chemical Compositions in Fu Zheng Fang with Different Dosage Forms Based on HPLC-Q-Exactive Orbitrap/MS Combined with Multivariate Statistical Analysis. *Curr. Pharm. Anal.* **2021**, *17*, 710–722. [CrossRef]
22. Yanfang, Y.; Guijun, Z.; Qiyu, S.; Liang, L.; Hui, P.; Jingjuan, W.; Li, X. Simultaneous Determination of 8 Compounds in Gancao-Ganjiang-Tang by HPLC-DAD and Analysis of the Relations between Compatibility, Dosage, and Contents of Medicines. *J. Evid.-Based Complement. Altern. Med.* **2017**, *2017*, 4703632. [CrossRef]
23. Xiaohui, F.; Zhengliang, Y.; Yiyu, C. A Computational Method Based on Information Fusion for Evaluating the Similarity of Multiple Chromatographic Fingerprints of TCM. *Gaodeng Xuexiao Huaxue Xuebao* **2006**, *27*, 26–29.
24. Qinglian, Y.; Tao, Q. Research of Digital Based on Network Model in the Fingerprint of Traditional Chinese Medicines (TCM). In Proceedings of the 3rd International Conference on Intelligent Computing and Cognitive Informatics (ICICCI), Mexico City, Mexico, 19–21 December 2018; p. 01005.
25. Xiaoyuan, L.; Wenwen, J.; Me, S.; Yue, S.; Hongming, L.; Lei, N.; Hengchang, Z. Quality Evaluation of Traditional Chinese Medicines Based on Fingerprinting. *J. Sep. Sci.* **2020**, *43*, 6–17. [CrossRef]
26. Xiangqin, W.; Fei, S.; Wei, Y.; Le, C.; Shumei, W.; Shengwang, L. Fingerprinting of Mineral Medicine Natrii Sulfas by Fourier Transform Infrared Spectroscopy. *Spectroscopy* **2021**, *36*, 38–43.
27. Xue, Z.; Hongwei, W.; Lina, L.; Shihan, T.; Huihui, L.; Hongjun, Y. Establishment and Application of Quality Evaluation Method for Xiaochaihu Granules Based on Calibrator Samples. *Zhongguo Zhongyao Zazhi* **2022**, *47*, 85–94. [CrossRef]
28. Aoxue, L.; Tongtong, X.; Yu, Y.; Qingyu, W.; Dandan, Z.; Yiwei, S. Study on Determination of Seven Components in Xiaochaihu Granules by QAMS. *Yaowu Pingjia Yanjiu* **2020**, *43*, 2217–2221. [CrossRef]
29. Guangzheng, X.; Hui, W.; Yingqian, D.; Keyi, X.; Weibo, Z.; Xingchu, G. Research Progress on Quality Control Methods for Xiaochaihu Preparations. *Separations* **2021**, *8*, 199. [CrossRef]

30. Jiaying, W.; Bingyong, M.; Jiayu, G.; Shumao, C.; Qiuxiang, Z. Isolation and Identification of the Intestinal Bacteria Capable of Utilizing Stachyose and its Utilization Characteristics. *Shipin Yu Fajiao Gongye* **2020**, *46*, 16–23. [CrossRef]
31. Wei, L.; Jiawei, W.; Lingyuan, M.; Xin, W.; Xia, X.; Baowei, Y. Application and Effects of Stachyose on Enterobacteria: Research Progress. *Chin. J. Microecol.* **2017**, *29*, 1110–1113, 1117. [CrossRef]
32. Jiaying, W.; Minxuan, C.; Tianci, J.; Shunhe, W.; Shumao, C.; Xin, T.; Bingyong, M. Utilization Characteristics of Stachyose by Bifidobacterium and Lactobacillus. *Shipin Yu Fajiao Gongye* **2021**, *47*, 13–20. [CrossRef]
33. Zeqi, C.; Wei, G.; Fei, L.; Chengyi, Z.; Fang, Z.; Wenzhu, L.; Jianyang, P.; Haibin, Q. Simultaneous Determination of Seven Saccharides in the Intermediates of Danshen Chuanxiongqin Injection by HPLC-ELSD. *Zhongguo Xiandai Yingyong Yaoxue* **2021**, *38*, 1349–1353. [CrossRef]
34. Yang, Z.; Desheng, X.; Li, L.; Furong, Q.; Jiiongliang, C.; Guanglin, X. A LC-MS/MS Method for the Determination of Stachyose in Rat Plasma and its Application to a Pharmacokinetic Study. *J. Pharm. Biomed. Anal.* **2016**, *123*, 24–30. [CrossRef]
35. Ghosh, R.; Kline, P. HPLC with Charged Aerosol Detector (CAD) as a Quality Control Platform for Analysis of Carbohydrate Polymers. *BMC Res. Notes* **2019**, *12*, 268. [CrossRef]
36. Linlin, W.; Shunnan, Z.; Lihong, Z.; Haoshu, X.; Xingchu, G.; Sijie, Z.; Jianyang, P.; Haibin, Q. Establishment and Validation of the Quantitative Analysis of Multi-Components by Single Marker for the Quality Control of Qishen Yiqi Dripping Pills by High-Performance Liquid Chromatography with Charged Aerosol Detection. *Phytochem. Anal.* **2021**, *32*, 942–956. [CrossRef]
37. Johanne, P.; Julian, W. Hydrophilic Interaction Chromatography Coupled with Charged Aerosol Detection for Simultaneous Quantitation of Carbohydrates, Polyols and Ions in Food and Beverages. *Molecules* **2019**, *24*, 4333. [CrossRef]
38. Aneta, S.; Ewa, J.-R.; Anna, S. High-Performance Liquid Chromatography Determination of Free Sugars and Mannitol in Mushrooms Using Corona Charged Aerosol Detection. *Food Anal. Method* **2021**, *14*, 209–216. [CrossRef]
39. Ying, W.; Yuanxi, L.; Hongshui, Y.; Weiyi, X.; Jianming, C.; Hongyu, J.; Shuangcheng, M. Comparison between Charged Aerosol Detector and Evaporative Light Scattering Detector for Analysis of Sugar in Zhusheyong Yiqi Fumai and Study on Accuracy of Methods. *Zhongguo Zhongyao Zazhi* **2020**, *45*, 5511–5517. [CrossRef]
40. Jingyuan, S.; Haibin, Q.; Xingchu, G. Comparison of Two Algorithms for Development of Design Space-Overlapping Method and Probability-Based Method. *Zhongguo Zhongyao Zazhi* **2018**, *43*, 2074–2080. [CrossRef]

Disclaimer/Publisher's Note: The statements, opinions and data contained in all publications are solely those of the individual author(s) and contributor(s) and not of MDPI and/or the editor(s). MDPI and/or the editor(s) disclaim responsibility for any injury to people or property resulting from any ideas, methods, instructions or products referred to in the content.

Article

In-Line Vis-NIR Spectral Analysis for the Column Chromatographic Processes of the *Ginkgo biloba* L. Leaves. Part II: Batch-to-Batch Consistency Evaluation of the Elution Process

Wenlong Li [1,2,3], Xi Wang [2,3], Houliu Chen [1], Xu Yan [1] and Haibin Qu [1,3,*]

1. Pharmaceutical Informatics Institute, College of Pharmaceutical Sciences, Zhejiang University, Hangzhou 310058, China
2. College of Pharmaceutical Engineering of Traditional Chinese Medicine, Tianjin University of Traditional Chinese Medicine, Tianjin 300193, China
3. State Key Laboratory of Component-Based Chinese Medicine, Tianjin University of Traditional Chinese Medicine, Tianjin 301617, China
* Correspondence: quhb@zju.edu.cn; Tel./Fax: +86-571-88208428

Abstract: An in-line monitoring method for the elution process of *Ginkgo biloba* L. leaves using visible and near-infrared spectroscopy in conjunction with multivariate statistical process control (MSPC) was established. Experiments, including normal operating batches and abnormal ones, were designed and carried out. The MSPC model for the elution process was developed and validated. The abnormalities were detected successfully by the control charts of principal component scores, Hotelling T^2, or DModX (distance to the model). The results suggested that the established method can be used for the in-line monitoring and batch-to-batch consistency evaluation of the elution process.

Keywords: Vis-NIR spectroscopy; *Ginkgo biloba* L. leaves; in-line analysis; column chromatographic elution; consistency evaluation; multivariate statistical process control (MSPC)

1. Introduction

Botanical drugs are widely used around the world. Different from chemical drugs, the raw materials of botanical drugs are from natural products with complicated components. Therefore, taking only one or a few indicators as the quality indices is not suitable during process monitoring for botanical drug production, and more comprehensive information of the in-process materials should be taken into consideration [1–3]. Visible-near infrared (Vis-NIR) spectroscopy is increasingly widely used in the food industry, chemical engineering, and process analysis of the pharmaceutical process [4–6]. Traditionally, the application of the spectral analysis technique includes the establishment, optimization, validation, and maintenance of the calibration models [7,8], which are complex, time-consuming, and require professionals to operate. To overcome these disadvantages, there is an urgent need to develop new process analytical technology for the pharmaceutical process.

Multivariate statistical process control (MSPC) is a process control technology tool for monitoring the process performance using multidimensional data collected from the processes. The data can be process parameters, such as pH, temperature, pressure, etc., and can also be spectra of the materials in process. In a previous study [9–11], we applied the technique to the manufacturing process of traditional Chinese medicine, which uses herbal medicines as raw materials. Several NIR spectroscopy-based process trajectories have been developed for monitoring pharmaceutical processes [12]. They can correctly identify the normal operation condition (NOC) batches and abnormal operation condition (AOC) batches and can also be used for the consistency evaluation of different manufacturing processes [13–16].

Ginkgo biloba L. preparation is widely used for its potential effects on memory and cognition, and in the treatment of many cardiovascular diseases [17–19]. In the production

process of *Ginkgo* preparation, column chromatography is a widely used technique for its separation and purification. However, due to the lack of on-line monitoring means, column chromatography techniques are always operated based on operator skills, and it is difficult to reduce the batch-to-batch variability, which ultimately affects the consistency of the product quality. This is a technical problem to be solved urgently in the column chromatography technique of *Ginkgo biloba* L. extract [20].

In the present study, Vis-NIR spectroscopy-based process trajectories were developed for the monitoring and control of the column chromatographic processes of *Ginkgo biloba* L. leaves extract. The in-line spectra of liquid materials were collected in the production process and, combined with the MSPC method, key information that can reflect the status of the production process was extracted. Principal component (PC) scores and Hotelling T^2 and DModX (distance to the model) control charts were quickly and effectively built and used for the real-time monitoring of the production process. The presented method provides a promising tool for the recognition of abnormal batches and the consistency evaluation of different batches for the column chromatographic processes of *Ginkgo biloba* L. leaves extract, and it can also be consulted for solving similar problems.

2. Materials and Methods

2.1. Experimental Design of NOC and AOC Batches

Experimental apparatus and experimental procedures were described in part I [21], and detailed operating conditions are listed in Table 1. Eleven batches of column chromatography processes of *Ginkgo biloba* L. leaves extracts were designed and executed, which included six NOC batches (batch 1–6) and five AOC batches (batch 7–11). Five NOC batches were used to develop the MSPC models and the AOC batches were used to investigate the prediction performance of the models. The five AOC batches covered common problems that often occurred in practical production. Batch 7 gives a low loading sample, whereas batch 8 gives a high one. Batches 9 and 10 were designed to simulate the case of abnormal elution solvent. In the elution solvent of batch 9, the concentration of ethanol is low, whereas it is high in batch 10. In batch 11, the elution flow rate is low, which is used to simulate the blocking of the chromatography column.

Table 1. Experimental design of the column chromatographic processes of *Ginkgo biloba*.

Batch No.	Volumes of Concentrated Solution Used in a Batch (L)	Flow Rate of Elution Solvent (mL·min^{-1})	Ethanol Concentrations in the Elution Solvent (v/v, %)
1–6	6.66	25	70
7	4.50	25	70
8	9.00	25	70
9	6.66	25	50
10	6.66	25	90
11	6.66	15	70

2.2. In-Line Spectra Acquisition

The in-line Vis-NIR spectra were collected as described in part I with a Sartorius X-One Vis-NIR spectrometer [22]. For every batch, a matrix composed of spectral data at different time-points was obtained. The original spectrogram of eluent of batch 1 is shown in Figure 1.

2.3. Data Processing

Five NOC batches were selected as calibration set randomly in order to establish the MSPC models. The remaining one batch of normal operation (batch 4) and other five AOC batches were used as a validation set for model evaluation.

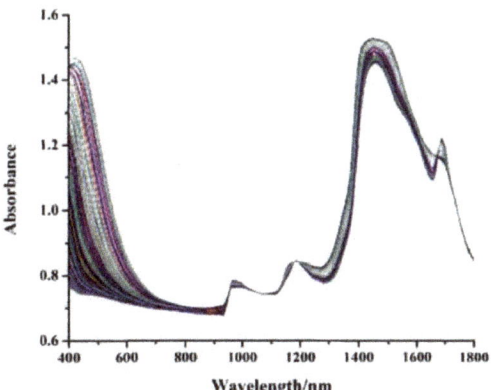

Figure 1. In-line original spectrogram of eluent of batch 1.

2.4. Unfolding of the Batch Data

Principal component analysis (PCA), partial least squares (PLS), and other multivariate projection algorithms were applied for the two-dimensional data analysis. However, the data of batch production process are composed of 3 dimensions: batch No., time, and process variables. Single-batch data are composed of a dimension of sampling points and a dimension of process variables, which are ordered by time to form an independent two-dimensional data block, whereas multi-batch process data make up a 3D data matrix X (I(nt), where I is the experimental batch, J represents the process time, and K represents the number of process variables (spectral variables) [23,24], as shown in Figure 2.

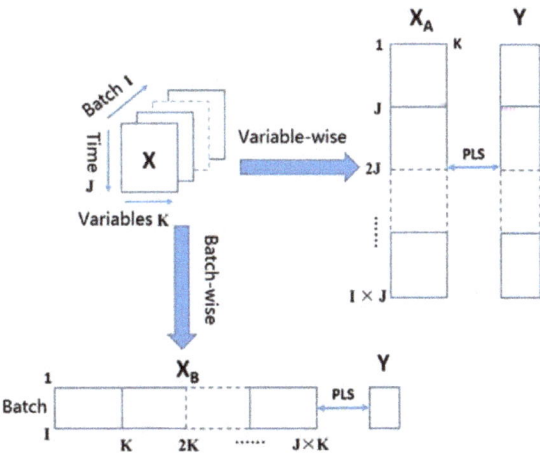

Figure 2. Two different ways to unfold 3-D spectral data.

The 3D data matrix should unfold in two ways, named "variable-wise" or "batchwise", before the multiway principal component analysis (MPCA) or multiway partial least squares (MPLS) analysis. In the batchwise method, the data of the time and process variable dimensions are combined as one, and the 3D data matrix is decomposed into 2D matrix XB (I(e), in which, every row includes all data of a batch. However, this method cannot monitor the technical process at each time point, and can only be used to analyze the whole batch, which is the limitation of the batchwise method [25]. On the other hand, in the variable-wise method, the data of time and batch dimensions are combined, and two-dimensional matrices XA (K(d) are constituted, in which, each time point corresponds to a Y value, respectively. In this way, the matrix can be used to monitor the whole process of a batch at each time-point.

2.5. Multivariate Statistical Process Control Charts

The control charts, including PC scores, Hotelling T^2, DModX, or SPE, are important tools for the MSPC model. The PC scores trajectory reflects the change trend of the main components with time. If the used PCs can explain the information of the process variables enough, the change trend of the process can be visualized with the PC scores trajectory [26]. However, if too many principal components are taken into account, the monitoring will be more complex. Therefore, in the actual application, only the first principal component is often used. In the MPLS and MPCA method, the control limits of PC scores are different from those of PCA and PLS, which are determined by the average value and standard deviation of the modeling data. In the MPLS and MPCA methods, the control limits of PC scores were determined according to the average values and standard deviations at the same time point of different batches [27].

Hotelling T^2 statistic is always used to determine how similar new observations are to the historical data collected under normal conditions. The control limit of Hotelling T^2 statistic can be determined using the F distribution [28]. Hotelling T^2 can be described as

$$T_K^2 = T_K \lambda_A^{-1} T_K^T, \qquad (1)$$

where A represents the number of the selected PCs; λ_A^{-1} and T_K represent the diagonal matrix composed of the eigenvalues corresponding to the first A PCs and the score vector of the Kth PC, respectively [29].

SPE statistic, also known as Q statistic, is the sum of squares of model residuals, representing changes in sampling points that are not explained by the model. There are N sampling points, Xn (1 × K), and the calculation formula of SPE statistics is

$$\text{SPE}n = e_n e_n^T = \sum_{k=1}^{K} \left(x_{nk} - \hat{x}_{nk} \right) \qquad (2)$$

where e_n is the residual term of sampling point x_n, x_{nk} is the observed value of the kth variable of sampling point x_n, and $\hat{x}_{nk}$ is the model predictive value of the kth variable of sampling point x_n.

In SIMCA software (Stockholm, Sweden), DModX statistics replace SPE statistics for analysis. DModX is the distance of a given observation to the model plane. DModX statistic and its control limit can be determined according to the formula found in reference [16]. DModX can be described as

$$\text{DModX}_n = \frac{e_n e_n^T}{K - A} = \sum_{K-1}^{K} \frac{(x_{nK} - X_{nK})^2}{K - A} \qquad (3)$$

where x_n, e_n, and X_{nK} represent the process variable, residual vector of x_n, and estimate of x_n. The control limit of DModX statistic is calculated as

$$\text{DModX}_{\text{lim}} = D_{cal} \pm 3SD, \qquad (4)$$

In Equation (4), D_{cal} and SD represent the mean value and the standard deviation of the standardized DModX values, respectively [29].

All of the computations were performed using SIMCA-P (V12.0.1, Umetrics, Umea, Sweden) software package.

3. Results and Discussion

3.1. Spectral Data Pretreatment

The PC score was performed on the original spectra of batch 2, and Figure 3A shows the trajectories of the first two PC scores, in which there are obvious spectral fluctuations at approximately 180 min, which may be caused by a baseline drift. In comparison with the first derivative, Savitzky–Golay (SG) smoothing, SNV, MSC, and other pretreatment methods, the 2nd derivative can eliminate the baseline drift effectively, which can be seen from the PC score diagram after pretreatment (Figure 3B).

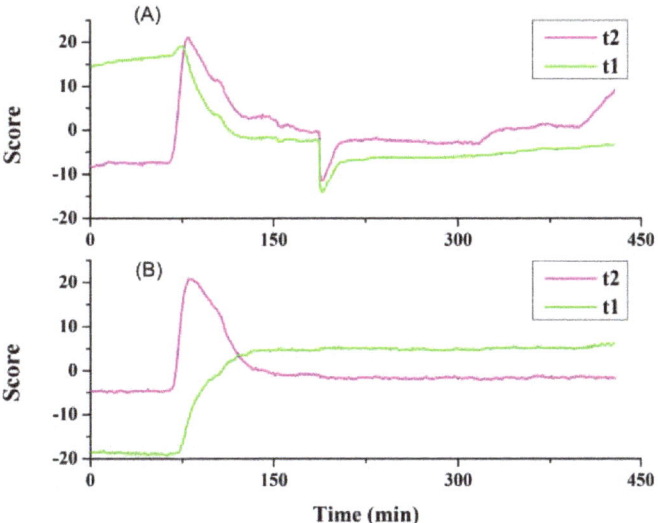

Figure 3. The first 2 PC scores variation trend during the chromatographic process of batch 2 raw spectral data, ((A) Track map of the first 2 PC scores; (B)—Second derivative spectral data).

3.2. PCA of the NOC Batches

Seen from the in-line spectra of the elution process in part I, the whole band spectra have obvious fluctuations in the elution process. To monitor the whole elution process more comprehensively, whole band spectra were selected for subsequent modeling. To understand the change in trend of the elution process and analyze the batch process consistency more clearly, 3D data of the five NOC batches were unfolded with the variable-wise method according to the description in the "*Unfolding of the batch data*" section, and the data were pretreated with the second derivative before the MPLS analysis.

The PCA results of the NOC batches are shown in Figure 4, where the blue points represent the starting points of every batch, the green points are the endpoints, and the color changes according to the time sequence of every batch. As seen from Figure 4, along with the process of elution, the PC scores of the in-line spectra showed an obvious trend, and different batches had similar trajectories. The elution process can be divided into four stages. At the beginning of the elution, the elution solvent (70% ethanol) needed a certain time from entering into the column to flow out of the column, so the spectra were in a relatively stable stage. After 60 min, the solvent began to gradually change from water to 70% ethanol, and a large amount of active components were eluted out simultaneously. The spectra in this stage changed greatly. After the elution solvent converted into 70% ethanol completely, the main changes in the spectra were caused by changes in the chemical substance contents in the elution liquid, which were relatively small; and when the elution process reached the end-point, the score was concentrated in a small space, and the spectra changed slightly.

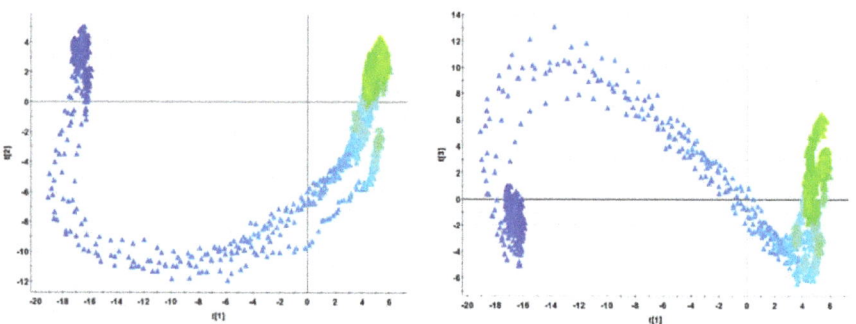

Figure 4. The spectral principal component score trajectories of the 5 NOC batches.

3.3. MPCA of the NOC Batches

To investigate the overall fluctuation in the batches, the process spectra cubes were unfolded in a batchwise way according to the description in the "*Unfolding of the batch data*" section, and the PCA model was established using the NOC batches. The cross-validation results indicated that the best PC number is 2, with which, the cumulative contribution rate of variance reached 60.5%. The PC score, Hotelling T^2, and DModX control charts are shown in Figure 5, from which, it can be concluded that all three statistics of the NOC batches were within the control limits. For the AOC batches, the DModX values were all beyond the control limit, whereas the PC scores and Hotelling T^2 statistics of batches 7, 8, and 11 were normal, which may be due to the model variance accumulation rate not being high enough to detect the abnormal sensitively. The results indicate that the DModX control chart can monitor the abnormal part, which was not explained by the model and is complementary to the PC scores and Hotelling T^2 control charts; therefore, a comprehensive analysis of the three control charts is needed for a new batch.

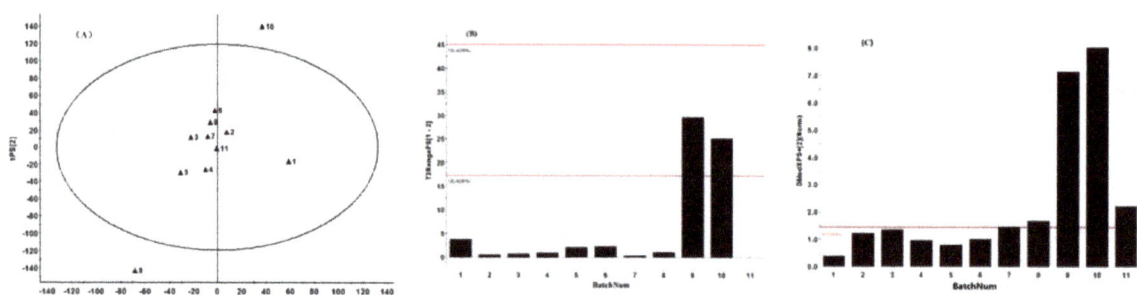

Figure 5. Three PCA model statistics of the 11 batches of column chromatographic processes ((**A**)—PC1; (**B**)—Hotelling T^2; (**C**)—DModX).

Although the PCA model can identify the AOC batches with abnormal initial materials or technical parameters, it can only be used to analyze the whole batch after it finishes and cannot be used to monitor the process synchronously [15]. For this reason, a more convenient and effective method should be developed.

3.4. Establishment of the MSPC Model

The modeling waveband (400–1800 nm) includes 141 wavelength points; therefore, the 3D matrix of the process spectral data (5 × 420 × 141) can be unfolded into a 2D matrix with 5 × 420 rows and 141 columns in the variable-wise way. The validation batches were also unfolded in the same way.

After unfolding the 3D matrix, the MPLS model of the five NOC batches, which take the spectral data matrix as X and the time vector as Y, was established. According to the cross-validation results, three latent variables were used in the model, which explains the 74.4% variance. The obtained PC1, Hotelling T^2, and DModX control charts (see Figure 6) can be used for the real-time monitoring of new batches.

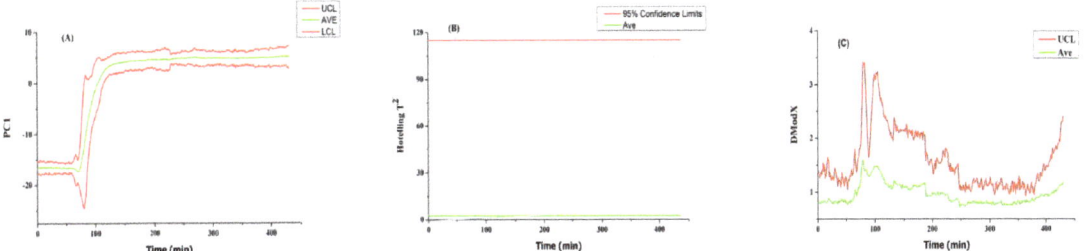

Figure 6. Three control charts of MSPC models for the column chromatographic processes of *Ginkgo biloba* ((**A**)—PC1; (**B**)—Hotelling T^2; (**C**)—DModX).

3.5. Validation of the Established MSPC Model

3.5.1. NOC Batch

To examine whether the process monitoring model will give a false alarm, Batch 4 (an NOC batch) was used for verification. The process control charts are shown in Figure 7, from which, it can be concluded that, for a randomly selected normal validation batch, in the three control charts, the trajectories are all within the control ranges, which indicates that the overall operation of the batch is in normal conditions, and the quality of intermediates can be considered in the acceptable range.

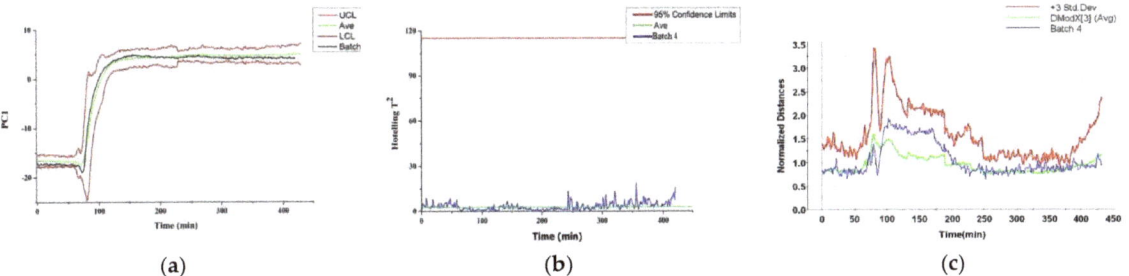

Figure 7. Three control charts of the batch 4 column chromatographic processes of *Ginkgo biloba* ((**a**)—PC1; (**b**)—Hotelling T^2; (**c**)—DModX).

3.5.2. AOC Batches with Abnormal Starting Materials

Batches 7 and 8 were used to simulate the AOC batches by changing the volume of the starting materials. As seen from the PC1 score trajectory (Figure 8), the two batches deviated from the normal range at the beginning stage of the elution. However, after 60 min, the elution solvent gradually changed from water to an ethanol solution, and the spectra fluctuated greatly, so the abnormality of the starting material was masked. Two hours later, the whole solvent system was replaced by 70% ethanol, and the anomalies in the elution were mainly caused by the different contents of the chemical substances. After 200 min, the abnormality could still be observed in the first PC score and DModX control charts. The Hotelling T^2 trajectories lie within the 95% confidence limits during the entire process. However, compared with the NOC batch, the values are significantly larger.

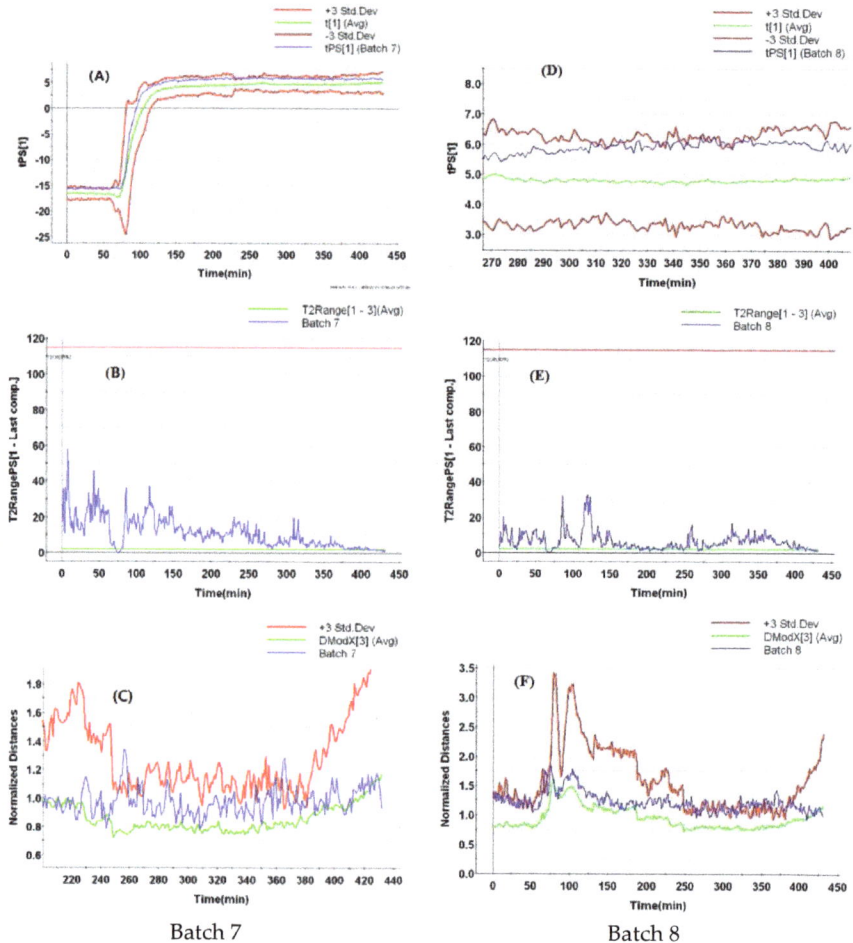

Batch 7   Batch 8

Figure 8. The control charts of batch 7 and batch 8 with abnormal initial materials ((**A**,**D**)—PC1; (**B**,**E**)—Hotelling T^2; (**C**,**F**)—DModX).

3.5.3. AOC Batches with Abnormal Ethanol Concentrations

Figure 9 shows the process control charts of two AOC batches with abnormal ethanol concentrations. From PC1 and Hotelling T^2 control charts, no obvious anomaly can be found. In the first 60 min, as the eluent liquid is the solvent of the starting materials, both trajectories laid in the normal range. However, the solvent changed after 60 min, which resulted in the process trajectories changing; this phenomenon is fully reflected in both PC1 and Hotelling T^2 control charts.

3.5.4. AOC Batches with Abnormal Flow Rates

The chromatography column will be blocked if the raw solution is insufficiently flitted, the concentration is too high, or the column is polluted by the irreversible adsorption of lipids, polysaccharides, and other impurities after repeated use. The blockage is a gradual process, in which, the elution rate will slow down. During this process, if the abnormal flow rate can be effectively monitored, complete blockage can be avoided through the refreshing of the resin, which will also increase the service life of the resin.

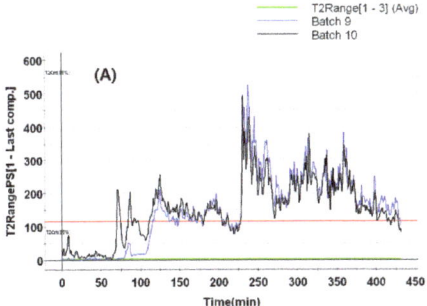
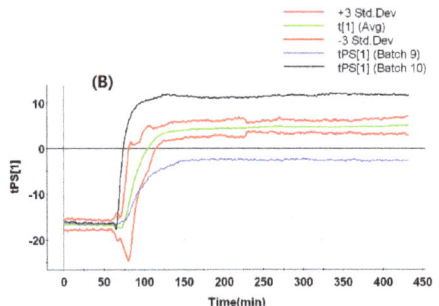

Figure 9. The control charts of batch 7 and batch 8 with abnormal ethanol concentrations ((**A**)—PC1; (**B**)—Hotelling T^2).

Batch 11 was used to simulate the blockage fault of column by reducing the flow rate of the elution solvent; the process monitoring charts of the batch are shown in Figure 10. As seen from the DModX chart, the effect of the flow rate on the process gradually increased with time. The anomaly was not obvious at the beginning of the elution, but it was quite obvious after 2–3 h. However, after 6 h, the elution process approached the end-point, and the chemical contents and the spectra were no longer changed; therefore, the DModX returned to a normal level. The Hotelling T^2 values are larger than average during the entire process, which indicates that the process is in an abnormal state.

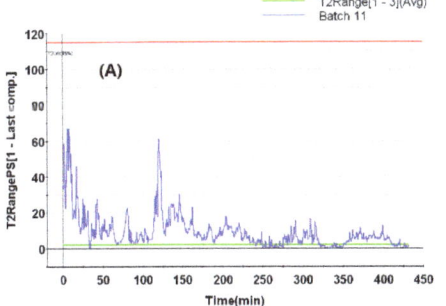
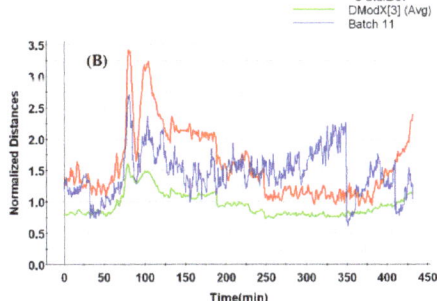

Figure 10. The control charts of batch 11 with abnormal flow rate of elution solvent ((**A**)—Hotelling T^2; (**B**)—DModX).

4. Discussion

The ingredients contained in plant medicines are complex. The target compound needs to be separated and purified by chemical means [30,31], and column chromatography technology is a traditional and very effective method for separating natural products [32]. However, because the slogan of green environmental protection goes hand in hand with the policy of eliminating outdated technologies, column chromatography has had to face a series of challenges in recent years, and the process of column chromatography needs to be monitored and controlled in order to avoid consuming too much time and solvent [33]. As one of the important means of process analysis technology (PAT), NIRs are often used to monitor the key quality and performance characteristics of raw materials, intermediates, and processes in real time [34].

5. Conclusions

In this study, a Vis-NIR spectroscopy-based MSPC method was developed for the in-line monitoring and control of the column chromatographic processes of *Ginkgo biloba*. The established model can provide effective supervision for the process and reflect the running state of the process accurately. The PC score, Hotelling T^2, and DModX trajectories control charts were mutually supplemented. A combined use of the three statistics will help to obtain more accurate results. It is undeniable that there were still problems that can be improved in this study: the number of batches used for calibration and verification was small, and the more data used to establish the calibration set, the higher the accuracy. This is where we need to improve.

Author Contributions: Conceptualization, W.L., X.W., X.Y. and H.Q.; methodology, X.Y. and X.W.; software, H.C., X.W. and X.Y.; validation, X.Y. and H.Q.; formal analysis, W.L. and H.C.; investigation, W.L. and H.Q; resources, W.L., X.Y. and H.Q.; data curation, W.L. and X.W.; writing—original draft preparation, W.L., H.C. and H.Q.; writing—review and editing, W.L., H.C. and X.W.; visualization, W.L. and H.Q.; supervision, W.L. and H.Q.; project administration, W.L., H.C., X.W., X.Y. and H.Q.; funding acquisition, W.L. and H.Q. All authors have read and agreed to the published version of the manuscript.

Funding: This study was supported by the Key R & D Project of the Zhejiang Science and Technology Program (2018C03075), Hebei Industrial Innovation and Entrepreneurship Team (No. 215A2501D), Wuhan Science and Technology Project (No. 2020020602012116), the Key Project from National Project for Standardization of Chinese Materia Medica (ZYBZH-C-JIN-43) and the National S&T Major Project of China (No. 2018ZX09201011).

Data Availability Statement: Data are contained within the article.

Conflicts of Interest: The authors declare no conflict of interest.

References

1. Food and Drug Administration. *Guidance for Industry: PAT—A Framework for Innovative Pharmaceutical Development, Manufacturing and Quality Assurance*; Food and Drug Administration: Washington, DC, USA, 2004.
2. Food and Drug Administration. *Guidance for Industry: Botanical Drug Products*; Food and Drug Administration: Washington, DC, USA, 2004.
3. Food and Drug Administration. *Botanical Drug Development Guidance for Industry*; Food and Drug Administration: Washington, DC, USA, 2015.
4. Biagi, D.; Nencioni, P.; Valleri, M.; Calamassi, N.; Mura, P. Development of a Near Infrared Spectroscopy method for the in-line quantitative bilastine drug determination during pharmaceutical powders blending. *J. Pharm. Biom. Anal.* **2021**, *204*, 114277. [CrossRef] [PubMed]
5. Khatiwada, B.P.; Subedi, P.P.; Hayes, C.; Jnr, L.C.C.C.; Walsh, K.B. Assessment of internal flesh browning in intact apple using visible-short wave near infrared spectroscopy. *Postharvest Biol. Technol.* **2016**, *120*, 103–111. [CrossRef]
6. Cecchini, C.; Antonucci, F.; Costa, C.; Marti, A.; Menesatti, P. Application of near-infrared handheld spectrometers to predict semolina quality. *J. Sci. Food. Agric.* **2021**, *101*, 151–157. [CrossRef]
7. Wang, P.; Zhang, H.; Yang, H. Rapid determination of major bioactive isoflavonoid compounds during the extraction process of kudzu (*Pueraria lobata*) by near-infrared transmission spectroscopy. *Spectrochim. Acta A* **2015**, *137*, 1403–1408. [CrossRef] [PubMed]
8. Mishra, P.; Herrmann, I.; Angileri, M. Improved prediction of potassium and nitrogen in dried bell pepper leaves with visible and near-infrared spectroscopy utilising wavelength selection techniques. *Talanta* **2021**, *225*, 121971. [CrossRef]
9. Giannuzzi, D.; Mota, L.F.M.; Pegolo, S.; Gallo, L.; Schiavon, S.; Tagliapietra, F.; Katz, G.; Fainboym, D.; Minuti, A.; Trevisi, E.; et al. In-line near-infrared analysis of milk coupled with machine learning methods for the daily prediction of blood metabolic profile in dairy cattle. *Sci. Rep.* **2022**, *12*, 8058. [CrossRef]
10. Assi, S.; Arafat, B.; Lawson-Wood, K.; Robertson, I. Authentication of Antibiotics Using Portable Near-Infrared Spectroscopy and Multivariate Data Analysis. *Appl. Spectrosc.* **2021**, *75*, 434–444. [CrossRef]
11. Li, W.; Han, H.; Cheng, Z.; Zhang, Y.; Liu, S.; Qu, H. A Feasibility Research on the Monitoring of Traditional Chinese medicine Production Process Using NIR-based Multivariate Process Trajectories. *Sens. Actuators B Chem.* **2016**, *231*, 313–323. [CrossRef]
12. Xue, J.T.; Yang, Q.W.; Li, C.Y.; Liu, X.L.; Niu, B.X. Rapid and simultaneous quality analysis of the three active components in Lonicerae Japonicae Flos by near-infrared spectroscopy. *Food. Chem.* **2021**, *342*, 128386.
13. Wu, H.L.; Chen, G.C.; Zhang, G.B.; Dai, M.H. Application of Multimodal Fusion Technology in Image Analysis of Pretreatment Examination of Patients with Spinal Injury. *J. Healthc. Eng.* **2022**, *2022*, 4326638. [CrossRef]
14. Mishra, V.; Thakur, S.; Patil, A.; Shukla, A. Quality by design (QbD) approaches in current pharmaceutical set-up. *Expert Opin. Drug Deliv.* **2018**, *15*, 737–758. [CrossRef] [PubMed]
15. Mercier, S.M.; Diepenbroek, B.; Wijffels, R.H.; Streefland, M. Multivariate PAT solutions for biopharmaceutical cultivation: Current progress and limitations. *Trends Biotechnol.* **2014**, *32*, 329–336. [CrossRef] [PubMed]
16. Kim, E.J.; Kim, J.H.; Kim, M.S.; Jeong, S.H.; Choi, D.H. Process Analytical Technology Tools for Monitoring Pharmaceutical Unit Operations: A Control Strategy for Continuous Process Verification. *Pharmaceutics* **2021**, *13*, 919. [CrossRef] [PubMed]

17. Liu, L.M.; Wang, Y.T.; Zhang, J.C.; Wang, S.F. Advances in the chemical constituents and chemical analysis of Ginkgo biloba leaf, extract, and phytopharmaceuticals. *J. Pharm. Biomed. Anal.* **2021**, *193*, 113704. [CrossRef]
18. Noor-E-Tabassum; Das, R.; Lami, M.S.; Chakraborty, A.J.; Mitra, S.; Tallei, T.E.; Idroes, R.; Mohamed, A.A.; Hossain, M.J.; Dhama, K.; et al. Ginkgo biloba: A Treasure of Functional Phytochemicals with Multimedicinal Applications. *Evid. Based Complement. Altern. Med.* **2022**, *2022*, 8288818. [CrossRef]
19. Liu, Y.X.; Xin, H.W.; Zhang, Y.C.; Che, F.; Shen, N.; Cui, Y. Leaves, seeds and exocarp of Ginkgo biloba L. (Ginkgoaceae): A Comprehensive Review of Traditional Uses, phytochemistry, pharmacology, resource utilization and toxicity. *J. Ethnopharmacol.* **2022**, *298*, 115645. [CrossRef]
20. Blumberg, L.M. Practical limits to column performance in liquid chromatography-Optimal operations. *J. Chromatogr. A* **2020**, *1629*, 461482. [CrossRef]
21. Li, W.; Yan, X.; Chen, H.; Qu, H. In-line Vis-NIR spectral analysis for the column chromatographic processes of Ginkgo biloba part I: End-point determination of the elution process. *Chemom. Intell. Lab. Syst.* **2018**, *172*, 159–166. [CrossRef]
22. Kim, B.; Woo, Y.A. Optimization of in-line near-infrared measurement for practical real time monitoring of coating weight gain using design of experiments. *Drug Dev. Ind. Pharm.* **2021**, *47*, 72–82. [CrossRef]
23. Zhang, D.J.; Sun, L.L.; Mao, B.B.; Zhao, D.S.; Cui, Y.L.; Sun, L.; Zhang, Y.X.; Zhao, X.; Zhao, P.; Zhang, X.L. Analysis of chemical variations between raw and wine-processed Ligustri Lucidi Fructus by ultra-high-performance liquid chromatography-Q-Exactive Orbitrap/MS combined with multivariate statistical analysis approach. *Biomed. Chromatogr.* **2021**, *35*, e5025. [CrossRef]
24. Eriksson, L.; Johansson, E.; Kettaneh-Wold, N. *Multi-and Megavariate Data Analysis: Principles and Applications*; Umetrics: Umea, Sweden, 2006.
25. Sheng, X.C.; Xiong, W.L. Soft sensor design based on phase partition ensemble of LSSVR models for nonlinear batch processes. *Math. Biosci. Eng.* **2019**, *17*, 1901–1921. [CrossRef] [PubMed]
26. Feng, H.M.; Li, S.N.; Hu, Y.F.; Zeng, X.Y.; Qiu, P.; Li, Y.X.; Li, W.L.; Li, Z. Quality assessment of Succus Bambusae oral liquids based on gas chromatography/mass spectrometry fingerprints and chemometrics. *Rapid Commun. Mass Spectrom.* **2021**, *35*, e9200. [CrossRef] [PubMed]
27. Clavaud, M.; Lema-Martinez, C.; Roggo, Y.; Bigalke, M.; Guillemain, A.; Hubert, P.; Ziemons, E.; Allmendinger, A. Near-Infrared Spectroscopy to Determine Residual Moisture in Freeze-Dried Products: Model Generation by Statistical Design of Experiments. *J. Pharm. Sci.* **2020**, *109*, 719–729. [CrossRef] [PubMed]
28. Oliveira, R.R.; Avila, C.; Bourne, R.; Muller, F.; Juan, A. Data fusion strategies to combine sensor and multivariate model outputs for multivariate statistical process control. *Anal. Bioanal. Chem.* **2020**, *412*, 2151–2163. [CrossRef] [PubMed]
29. Wu, S.J.; Cui, T.C.; Zhang, Z.Y.; Li, Z.; Yang, M.; Zang, Z.Z.; Li, W.L. Real-time monitoring of the column chromatographic process of Phellodendri Chinensis Cortex part II: Multivariate statistical process control based on nearinfrared spectroscopy. *New J. Chem.* **2022**, *46*, 10690–10699. [CrossRef]
30. Wang, X.J.; Xie, Q.; Liu, Y.; Jiang, S.; Li, W.; Li, B.; Wang, W.; Liu, C.X. Panax japonicus and chikusetsusaponins: A review of diverse biological activities and pharmacology mechanism. *Chin. Herb. Med.* **2021**, *13*, 64–77. [CrossRef] [PubMed]
31. Zhang, L.; Jia, Y.Z.; Li, B.; Peng, C.Y.; Yang, Y.P.; Wang, W.; Liu, C.X. A review of lignans from genus Kadsura and their spectrum characteristics. *Chin. Herb. Med.* **2021**, *13*, 157–166. [CrossRef]
32. Wang, N.N.; Chen, T.; Yang, X.; Shen, C.; Li, H.M.; Wang, S.; Zhao, J.Y.; Chen, J.L.; Chen, Z.; Li, Y.L. A practicable strategy for enrichment and separation of four minor flavonoids including two isomers from barley seedlings by macroporous resin column chromatography, medium-pressure LC, and high-speed countercurrent chromatography. *J. Sep. Sci.* **2019**, *42*, 1717–1724. [CrossRef]
33. Wang, R.; Li, W.; Chen, Z. Solid phase microextraction with poly(deep eutectic solvent) monolithic column online coupled to HPLC for determination of non-steroidal anti-inflammatory drugs. *Anal. Chim. Acta* **2018**, *1018*, 111–118. [CrossRef]
34. Choudhary, S.; Herdt, D.; Spoor, E.; García Molina, J.F.; Nachtmann, M.; Rädle, M. Incremental Learning in Modelling Process Analysis Technology (PAT)—An Important Tool in the Measuring and Control Circuit on the Way to the Smart Factory. *Sensors* **2021**, *21*, 3144. [CrossRef]

Article

Development and Validation of a Near-Infrared Spectroscopy Method for Multicomponent Quantification during the Second Alcohol Precipitation Process of *Astragali radix*

Wenlong Li [1,2,3], Yu Luo [1], Xi Wang [2], Xingchu Gong [1], Wenhua Huang [4], Guoxiang Wang [4] and Haibin Qu [1,*]

1. Pharmaceutical Informatics Institute, College of Pharmaceutical Sciences, Zhejiang University, Hangzhou 310058, China
2. College of Pharmaceutical Engineering of Traditional Chinese Medicine, Tianjin University of Traditional Chinese Medicine, Tianjin 300193, China
3. Haihe Laboratory of Modern Chinese Medicine, Tianjin 301617, China
4. Livzon (Group) Limin Pharmaceutical Factory, Shaoguan 512028, China
* Correspondence: quhb@zju.edu.cn; Tel./Fax: +86-571-88208428

Abstract: The objective of this study was to develop and validate a near-infrared (NIR) spectroscopy based method for in-line quantification during the second alcohol precipitation process of *Astragali radix*. In total, 22 calibration experiments were carefully arranged using a Box–Behnken design. Variations in the raw materials, critical process parameters, and environmental temperature were all included in the experimental design. Two independent validation sets were built for method evaluation. Validation set 1 was used for optimization. Different spectral pretreatments were compared using a "trial-and-error" approach. To reduce the calculation times, the full-factorial design was applied to determine the potential optimal combinations. Then, the best parameters for the pretreatment algorithms were compared and selected. Partial least squares (PLS) regression models were obtained with low complexity and good predictive performance. Validation set 2 was used for a thorough validation of the NIR spectroscopy method. Based on the same validation set, traditional chemometric validation and validation using accuracy profiles were conducted and compared. Conventional chemometric parameters were used to obtain the overall predictive capability of the established models; however, these parameters were insufficient for pharmaceutical regulatory requirements. Then, the method was fully validated according to the ICH Q2(R1) guideline and using the accuracy profile approach, which enabled visual and reliable representation of the future performances of the analytical method. The developed method was able to determine content ranges of 8.44–39.8% at 0.541–2.26 mg/mL, 0.118–0.502 mg/mL, 0.220–0.940 mg/mL, 0.106–0.167 mg/mL, 0.484–0.879 mg/mL, and 0.137–0.320 mg/mL for total solid, calycosin glucoside, formononetin glucoside, 9,10-dimethoxypterocarpan glucopyranoside, 2′-dihydroxy -3′, 4′-dimethoxyisoflavan glucopyranoside, astragloside II, and astragloside IV, respectively. These ranges were specific to the early and middle stages of the second alcohol precipitation process. The method was confirmed to be capable of achieving an in-line prediction with a very acceptable accuracy. The present study demonstrates that accuracy profiles offer a potential approach for the standardization of NIR spectroscopy method validation for traditional Chinese medicines (TCMs).

Keywords: near-infrared spectroscopy; *Astragali radix*; alcohol precipitation; validation; accuracy profile

1. Introduction

Astragali radix is one of the most extensively used Chinese herbal medicines because of its effect of increasing the overall vitality of the system, and it has been prescribed for general debility and chronic illnesses for centuries [1]. In recent years, it has been used clinically for spinal cord injury [2], tissue fibrosis, and other diseases [3]. Alcohol precipitation is a vital separation unit that is widely used in the manufacture of botanical medicines to purify the water extracts of medicinal plants [4]. In the manufacturing of Chinese patented drugs derived from *Astragali radix*, a second ethanol precipitation is performed after the first ethanol precipitation to remove more impurities, such as saccharides or proteins. The quality of the intermediates in the following processes and the finished products is thought to be affected by the second ethanol precipitation. To ensure a precise and reproducible alcohol precipitation process, the composition of the alcohol precipitation liquid should be closely supervised during the process.

Due to the complicated ingredients in *Astragali radix*, it is insufficient to realize quality control using a single indicator. The total solid (TS) content represents the total soluble solids (mostly saccharides) that are partially removed during the precipitation process. Flavonoids and saponins are bioactive components that are responsible for pharmacological activities and therapeutic efficacy [5,6]. In this study, the contents of TS, four flavonoid compounds, and two saponin compounds were taken as critical quality attributes (CQAs) of the alcohol precipitation liquids. However, these seven quality indices are often determined by the time-consuming, loss-on-drying method or by high-performance liquid chromatography (HPLC), which fail to satisfy the need for real-time monitoring.

With its advantages of nondestructive and high-speed acquisition, near-infrared (NIR) spectroscopy is a good process analytical technology (PAT) tool that has long been used in the pharmaceutical industry. To develop a sound NIR spectroscopy method, representative samples should be carefully selected, which need to be robust with the expected variation [7]. In the modeling, spectral pretreatments should also be properly selected. Among the different types of selection approaches, the "trial-and-error" approach [8] is a fit-for-use oriented approach, which applies all the possible pretreatments to the data set and selects the optimal one according to the goal of the analysis. However, this approach may be computationally intensive.

Prior to routine analysis, the established NIR spectroscopy method should be validated to demonstrate that it is suitable for its intended purpose [9]. However, validation of the chemometric method is not straightforward compared with the validation of conventional analytical techniques, such as chromatography or titrimetry. In pharmaceutical applications, validation based on traditional chemometric parameters is widely used to assess the performance of the developed NIR spectroscopy method [10]. Such parameters include the correlation coefficient (R), root mean square error of calibration (RMSEC), root mean square error of cross-validation (RMSECV), and root mean square error of prediction (RMSEP). However, the model performance evaluation is an area that has not been fully explored. Some studies have demonstrated that all these parameters provide insufficient information to guarantee the suitability of the method for the intended purpose [11–13]. For example, R is an index affected by unwanted factors, such as data distribution. The more centralized the data distribution is, the more difficult it is to obtain a higher R, while the more dispersed the data distribution is, the more likely it is to obtain a higher R.

The validation strategy of an accuracy profile, which was introduced by the commission of the Société Française des Sciences et Techniques Pharmaceutiques (SFSTP) [14–16], involves acquiring the content ranges over which future measurements will be sufficiently accurate. The accuracy profile is based on β-expectation tolerance intervals, which reflect the total measurement error.

A method is considered to be valid when the β-expectation tolerance intervals are fully included within the predefined acceptance limits. As described by De Bleye et al., accuracy profiles have been used for many NIR spectroscopic methods in pharmaceutical applications [10].

In the manufacturing of traditional Chinese medicine (TCM) preparations, NIR spectroscopy has been extensively applied [17]. However, most NIR spectroscopy methods have been considered valid when satisfactory traditional chemometric parameters were obtained, and few of the methods were further validated according to the ICH Q2(R1) guideline. In reported studies, there have been some cases of using the accuracy profile approach during the validation stage, such as methods for the determination of baicalin in Yinhuang oral solution [18], chlorogenic acid in the ethanol precipitation solution of *Lonicera japonica* [19], and licorice acid in a blending process [20]. However, to our knowledge, there have been few studies that have focused on a thorough validation of an in-line NIR spectroscopy method for multicomponent quantification.

The objective of this study was first to develop an NIR spectroscopy method for multicomponent quantification during the second alcohol precipitation process of *Astragali radix*, meanwhile during which a selection method based on the design of experiments (DOE) was used to reduce the calculation times for the selection of pretreatments. The second aim was to fully validate the in-line method for the seven analytes and to compare the traditional chemometric validation with the accurate profile approach.

2. Materials and Methods

2.1. Materials

The concentrated supernatants of the first ethanol precipitation of *Astragali radix* were supplied by Livzon (Group), Limin Pharmaceutical Factory (Shaoguan, China). Anhydrous alcohol was purchased from Changqin Chemical Co., Ltd. (Hangzhou, China). Standard substances of calycosin-7-O-β-D-glucoside (CG), formononetin–7–O–β–D-glucoside (FG), 9,10-dimethoxypterocarpan–3–O-β-D-glucopyranoside (DPGP), 2′-dihydroxy-3′,4′-dimethoxyisoflavan-7-O-β-D- glucopyranoside (DDIFGP), astragloside II (AG II), and astragloside IV (AG IV) were purchased from Shanghai Winherb Medical Technology Co., Ltd. (Shanghai, China).

2.2. Alcohol Precipitation Process and Experimental Setup

Typical operating conditions for the second alcohol precipitation process were as follows. First, 300 g of concentrated supernatant of the first ethanol precipitation of *Astragali radix* (TS content was 40%) was placed into a 2 L jacketed glass container. The solution was maintained at a constant temperature of 25 °C using a circulation bath. A 95% (v/v) alcohol solution that was 900 g was added into the glass container at a constant speed using a peristaltic pump, and the mixed solution was stirred using a mechanical stirrer at a speed of approximately 350 rpm. The alcohol adding time was 20 min, and the mixed solution was allowed to stand for 10 min without stirring. The supernatant was obtained as the second alcohol precipitation liquid.

The experimental setup is shown schematically in Figure 1. An NIR immersion transflectance probe with a 2 mm optical path length (Hellma, Müllheim, Germany) was directly inserted into the glass container and connected to the spectrometer by optic fibers. During the precipitation process, a sample of approximately 5 mL was collected near the probe every 5 to 10 min for reference assays.

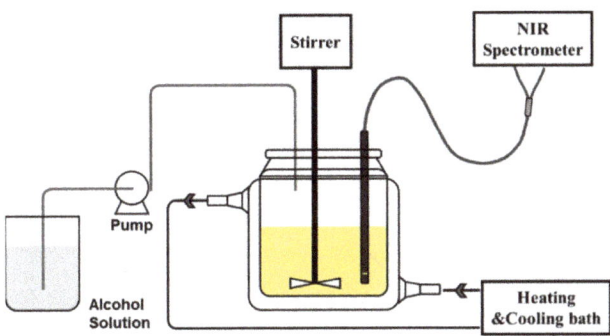

Figure 1. Schematic of the experimental setup for the second alcohol precipitation process.

2.3. NIR Spectral Acquisition

NIR spectra were collected in transflectance mode using an Antaris FT-NIR spectrophotometer (Thermo Nicolet Corporation, Waltham, MA, USA) fitted with an InGaAs detector and a 2 mm optical path length immersion probe. The instrument resolution was specified at 8 cm^{-1}. Each spectrum was acquired by averaging 32 scans over the wavenumber range of 4500–10,000 cm^{-1}, and background spectra were obtained in air.

2.4. Reference Assays

The seven quality indicators were measured by corresponding standard assays. A gravimetric-based loss-on-drying method was run to determine the contents of TS via hot air dying at 105 °C for 3 h. Quantitative analyses of CG, FG, DPGP, DDIFGP, AG II, and AG IV were performed using the HPLC-UV-ELSD method [19]. The flavonoids of CG, FG, DPGP, and DDIFGP were detected using UV. The saponins of AG II and AG IV were detected using ELSD.

2.5. Calibration Protocol

Calibration models should be developed using carefully selected and representative samples, and method development procedures need to be robust with respect to the expected variability of the products to be analyzed and the manufacturing processes used to prepare them [7]. Variations of raw materials, process parameters, and environmental temperature were introduced into the calibration sample set using a Box–Behnken design. As shown in Table 1, experiments of 29 runs (the central operating conditions were repeated 5 times) were performed to cover the different sources of variability. The concentrated supernatants of the first ethanol precipitation with TS contents of 45%, 40%, and 35% were prepared to obtain different raw materials for the second precipitation process. As illustrated in our previous study [20,21], the concentration of ethanol and the mass ratio of ethanol to concentrated raw materials were considered to be the two critical process parameters for the alcohol precipitation process. Additionally, the spectral acquisition position of each experiment was randomly set to include the different positions of the samples in the container. In the meantime of spectral acquisition, samples were collected for reference assays. Seven samples were collected for each experiment for batches 1 to 22, as shown in Table 1, and in total, 154 samples were included in the calibration set.

Table 1. Conditions for the alcohol precipitation experiments according to the Box–Behnken design.

Batch	TS Contents of the Concentrated Raw Materials (%)	Ethanol Concentration (%)	Mass Ratio of Ethanol to the Concentrated Raw Materials (g/g)	Temperature (°C)	Usage
1	35	93	3.0	25	Calibration
2	45	93	3.0	25	Calibration
3	35	97	3.0	25	Calibration
4	45	97	3.0	25	Calibration
5	40	97	2.5	25	Calibration
6	40	93	3.5	25	Calibration
7	35	95	2.5	25	Calibration
8	45	95	2.5	25	Calibration
9	35	95	3.5	25	Calibration
10	45	95	3.5	25	Calibration
11	40	95	2.5	20	Calibration
12	40	93	3.5	25	Calibration
13	35	95	3.0	20	Calibration
14	45	95	3.0	20	Calibration
15	40	93	3.0	20	Calibration
16	40	97	3.0	20	Calibration
17	40	95	2.5	30	Calibration
18	40	95	3.5	30	Calibration
19	35	95	3.0	30	Calibration
20	45	95	3.0	30	Calibration
21	40	97	3.5	25	Calibration
22	40	97	3.0	30	Calibration

Table 1. Cont.

Batch	TS Contents of the Concentrated Raw Materials (%)	Ethanol Concentration (%)	Mass Ratio of Ethanol to the Concentrated Raw Materials (g/g)	Temperature (°C)	Usage
23	40	95	3.5	20	Validation and robustness evaluation
24	40	93	3.0	30	Validation and robustness evaluation
25	40	95	3.0	25	Robustness evaluation
26	40	95	3.0	25	Validation
27	40	95	3.0	25	Validation
28	40	95	3.0	25	Validation
29	40	95	3.0	25	Validation

2.6. Validation Protocol

Validation set 1 was constructed to optimize the calibration model. It encompassed samples collected from batches 23 to 29. Independent raw material with TS contents of 40% were used in the experiments, and 3 to 7 samples were randomly collected for each batch. Finally, 38 samples were included in external validation set 1.

Validation set 2 was constructed for a thorough validation of the in-line NIR method. Three batches with normal operating conditions for the second alcohol precipitation process were repeated and performed over three days. The position of the optical probe was fixed. One spectrum was collected every 1 min to provide real-time spectral information. Samples for reference assays were collected at 6 time points (0, 4, 8, 13, 18, and 30 min) near the probe in each validation batch to obtain 6 different content levels. Samples collected at the same time point during the 9 repeated batches were considered to have the same content levels. Finally, 54 samples (9 batches × 6 content levels) were included in validation set 2.

2.7. Multivariate Data Treatment

Partial least square regression (PLSR) was used to build the prediction models based on the calibration set. Appropriate pretreatments of the spectra were selected via a "trial-and-error" approach. To reduce the computational complexity, an experimental design was first set up to determine the optimal directions. The parameters for the improved algorithms were adjusted to select the best pretreatments.

Model validation consisted of traditional chemometric validation and validation using accuracy profiles. Conventional chemometric parameters were first used to obtain the global predictive capability of the established models based on validation sets 1 and 2. Then, the accuracy profiles computed based on validation set 2 were used for model assessment and thorough validation. Conventional statistical parameters, such as R, RMSEC, RMSECV, RMSEP, and the relative standard error of prediction (RSEP), were calculated to evaluate the model performance.

In validation set 2, for each content level, the average content values of the 9 samples were used as the true reference values. Due to process variation, it was not possible to obtain exactly the same contents when we repeated the 9 validation batches. Therefore, all prediction values were normalized in Equation (1) when calculating the accuracy profile for each content level [13].

$$y_{ij \text{ nor}} = \frac{\widehat{y}_{ij}}{y_{ij}} \cdot \overline{y}_j \quad (1)$$

In the ith validation batch, for a sample of the jth content level, $y_{ij \text{ norm}}$ is the normalized NIR predicted value, y_{ij} is the reference value, $\widehat{y}_{ij}$ is the NIR predicted value, and $\overline{y}_j$ is the true reference value of the jth content level.

The TQ analyst software package (Thermo Fisher scientific, Madison, WI, USA) and ChemDataSolution chemometrics software (Dalian ChemDataSolution Information Technology Co. Ltd., Dalian, China) were used for spectral data treatments. The accuracy profiles were computed with e.noval V3.0b demo (Arlenda, Liège, Belgium).

3. Results and Discussion

3.1. The Second Alcohol Precipitation Process and NIR Spectra

Due to the addition of ethanol, the system changed as alcohol-insoluble impurities emerged. The NIR spectra obtained under the main operating conditions are shown in Figure 2. The peak at 6900 cm^{-1} is attributed to water and ethanol. The peaks at 5200–5500 cm^{-1} and 8500 cm^{-1} appeared after ethanol addition, which correspond to the 1st and 2nd overtones of C-H stretching in ethanol. [22] The system was clear before the critical point when the alcohol-insoluble impurities emerged. The whole system became turbid, and the spectra shifted after the critical point (at approximately 12 min). During the standing period (20–30 min) after the completion of ethanol addition, the impurities gathered and began to precipitate. The upper part of the system was clear again, and the precipitate was in the bottom part. The corresponding NIR spectra for this period became smooth, and the spectral shifts decreased. Therefore, the NIR spectra reflected the changing process state. The regions of 5000–5095 cm^{-1} and 5300–10,000 cm^{-1} were selected for model construction after removing the saturated absorption regions.

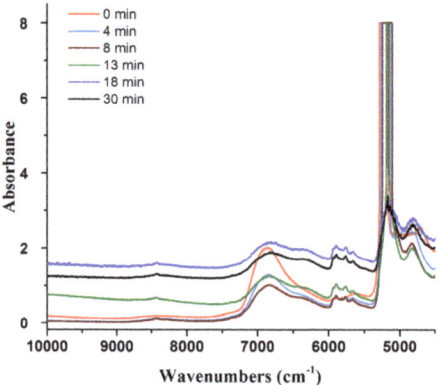

Figure 2. The NIR spectra during the process under normal operating conditions.

3.2. Spectral Pretreatments

Different spectral pretreatments were employed to remove unwanted artifacts caused by small particles and other interference factors during the process. The typical methods used to preprocess the raw spectral data include: baseline correction, scatter correction, smoothing, and scaling [23,24]. Table 2 lists some pretreatment algorithms that are often used. Appropriate combinations of different pretreatments were selected via a "trial-and-error" approach; however, they may be computationally intensive. To reduce the number of calculations, DOE was used to determine the optimization directions. The four pretreatment steps in Table 2 were used as factors, and the full factorial design was applied to obtain 48 combinations of different pretreatments. The Norris–Williams derivation was only used to process the first or second derivation spectra. The combinations were used to process the spectra, and 48 PLSR models were obtained. The models with better performance were selected compared to the model constructed from the raw data without any pretreatments. Using the TS contents as an example, the potential optimal combinations are listed in Table 3.

Next, predictive performance and model complexity were used as evaluation indices to further optimize the parameters of the pretreatment algorithms shown in Table 3. As a result, there were a total of 80 models constructed, as included in Figure 3. Take the red triangle as reference, which represents the model built from raw data. The performances of the models in the lower left part were improved by the pretreatments. Finally, the best pretreatments that yielded a simple model with the best predictive performance were chosen. For the seven analytes, the best pretreatment combinations are listed in Table 4.

Table 2. Overview of some commonly used algorithms for each pretreatment method.

Baseline Correction	Scatter Correction	Smoothing	Scaling
-	-	-	Mean centering
1st D [a]	MSC [c]	SG [e]	Auto scaling
2nd D [b]	SNV [d]	NW [f]	

[a] First derivation (1st D). [b] Second derivation (2nd D). [c] Multiplicative scatter correction (MSC). [d] Standard normal variate (SNV). [e] Savitzky–Golay polynomial derivative filter (SG) using a 9-point window and a third-order polynomial as the initial default parameters. [f] Norris–Williams derivation (NW) using 5-point smoothing and a gap size of 5 as the initial default parameters.

Table 3. Potential optimal pretreatment combinations for the TS content models.

Baseline Correction	Scatter Correction	Smoothing	Scaling	LVs [a]	Calibration Set		Validation Set 1	
					R_C	RMSEC	R_P	RMSEP
-	-	-	Auto scaling	6	0.9783	2.28	0.9954	1.07
-	MSC	-	Auto scaling	6	0.9692	2.71	0.9954	1.03
-	MSC	SG	Auto scaling	5	0.9622	3.00	0.9939	1.18
1st D	-	NW	Auto scaling	3	0.9651	2.88	0.9950	1.06
1st D	-	SG	Auto scaling	7	0.9861	1.83	0.9950	1.23
2nd D	-	NW	Auto scaling	4	0.9768	2.36	0.9963	0.83

[a] The number of latent variables (LVs) reveals the complexity of the models.

Table 4. Performance parameters of the seven calibration models.

Analytes	Pretreatment Combinations	LVs	Calibration Set		Cross-Validation		Validation Set 1	
			R_C	RMSEC	R_{CV}	RMSECV	R_P	RMSEP
TS	NW [a] + 2nd D + auto scaling	3	0.9711	2.63%	0.9602	3.11%	0.9974	0.74%
CG	SG [b] + 1st D + auto scaling	3	0.9614	0.169 mg/mL	0.9504	0.192 mg/mL	0.9963	0.0460 mg/mL
FG	NW [a] + 2nd D + auto scaling	3	0.971	0.0316 mg/mL	0.9549	0.0395 mg/mL	0.9924	0.0142 mg/mL
DPGP	NW [a] + 2nd D + auto scaling	3	0.9691	0.0629 mg/mL	0.9522	0.0784 mg/mL	0.9967	0.0197 mg/mL
DDIFGP	NW [c] + 1st D + auto scaling	3	0.9621	0.0256 mg/mL	0.9517	0.0289 mg/mL	0.9941	0.00865 mg/mL
AG II	SG [e] + mean centering	6	0.9675	0.0601 mg/mL	0.9618	0.0652 mg/mL	0.9656	0.0841 mg/mL
AG IV	NW [d] + 1st D + auto scaling	3	0.9416	0.0590 mg/mL	0.9268	0.0659 mg/mL	0.9762	0.0325 mg/mL

[a] NW using a 7-point smoothing and a gap size of 7. [b] SG using an 11-point window and a second-order polynomial. [c] NW using a 3-point smoothing and a gap size of 5. [d] NW using a 5-point smoothing and a gap size of 3. [e] SG using a 9-point window and a third-order polynomial.

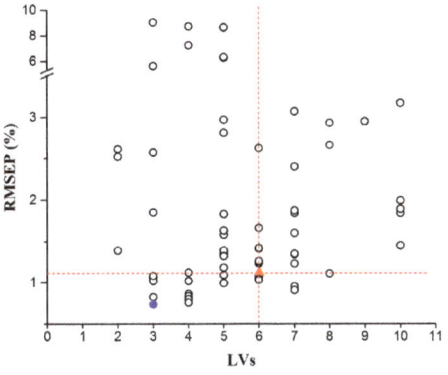

Figure 3. Overview of the TS content models for all pretreatment combinations. (The red triangle represents the model built from raw data, and the blue dot represents the final model built from the best pretreatments).

3.3. Development of Calibration Models

A four-fold cross-validation was used to choose the number of LVs for each PLSR model. The quantitative models were constructed for NIR spectra and the seven analytes. The conventional statistical parameters, such as R of calibration (R_C), R of cross-validation (R_{CV}), R of prediction (R_P), RMSEC, RMSECV, and RMSEP, were calculated to evaluate the model performance. The detailed performance parameters of the seven models are summarized in Table 4. The obtained models have a small number of LVs, which limit the risk of overfitting. Promising results in terms of high correlation coefficients and low prediction errors were obtained. Figure 4 shows the correlation plots of reference values versus the NIR predictions for the seven analytes. The plots present the fitting and predictive ability of the seven models for the entire content range. The models demonstrate good global predictive performance for the seven analytes of the target process samples in the external validation set 1.

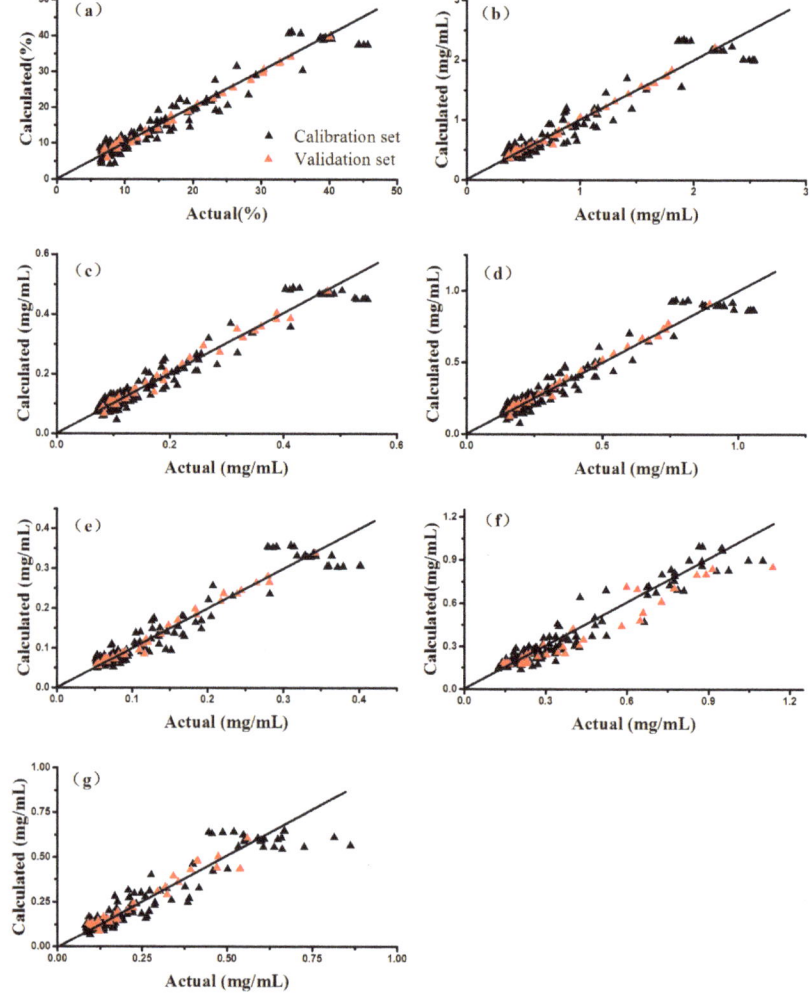

Figure 4. Correlation plots for reference values versus NIR predictions for the PLSR models of TS (**a**), CG (**b**), FG (**c**), DPGP (**d**), DDIFGP (**e**), AG II (**f**), and AG IV (**g**).

3.4. In-Line Monitoring of the Second Alcohol Precipitation Process and Chemometric Validation

The established calibration models were used to predict the contents of the seven analytes during the second alcohol precipitation process under normal operating conditions. Figure 5 shows the in-line monitoring results of nine alcohol precipitation batches. To further evaluate the model performance, 54 samples in validation set 2 were collected for reference assays from the nine batches. The newly added parameter of RSEP together with RMSEP and R were used to evaluate the different aspects of the model quality. Table 5 shows the chemometric validation parameters of the quantitative models for the seven analytes. The results demonstrate that the models still present a good overall predictive performance for the samples in the independent new batches. Compared with the parameters in Table 4, the difference in RMSEP values is low. In addition, the first four models for TS, CG, FG, and DPGP perform better than the models for DDIFGP, AG II, and AG IV regarding the RSEP values.

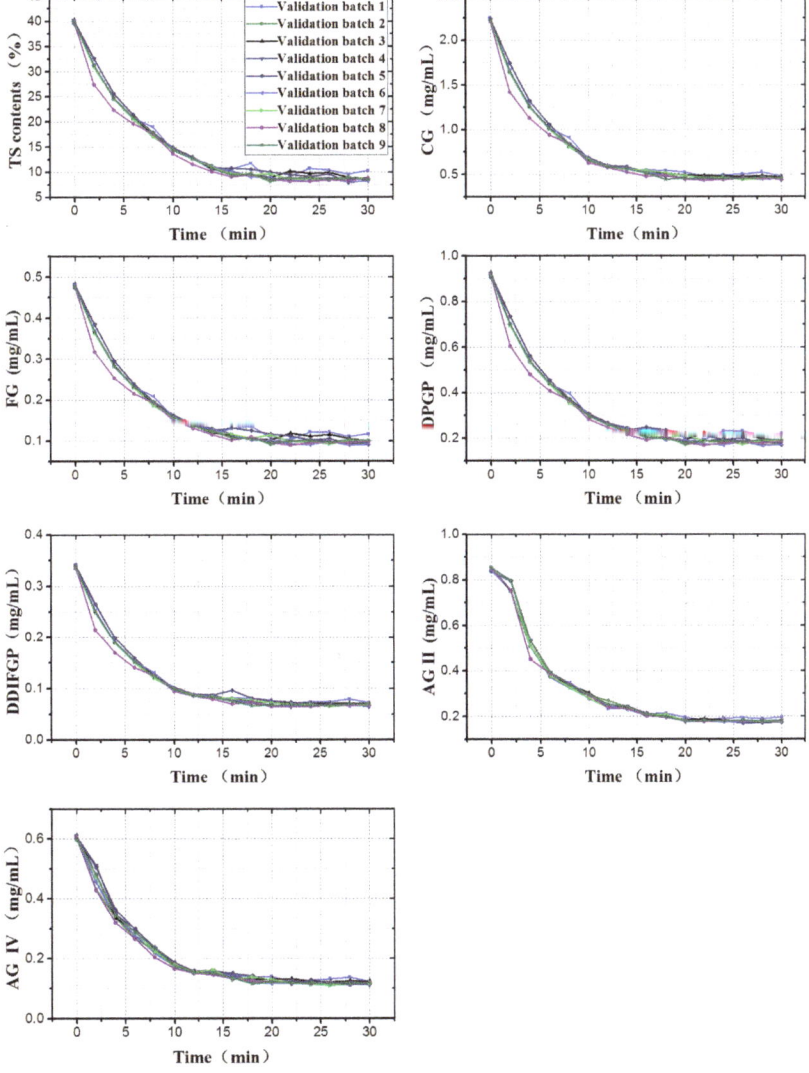

Figure 5. In-line monitoring results of 9 validation batches of the second alcohol precipitation.

Table 5. Chemometric validation parameters for the 7 quantitative models.

Analytes	Validation Set 2		
	R	RMSEP	RSEP
TS	0.9988	0.53%	2.41%
CG	0.9977	0.0437 mg/mL	3.73%
FG	0.9961	0.0126 mg/mL	4.79%
DPGP	0.9972	0.0199 mg/mL	4.10%
DDIFGP	0.9748	0.0253 mg/mL	13.4%
AG II	0.9331	0.0766 mg/mL	15.6%
AG IV	0.9864	0.0256 mg/mL	8.58%

3.5. Validation Based on Accuracy Profiles

The accuracy profile approach using the concept of total error (bias and standard deviation) is fully compliant with the ICH Q2(R1) requirements in which different validation characteristics should be considered. The following sections discuss in detail the validation criteria for the established models.

3.5.1. Trueness

Trueness represents the systematic error in the measurements. It refers to the closeness in agreement between the average of the measured results and the accepted reference value [25,26]. Trueness is generally expressed in terms of relative bias and recovery.

The calculated results of the six different content levels for each analyte are listed in Tables 6 and 7. All seven models exhibited a higher relative bias as a function of increasing content. The relative bias became much higher for the lower content range of 0.246–0.386 mg/mL for the AG II model, and for the other six models shown in Table 8, all values were within 12.5%. Most values fell within 5%.

Table 6. Validation criteria of trueness, precision, and accuracy of the models for the 7 analytes.

Analytes	Content Level	Trueness		Precision		Accuracy
		Relative Bias (%)	Relative Bias (%)	Repeatability (RSD%)	Intermediate Precision (RSD%)	Relative β-Expectation Tolerance Limits (%)
TS (%)	39.8	0.43	100.4	0.34	0.92	[−3.40, 4.25]
	25.0	−1.66	98.34	0.81	0.81	[−3.64, 0.33]
	17.9	−0.84	99.16	3.2	3.2	[−8.65, 6.97]
	12.7	−6.17	93.83	2.2	2.5	[−12.86, 0.53]
	9.4	1.65	101.7	5.4	5.4	[−11.58, 14.88]
	8.4	3.81	103.8	2.1	3.2	[−9.04, 10.99]
CG (mg/mL)	2.26	−1.47	98.53	2.2	2.8	[−9.63, 6.68]
	1.27	−0.85	99.15	1.4	1.4	[−4.29, 2.59]
	0.853	−2.18	97.82	3.9	4.0	[−12.02, 7.67]
	0.591	−1.37	98.63	1.4	2.7	[−11.31, 8.58]
	0.461	8.95	109.0	5.9	6.3	[−7.32, 25.23]
	0.408	12.31	112.3	2.2	3.2	[2.35, 22.27]
FG (mg/mL)	0.502	−4.51	95.49	0.86	1.5	[−9.57, 0.55]
	0.291	−3.03	96.97	1.5	1.5	[−6.60, 0.53]
	0.195	−0.37	99.63	3.6	3.9	[−10.33, 9.58]
	0.137	−3.18	96.82	1.9	2.0	[−8.26, 1.91]
	0.104	6.16	106.2	7.7	7.7	[−12.61, 24.92]
	0.091	8.59	108.6	7.7	8.9	[−14.86, 32.04]
DPGP (mg/mL)	0.939	−2.27	97.73	2.1	2.1	[−7.29, 2.74]
	0.525	2.36	102.4	1.7	2.4	[−5.09, 9.80]
	0.360	1.93	101.9	3.8	4.0	[−8.09, 11.96]
	0.246	−0.46	99.54	2.1	2.1	[−5.68, 4.75]
	0.189	8.55	108.6	7.5	7.5	[−9.75, 26.85]
	0.167	11.14	111.1	8.6	10	[−15.94, 38.22]

Table 7. Validation criteria of trueness, precision, and accuracy of the models for the 7 analytes.

Analytes	Content Level	Trueness		Precision		Accuracy
		Relative Bias (%)	Relative Bias (%)	Repeatability (RSD%)	Intermediate Precision (RSD%)	Relative β-Expectation Tolerance limits (%)
DDIFGP (mg/mL)	0.384	−10.55	89.45	9.2	9.2	[−33.10, 12.01]
	0.188	1.42	101.42	2.4	4.3	[−14.01, 16.85]
	0.121	2.51	102.5	3.0	3.3	[−5.83, 10.86]
	0.0847	3.28	103.3	3.6	5.6	[−14.69, 21.24]
	0.0701	7.53	107.5	8.9	12	[−29.41, 44.48]
	0.0633	8.01	108.0	4.2	6.6	[−13.49, 29.51]
AG II (mg/mL)	0.879	−3.15	96.85	2.0	3.4	[−14.60, 8.29]
	0.509	0.17	100.2	2.2	2.8	[−7.75, 8.08]
	0.346	−2.77	97.23	2.3	2.3	[−8.47, 2.92]
	0.386	−38.53	61.47	1.8	1.8	[−42.95, 34.11]
	0.284	−29.14	70.86	2.8	2.8	[−35.98, −22.30]
	0.246	−26.16	73.84	2.1	2.8	[−34.13, −18.19]
AG IV (mg/mL)	0.556	9.53	109.5	4.7	4.7	[−1.867, 20.93]
	0.336	1.72	101.7	3.4	4.9	[−13.44, 16.88]
	0.235	−2.14	97.9	2.3	2.7	[−9.11, 4.84]
	0.167	−6.37	93.6	2.8	2.8	[−13.21, 0.48]
	0.124	4.62	104.6	6.7	6.7	[−11.75, 20.99]
	0.106	12.08	112.1	4.0	8.3	[−19.33, 43.49]

Table 8. The valid range for each analyte and its proportion over the studied content range.

Analytes	LLOQ–ULOQ	Proportion (%)
TS	8.44–39.8%	100
CG	0.541–2.26 mg/mL	93.1
FG	0.118–0.502 mg/mL	93.5
DPGP	0.220–0.940 mg/mL	93.3
DDIFGP	0.106–0.167 mg/mL	18.9
AG II	0.484–0.879 mg/mL	62.4
AG IV	0.137–0.320 mg/mL	40.8

3.5.2. Precision

Precision represents the random error in the measurements. It refers to the closeness in agreement between a series of measurements of the same homogeneous sample obtained under various conditions [25]. Precision is evaluated at two levels: repeatability and intermediate precision, and the results are listed in Tables 6 and 7. The relative standard deviation (RSD%) shows good precision at high content levels for the seven analytes, whereas at low content levels, some random errors were observed. Of the seven models, the AG II model was the most precise model.

3.5.3. Accuracy

Accuracy expresses the closeness in agreement between a single measured result and the accepted reference value [25], and it represents the total measurement error, which is the sum of the trueness and precision. The accuracy at different content levels for the seven analytes calculated at the relative 95% β-expectation tolerance limits are shown in Tables 6 and 7. The obtained intervals produced error ranges that suggested that the future NIR predicted results will fall within a 95% probability. The accuracy profile was built by integrating the total error and the calculated tolerance limits at each content level in one plot, as shown in Figure 6. These profiles constitute a visual decision tool when compared with the predefined acceptance limits. The acceptance limits for the in-line determination of the seven analytes were fixed at 15%, and the NIR quantitative method was considered to be valid when the relative errors of the predicted values were within 15% of the studied content range. As shown in Figure 6, for the first four analytes, the relative β-expectation tolerance limits for most content levels were included within the acceptance limit of ±15%. For the last three analytes, the accuracy did not fulfill the acceptance limits for some content levels, especially for lower levels.

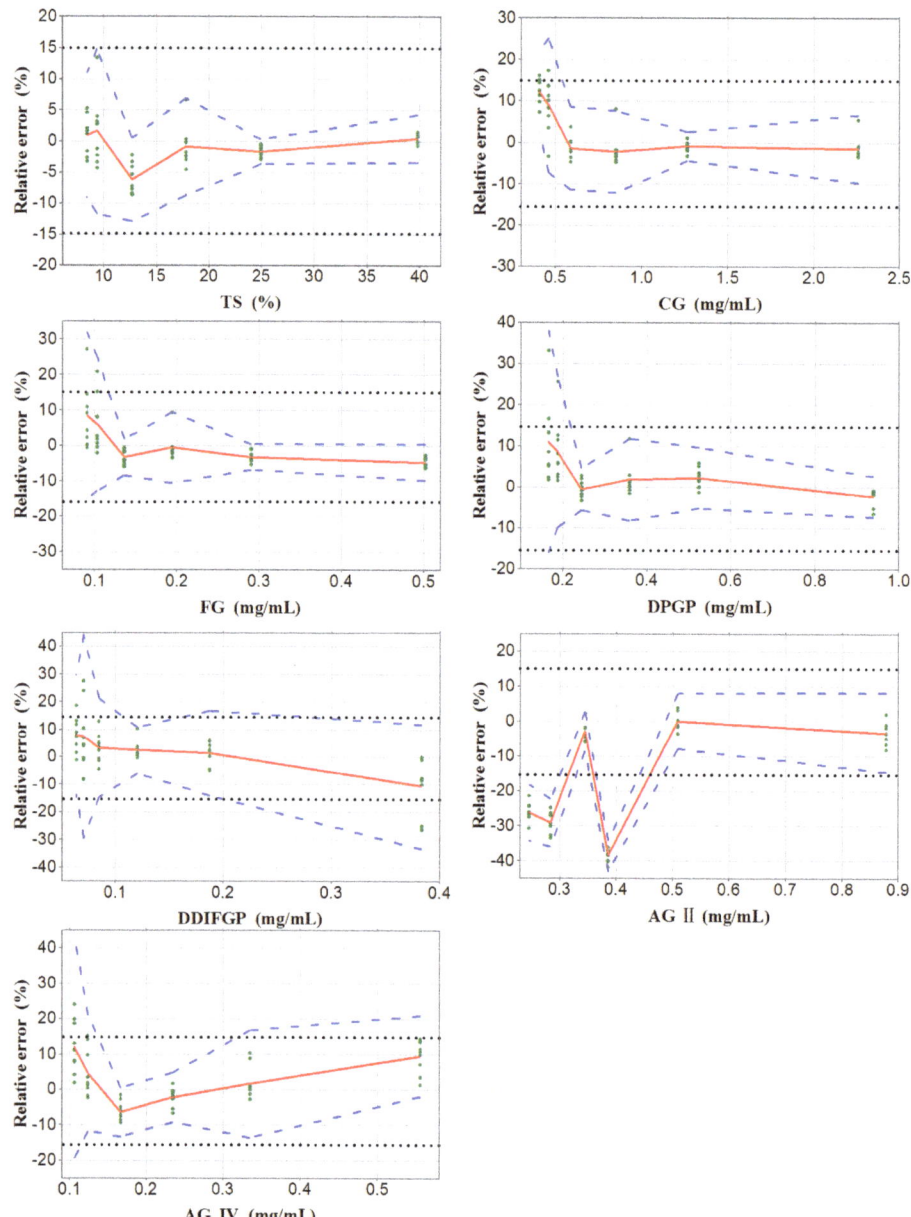

Figure 6. Accuracy profiles for the 7 PLSR models. (The plain red line represents the relative bias, the dashed blue lines represent the β-expectations tolerance limits (β = 95%), and the dotted black lines represent the acceptance limits (±15%)).

3.5.4. Linearity

The linearity of an analytical procedure is its ability within a definite range to obtain results that are directly proportional to the amount of the analyte in the sample [24]. Figure 7 presents the linear profiles of the seven models with R^2 values and the linear equations. The R^2 values are larger than 0.95, which indicate the overall high linearity of the models. The intercepts in the equations are close to 0, confirming the absence of a constant systematic error, and all slopes are close to 1 in the equations except the slope for DDIFGP, which indicates a certain proportional systematic error in that quantification model. However, the method can be considered linear within the content range where the β-expectation tolerance limits are within the absolute acceptance limits.

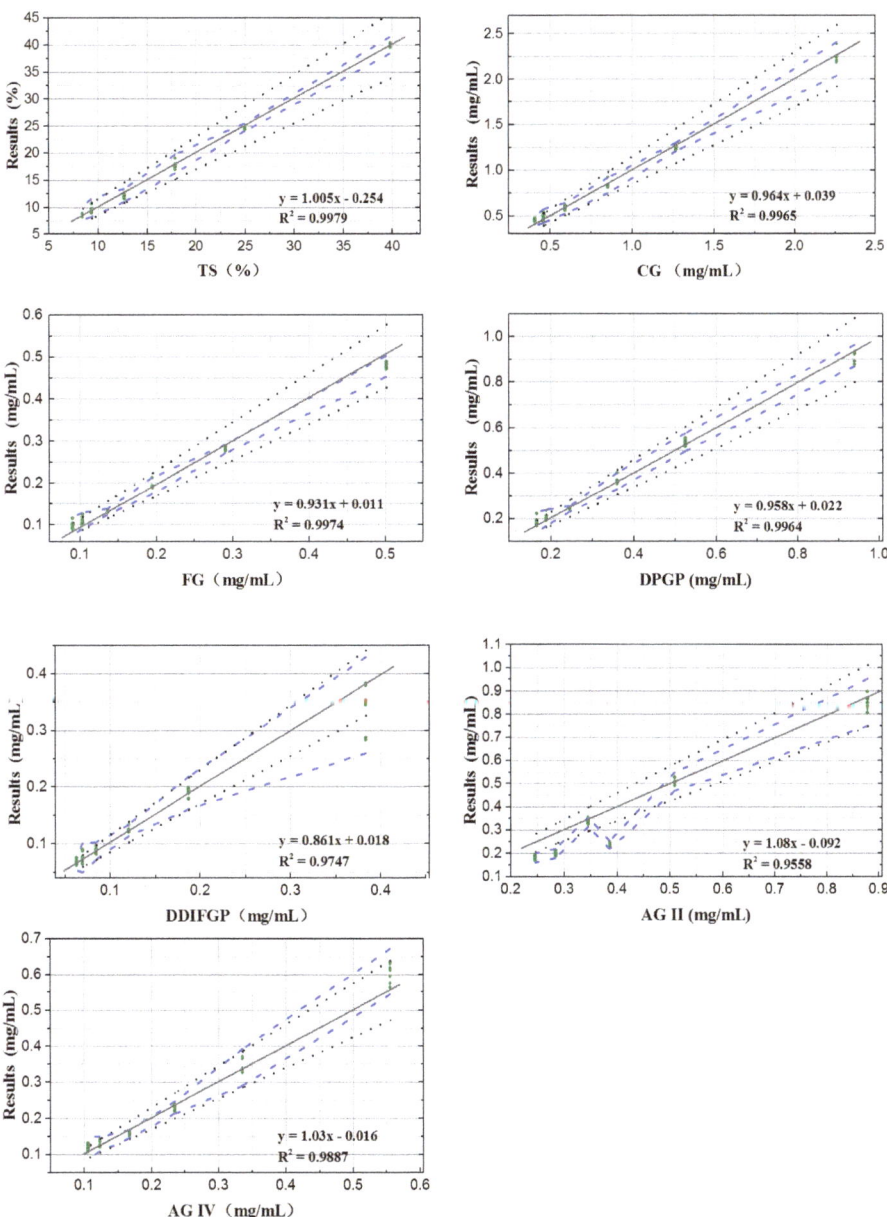

Figure 7. Linear profiles for the 7 PLSR models. (The dashed blue lines on this graph correspond to the accuracy profiles, i.e., the β-expectations tolerance limits expressed in absolute values. The dotted black lines represent the acceptance limits at ±15% expressed in concentration units. The solid line is the identity line for y = x).

3.5.5. Range

The range of an analytical procedure is the interval between the upper and lower amounts of analyte in a sample for which it has been demonstrated that the analytical procedure has a suitable level of precision, accuracy, and linearity [26]. According to Figure 6, the range between the lower and upper limits of quantification can be obtained where the relative β-expectation tolerance limits are included within the acceptance limits. The range for each quantification model is listed in Table 8. For the first four analytes, the established NIR models were able to predict accurate results over more than 90% of the content range for the second alcohol precipitation process. However, for the last three models, accurate results could be guaranteed within the higher content ranges for the process. Of the seven analytes, the content of DDIFGP was the lowest in the samples, which may be close to the sensitivity limit of the NIR spectrometer, which may be prone to large prediction errors. For AG II and AG IV, as the HPLC-ELSD method was applied as the reference quantitative method, which is less accurate than the HPLC-UV method, the prediction models yielded narrower valid ranges than the three models for CG, FG, and DPGP. Moreover, the complexity of the changing system during the in-line analysis and the existence of the alcohol-insoluble impurities made it more difficult to obtain more accurate models.

3.5.6. Robustness

The robustness of an analytical procedure is a measure of its capacity to remain unaffected by small but deliberate variations in the method parameters during normal usage [26]. At the stage of method development, the expected variability was built into the calibration set by DOE, and the obtained models had a small number of LVs, which characterized their robustness. At the stage of validation, stable RMSEP values calculated from two independent validation sets were obtained for the seven models, which indicated the robustness. To further evaluate method robustness, critical process parameters, and temperature were deliberately altered. As listed in Table 1, batch 23 and 24 were selected, and seven samples were randomly collected from each batch. The content results of the seven analytes in the samples were compared and are shown in Figure 8, which were obtained via the off-line reference methods and in-line NIR analysis. The established NIR spectroscopy method retained good in-process performance even with deliberate process variations.

3.5.7. Specificity

The specificity refers to the ability to unequivocally assess an analyte in the presence of components that are expected to be present [26]. The NIR spectrum is characterized by wide and overlapping absorption bands, and it is quite difficult to assign a value to a specific chemical component because of the complexity of TCMs. The specificity of the models was demonstrated by the variance in the reference data that was covered by the LVs [27]. Three LVs were used for quantitative models of TS, CG, FG, DPGP, DDIFGP, and AG IV, explaining 94.3%, 92.9%, 94.3%, 93.9%, 92.6%, and 88.7% of the total variance in the data, respectively. Six LVs were used for quantitative models of AG II, explaining 93.6% of the total variance in the data. This result indicated that each model contained enough content information of the target analyte and demonstrated the specificity [28].

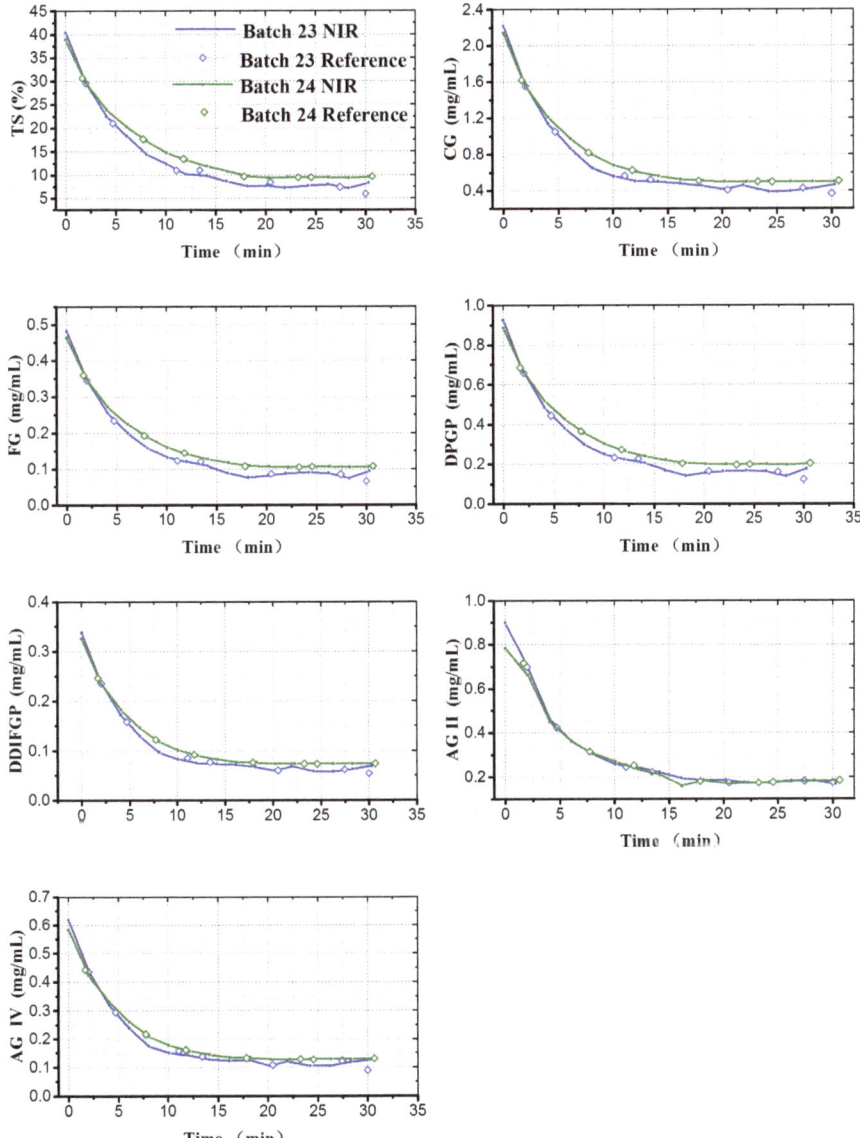

Figure 8. Robustness of the in-line NIR spectroscopy method. (In-line monitoring results are shown as the lines, where the results of the randomly collected 7 samples were obtained via off-line reference methods).

3.6. Method Uncertainty Assessment

The uncertainty characterizes the dispersion of the values that can reasonably be attributed to the measurement [29]. The uncertainties in the bias of the NIR spectroscopy method at each content level for the seven analytes are displayed in Tables 9 and 10. The expanded uncertainty refers to an interval around the results where an unknown true value can be observed with a confidence level of 95% [30], and the relative expanded uncertainties obtained by dividing the expanded uncertainties with the corresponding true reference content values were not higher than 11.5% over the entire TC validated range, which means that at a confidence level of 95%, the unknown true value is located at a maximum of ±

11.5% around the NIR result. For the other six analytes, the relative expanded uncertainties did not exceed 8.5%, 8.6%, 8.6%, 13.0%, 11.0%, or 7.6% over the respective valid ranges between the lower and upper limits of quantification.

Table 9. NIR method uncertainties for the 7 analytes.

Analytes	Content Level	Uncertainty	Expanded Uncertainty	Relative Expanded Uncertainty (%)
TS (%)	39.8	0.42	0.83	2.1
	25.0	0.21	0.43	1.7
	17.9	0.6	1.2	6.7
	12.7	0.35	0.70	5.5
	9.4	0.54	1.1	11.4
	8.4	0.30	0.60	7.1
CG (mg/mL)	2.26	0.071	0.14	6.3
	1.27	0.019	0.038	3.0
	0.853	0.036	0.072	8.4
	0.591	0.018	0.036	6.2
	0.461	0.031	0.063	13.6
	0.408	0.015	0.029	7.1
FG (mg/mL)	0.502	0.0083	0.017	3.32
	0.291	0.0045	0.0089	3.1
	0.195	0.0082	0.016	8.4
	0.137	0.003	0.0059	4.3
	0.104	0.0084	0.017	16.2
	0.091	0.0087	0.017	19.2
DPGP (mg/mL)	0.939	0.020	0.041	4.3
	0.525	0.014	0.028	5.4
	0.360	0.015	0.031	8.52
	0.246	0.0055	0.011	4.5
	0.189	0.015	0.030	15.8
	0.167	0.018	0.037	21.9

Table 10. NIR method uncertainties for the 7 analytes.

Analytes	Content Level	Uncertainty	Expanded Uncertainty	Relative Expanded Uncertainty (%)
DDIFGP (mg/mL)	0.384	0.037	0.075	19.4
	0.188	0.0092	0.018	9.8
	0.121	0.0042	0.0085	7.0
	0.0847	0.0053	0.011	12.4
	0.0701	0.0096	0.019	27.3
	0.0633	0.0047	0.0093	14.7
AG II (mg/mL)	0.879	0.034	0.067	7.6
	0.509	0.016	0.031	6.1
	0.346	0.0085	0.017	4.9
	0.386	0.0074	0.015	3.8
	0.284	0.0084	0.017	5.9
	0.246	0.0075	0.015	6.1
AG IV (mg/mL)	0.556	0.027	0.055	9.8
	0.336	0.018	0.037	11.0
	0.235	0.0067	0.013	5.7
	0.167	0.0049	0.010	5.9
	0.124	0.0087	0.017	14.1
	0.106	0.010	0.020	18.8

4. Conclusions

This study explored an in-line NIR spectroscopy method for multicomponent quantification during the second alcohol precipitation process of *Astragali radix*.

At the stage of method development, a calibration set that encompassed enough variation was built, and models were optimized using the DOEs, reducing the calculation times via a trial-and error approach. Finally, the established PLS models had a small number of LVs, and promising results in terms of high correlation coefficients and low prediction errors were obtained.

At the validation stage, traditional chemometric validation and the accurate profile approach were compared. The general good predictive capability of the seven models was demonstrated using the conventional statistical parameters, and the models were further validated using accuracy profiles. According to the predefined acceptance limits, the accuracy profiles produced a reliable representation of the future performances of the NIR spectroscopy method. A visual decision tool to select valid content ranges showed the following results: 8.44–39.8%, 0.541–2.26 mg/mL, 0.118–0.502 mg/mL, 0.220–0.940 mg/mL, 0.106–0.167 mg/mL, 0.484–0.879 mg/mL, and 0.137–0.320 mg/mL for TS, CG, FG, DPGP, DDIFGP, AG II, and AG IV, respectively. Generally, the developed NIR spectroscopy method can be applied for in-line prediction of the early and middle stage of the second alcohol precipitation process. Additionally, the validation results demonstrated acceptable trueness, precision, accuracy, linearity, specificity, and robustness over the ranges, which were in compliance with the ICH Q2(R1) guideline.

Author Contributions: Conceptualization, W.L., Y.L. and X.W.; methodology, Y.L. and X.W.; software, X.W. and Y.L.; validation, Y.L., X.W. and X.G.; formal analysis, W.L., X.G. and H.Q.; investigation, W.L. and H.Q.; resources, W.H. and G.W.; data curation, W.L., Y.L. and X.W.; writing—original draft preparation, W.L., Y.L. and H.Q.; writing—review and editing, W.L., Y.L., W.H., G.W. and H.Q.; visualization, W.L., X.G. and W.H.; supervision, W.L. and H.Q.; project administration, W.H., G.W. and H.Q.; funding acquisition, W.H. and G.W. All authors have read and agreed to the published version of the manuscript.

Funding: This research was funded by [Xingchu Gong] the National Project for Standardization of Chinese Materia Medica (ZYBZH-C-GD-04), and [Wenlong Li] the Key Project from the National Project for Standardization of Chinese Materia Medica (ZYBZH-C-JIN-43).

Data Availability Statement: Data are contained within the article.

Conflicts of Interest: The authors declare no conflict of interest.

References

1. Li, S.Y.; Wang, D.; Li, X.R.; Qin, X.M.; Du, Y.G.; Li, K. Identification and activity evaluation of Astragalus Radix from different germplasm resources based on specific oligosaccharide fragments. *Chin. Herb. Med.* **2021**, *13*, 33–42. [CrossRef] [PubMed]
2. Cao, S.N.; Hou, G.J.; Meng, Y.; Chen, Y.; Xie, L.; Shi, B. Network Pharmacology and Molecular Docking-Based Investigation of Potential Targets of Astragalus membranaceus and Angelica sinensis Compound Acting on Spinal Cord Injury. *Dis. Mark.* **2022**, *2022*, 2141882. [CrossRef] [PubMed]
3. Gong, F.Y.; Qu, R.M.; Li, Y.C.; Lv, Y.; Dai, J. Astragalus Mongholicus: A review of its anti-fibrosis properties. *Front. Pharm.* **2022**, *13*, 976561. [CrossRef] [PubMed]
4. Gong, X.C.; Wang, S.; Li, Y.; Qu, H.B. Separation characteristics of ethanol precipitation for the purification of the water extract of medicinal plants. *Sep. Purif. Technol.* **2013**, *107*, 273–280. [CrossRef]
5. Liang, X.L.; Ji, M.M.; Chen, L.; Liao, Y.; Kong, X.Q.; Xu, X.Q.; Liao, Z.G.; Wilson, W.D. Traditional Chinese herbal medicine Astragalus Radix and its effects on intestinal absorption of aconite alkaloids in rats. *Chin. Herb. Med.* **2021**, *13*, 235–242. [CrossRef] [PubMed]
6. Zheng, H.; Dong, Z.; Shr, Q. *Modern Study of Traditional Chinese Medicine*; Xue Yuan Press: Beijing, China, 1998; Volume 4, p. 3886.
7. European Medicines Agency (EMA). *Guideline on the Use of Near Infrared Spectroscopy by the Pharmaceutical Industry and the Data Requirements for New Submissions and Variations*; EMA: Amsterdam, The Netherlands, 2014.
8. Engel, J.; Gerretzen, J.; Szymanska, E.; Jansen, J.J.; Downey, G.; Blanchet, L.; Buydens, L.M.C. Breaking with trends in pre-processing? *TrAC Trends Anal. Chem.* **2013**, *50*, 96–106. [CrossRef]
9. *International Conference on Harmonisation (ICH) of Technical Requirements for Registration of Pharmaceuticals for Human Use, Tpoic Q2 (A), Test on Validation of Analytical Procedures*; ICH: Geneva, Switzerland, 1994.
10. Bleye, C.; Chavez, P.F.; Mantanus, J.; Marini, R.; Hubert, P.; Rozet, E.; Ziemons, E. Critical review of near-infrared spectroscopic methods validations in pharmaceutical applications. *J. Pharm. Biomed.* **2012**, *69*, 125–132. [CrossRef] [PubMed]

11. Kauppinen, A.; Toiviainen, M.; Lehtonen, M.; Jarvinen, K.; Paaso, J.; Juuti, M.; Ketolainen, J. Validation of a multipoint near-infrared spectroscopy method for in-line moisture content analysis during freeze-drying. *J. Pharm. Biomed.* **2014**, *95*, 229–237. [CrossRef] [PubMed]
12. Schaefer, C.; Clicq, D.; Lecomte, C.; Merschaert, A.; Norrant, E.; Fotiad, F. A Process Analytical Technology (PAT) approach to control a new API manufacturing process: Development, validation and implementation. *Talanta* **2014**, *120*, 114–125. [CrossRef] [PubMed]
13. Fonteyne, M.; Arruabarrena, J.; Beer, J.; Hellings, M.; Kerkhof, T.V.; Burggraeve, A.; Vervaet, C.; Remon, J.P.; Beer, T. NIR spectroscopic method for the in-line moisture assessment during drying in a six-segmented fluid bed dryer of a continuous tablet production line: Validation of quantifying abilities and uncertainty assessment. *J. Pharm. Biomed.* **2014**, *100*, 21–27. [CrossRef] [PubMed]
14. Sun, M.F.; Yang, J.Y.; Cao, W.; Shao, J.Y.; Wang, G.X.; Qu, H.B.; Huang, W.H.; Gong, X.C. Critical process parameter identification of manufacturing processes of *Astragali radix* extract with a weighted determination coefficient method. *Chin. Herb. Med.* **2020**, *12*, 125–132. [CrossRef] [PubMed]
15. Hubert, P.; Nguyen-Huu, J.J.; Boulangerc, B.; Chapuzet, E.; Chiap, P.; Cohen, N.; Compagnon, P.A.; Dewe, W.; Feinberg, M.; Lallier, M.; et al. Harmonization of strategies for the validation of quantitative analytical procedures—A SFSTP proposal—Part II. *J. Pharm. Biomed.* **2007**, *45*, 70–81. [CrossRef] [PubMed]
16. Hubert, P.; Nguyen-Huu, J.J.; Boulangerc, B.; Chapuzet, E.; Cohen, N.; Compagnon, P.A.; Dewe, W.; Feinberg, M.; Laurentie, M.; Mercier, N.; et al. Harmonization of strategies for the validation of quantitative analytical procedures—A SFSTP proposal—Part III. *J. Pharm. Biomed.* **2007**, *45*, 82–96. [CrossRef] [PubMed]
17. Hubert, P.; Nguyen-Huu, J.J.; Boulanger, B.; Chapuzet, E.; Cohen, N.; Compagnon, P.A.; Dewe, W.; Feinberg, M.; Laurentie, M.; Mercier, N.; et al. Harmonization of strategies for the validation of quantitative analytical procedures: A SFSTP proposal Part IV. Examples of application. *J. Pharm. Biomed.* **2008**, *48*, 760–771. [CrossRef]
18. Wu, Z.; Ma, Q.; Lin, Z.; Peng, Y.; Ai, L.; Shi, X.; Qiao, Y. A novel model selection strategy using total error concept. *Talanta* **2013**, *107*, 248–254. [CrossRef]
19. Wu, Z.; Xu, B.; Du, M.; Sui, C.; Shi, X.; Qiao, Y. Validation of a NIR quantification method for the determination of chlorogenic acid in *Lonicera japonica* solution in ethanol precipitation process. *J. Pharm. Biomed.* **2012**, *62*, 1–6. [CrossRef] [PubMed]
20. Xue, Z.; Xu, B.; Yang, C.; Cui, X.; Li, J.; Shi, X.; Qiao, Y. Method validation for the analysis of licorice acid in the blending process by near infrared diffuse reflectance spectroscopy. *Anal. Meth.* **2015**, *7*, 5830–5837. [CrossRef]
21. Luo, Y.; Li, W.; Huang, W.; Liu, X.; Song, Y.; Qu, H. Simultaneous assay of six components in water extract and alcohol precipitation liquid in *Astragali radix* by HPLC-UV-ELSD. *J. Chin. Mater. Med.* **2016**, *41*, 850–858.
22. Xu, Z.; Huang, W.; Gong, X.; Ye, T.; Qu, H.; Song, Y.; Liu, D.; Wang, G. Design space approach to optimize first ethanol precipitation process of Dangshen. *J. Chin. Mater. Med.* **2015**, *40*, 4411–4416.
23. Wang, K.Y.; Bian, X.H.; Tan, X.Y.; Wang, H.T.; Li, Y.K. A new ensemble modeling method for multivariate calibration of near infrared spectra. *Anal. Meth.* **2021**, *13*, 1374–1380. [CrossRef]
24. Gerretzen, J.; Szymanska, E.; Jansen, J.J.; Bart, J.; Manen, H.; Heuvel, E.R.; Buydens, L.M.C. Simple and Effective Way for Data Preprocessing Selection Based on Design of Experiments. *Anal. Chem. Acta* **2015**, *87*, 12096–12103. [CrossRef] [PubMed]
25. *ISO 5725-1:1994; Accuracy (Trueness and Precision) of Measurement Methods and Results—Part 1: General Principles and Definitions*. International Organization for Standardization: Geneva, Switzerland, 1994.
26. *International Conference on Harmonisation (ICH) of Technical Requirements for Registration of Pharmaceuticals for Human Use, Topic Q2(R1): Validation of Analytical Methods: Text and Methodology*; ICH: Geneva, Switzerland, 2005.
27. The European Agency for the Evaluation of Medicinal Products. *Note for Guideline on the Use of Near-Infrared Spectroscopy by the Pharmaceutical Industry and the Data Requirements for News Submissions and Variations*; European Agency for the Evaluation of Medicinal Products: London, UK, 2003.
28. Li, T.T.; Hu, T.; Nie, L.; Zang, L.X.; Zeng, Y.Z.; Zang, H.C. Rapid monitoring five components of ethanol precipitation process of Shenzhiling oral solution using near infrared spectroscopy. *Chin. J. Trad. Chin. Med.* **2016**, *41*, 3543–3550.
29. *Guide to the Expression of Uncertainty in Measurement*; ISO: Geneva, Switzerland, 1993.
30. Zhang, C.; Liu, F.; Qiu, Z.J.; He, Y. Application of Deep Learning in Food: A review. *Compr. Rev. Food Sci. Food Saf.* **2019**, *18*, 1793–1811.

Article

Inhibitory Effect and Mechanism of Dill Seed Essential Oil on *Neofusicoccum parvum* in Chinese Chestnut

Tian-Tian Liu [1,2], Lin-Jing Gou [1,2], Hong Zeng [3], Gao Zhou [1,2], Wan-Rong Dong [1], Yu Cui [1], Qiang Cai [4,*] and Yu-Xin Chen [1,2,*]

[1] Hubei Key Laboratory of Industrial Microbiology, Key Laboratory of Fermentation Engineering (Ministry of Education), Cooperative Innovation Center of Industrial Fermentation (Ministry of Education & Hubei Province), Hubei University of Technology, Wuhan 430068, China
[2] National "111" Center for Cellular Regulation and Molecular Pharmaceutics, School of Food and Biological Engineering, Hubei University of Technology, Wuhan 430068, China
[3] Xinjiang Production & Construction Corps, Key Laboratory of Protection and Utilization of Biological Resources in Tarim Basin, College of Life Sciences, Tarim University, Alaer 843300, China
[4] Department of Plant Science, School of Life Sciences, Wuhan University, Wuhan 430072, China
* Correspondence: qiang.cai@whu.edu.cn (Q.C.); yuxinc@hbut.edu.cn (Y.-X.C.)

Abstract: The chestnut postharvest pathogen *Neofusicoccum* parvum (*N. parvum*) is an important postharvest pathogen that causes chestnut rot. Chestnut rot in postharvest reduces food quality and causes huge economic losses. This study aimed to evaluate the inhibitory effect of dill seed essential oil (DSEO) on *N. parvum* and its mechanism of action. The chemical characterization of DSEO by gas chromatography/mass spectrometry (GC/MS) showed that the main components of DSEO were apiole, carvone, dihydrocarvone, and limonene. DSEO inhibited the growth of mycelium in a dose-dependent manner. The antifungal effects are associated with destroying the fungal cell wall (cytoskeleton) and cell membrane. In addition, DSEO can induce oxidative damage and intracellular redox imbalance to damage cell function. Transcriptomics analysis showed DSEO treatment induced differently expressed genes most related to replication, transcription, translation, and lipid, DNA metabolic process. Furthermore, in vivo experiments showed that DSEO and DSEO emulsion can inhibit the growth of fungi and prolong the storage period of chestnuts. These results suggest that DSEO can be used as a potential antifungal preservative in food storage.

Keywords: dill; *Anethum graveolens* L.; essential oil; *Neofusicoccum parvum*; antifungal mechanism

1. Introduction

Chestnuts (Castanea) belonging to the family Fagaceae are distributed mainly in eastern Asia, southwestern Asia, southern Europe, and North America [1]. The world's economic cultivation of chestnut plants includes mainly the Chinese chestnut (*Castanea mollissima* Bl.), Japanese chestnut (*C. crenata* Si.), and European chestnut (*C. sativa* Mill.). As the origin of chestnut, China has a history of planting for more than 3000 years [2]. Chestnut is an important food resource, which is rich in nutritional value and has the laudatory name of "king of a thousand fruits". Chestnuts have considerable potential as functional food or food ingredients [1].

Nut rot during storage is a severe problem for chestnuts, which not only causes serious economic losses to the industry but also reduces the supply of nutrients and deteriorates food quality, affecting human health [3]. Italian researchers isolated *Neofusicoccum parvum* (family Botryosphaeriaceae) for the first time from a rot-affected chestnut and showed *N. parvum* caused dark necrosis of the kernel, which is congruent with the nut rot symptoms that occurred in nature [4]. It has been shown that *N. parvum* is one of three major pathogens causing chestnut rot during the chestnut harvest [5]. In addition to chestnut, *N. parvum* has a wide variety of hosts, such as woody plants *Eucalyptus* spp., Pear (*Pyrus* spp.), *Citrus limon*, and grapes (*Vitis vinifera*), and it has become a growing threat to agricultural and forest ecosystems [4,5]. There is currently no fungicide-specific control *N. parvum*-caused disease of plants. Broad-spectrum fungicides, such as carbendazim and flusilazole, have been used globally. However, it is wildly concerned that the long term use of fungicides causes fungal resistance, environmental pollution, and increasing human health risks [6]. It is worth mentioning that rotten nuts may appear healthy on the outside and are hard to be spotted. Thus, it is required to the development of effective strategies against chestnut rot to ensure that consumers buy good quality nuts. Plant essential oils have been widely used in food preservation and storage and are currently considered new and effective antimicrobial agents.

Dill (*Anethum graveolens* L.) is an annual aromatic herb belonging to the family Umbelliferae. Dill is a traditional spice chopped into soups, lettuce salads, and seafood to enhance the flavor of dishes. Dill seeds are also a traditional spice used primarily in the pickling industry due to their strong smell, such as pickled cucumbers, which can be eaten to repel insects [7,8]. Dill seed essential oil (DSEO) is a potent inhibitor of fungal growth, including *Aspergillus flavus*, *Aspergillus niger*, *Fusarium* sp., and *Alternaria alternata* [9,10]. Studies have confirmed that DSEO can be used in the storage and preservation of several foods, such as chickpea food seed, mayonnaise, and corn [9,11,12]. Although extensive investigations have been carried out to study its effective application, there is no consistent finding concerning the DSEO against *N. parvum* in chestnuts. Accordingly, this study aims to determine the inhibitory effect and mechanism of action of DSEO on *N. parvum*. The reliable contact method, liquid shaker method, and gas diffusion method were used to study the inhibitory effect of DSEO on *N. parvum* in vitro. In addition, the effect of DSEO on the fungal cell wall, cell membrane, and oxidative stress were also explored. At the same time, the molecular mechanism of the antifungal activity of DSEO was explored by transcriptomic analysis. Finally, we evaluated the efficacy of DSEO as an antifungal agent in chestnut storage and further expanded the practical application of DSEO in food storage.

2. Materials and Methods

2.1. Plant Materials and Fungal Pathogens

Dill seeds were purchased from the Chinese herbal medicine market in Hotan, Xinjiang Uyghur Autonomous Region of China, and identified by Dr. Gao Zhou. The sample's voucher specimen (No. HM-2020-001) is kept in the Natural Products and Chemical Drug Research and Manufacturing Laboratory, School of Bioengineering and Food, Hubei University of Technology. The mature fruits of chestnut were harvested in Yutoushan Village, Yantianhe Town, Macheng City, Hubei Province. The *N. parvum* used in this study was isolated from rotten chestnut fruit and identified by morphological and molecular biology methods by Dr. Gao Zhou.

2.2. Extraction and Characterization of DSEO

Dried dill seeds (200 g) were smashed, and the EO was obtained via hydrodistillation in 1000 mL H_2O for 5 h using a Clevenger apparatus [7]. After that, the DSEO was kept at 4 °C for further experiments. The chemical characterization of DSEO was performed by gas chromatography/mass spectrometry (GC/MS) [13]. After filtering the sample, 1 µL of the sample was injected into a GC/MS system (Agilent 7890/5975, Agilent Technologies, Santa Clara, CA, USA) and separated on an HP-5ms column (30 m × 0.25 mm × 0.25 µm) (Agilent Technologies). Helium was used as the carrier gas set up at a flow of 1 mL/min. The GC oven temperature increased from 80 °C to 180 °C at a rate of 5 °C/min, and then increased to 260 °C at a 10 °C/min rate. The inlet, interface, ion source, and quadrupole temperatures were set at 250 °C, 250 °C, 230 °C, and 150 °C, respectively. MS data were acquired at a mass range of 20–500 in full scan mode and a solvent delay of 2.5 min. Then, the linear retention index's corresponding relationship was calculated using a mixture of C8-C30 n-alkane (Shanghai Yuanye Biotechnology Co., Ltd., Shanghai, China) according to the Van den Dool and Kratz formula. Essential oil compounds were identified by comparing their mass spectra with those provided by the National Institute of Standards and Technology (NIST17) database.

2.3. Effect of DSEO on Mycelial Growth

Direct contact assays were used to determine the effect of DSEO on the mycelial growth of N. parvum by a previously published method [14]. DSEO was added to PDA to obtain different final concentrations (0.2, 0.4, 0.6, 0.8, and 1 µL/mL). After pouring PDA with or without DSEO into sterilized Petri dishes, the N. parvum mycelia blocks (5 mm in diameter) were introduced at the center of the Petri dishes and incubated at 28 ± 2 °C for 4 days.

To inhibit N. parvum by vapor contact assays, sterile filter papers with different concentrations of DSEO were placed on the medium-free side of Petri dishes to obtain specific concentrations of DSEO in the air. The mycelia blocks were sealed with parafilm on one side of the medium to prevent vapor leakage from DSEO [15]. The formula for calculating the concentration of DSEO in Petri dishes is as follows:

$$C = V1/V2.$$

V1: Volume of DSEO in filter paper; V2: Volume of Petri dishes.

2.4. Effect of DSEO on Fungal Biomass

Different concentrations of DSEO solution (0, 0.2, 0.4, 0.6, 0.8, 1 µL/mL) were aseptically obtained by diluting DSEO dissolved in 0.1% Tween 20 in a final volume of 25 mL of potato dextrose broth (PDB). Then, the N. parvum mycelial block was added to each DSEO solution and incubated in a shaker at 200 r/min at 28 ± 2 °C. After culturing for 3 days, the mycelia were filtered, dried, and weighed [16].

2.5. SEM Observation of the Effect of DSEO on Mycelial Morphology

Studies on the microstructure of mycelia were conducted following a previously described procedure [17]. The N. parvum mycelial block was cultured in a PDB medium containing 0.8 µL/mL DSEO (minimum inhibitory concentration) for 3 h. The sample without DSEO treatment acted as a control. All samples were fixed with 2.0% v/v glutaraldehyde for 24 h at 4 °C and then washed with 100 mM phosphate buffer (pH = 7.4). The samples were then dehydrated with different concentrations (30, 50, 70, 80, 90, 100%) of ethanol. Fixed samples were critical points dried under carbon dioxide and sputter-coated with gold. Then, the changes in mycelium were observed under a scanning electron microscope (SEM) (S-3400, Hitachi, Tokyo, Japan) [15].

2.6. Effect of DSEO on the Cell Wall Integrity

According to the research, the N. parvum mycelia suspension was stained with calcofluor white (CFW) [18,19]. After culturing the mycelial suspension for 12 h, DSEO solutions of different concentrations were added to final concentrations of 0.1, 0.2, and 0.4 µL/mL. The culture was continued for 3 h, and 1 mL of mycelial suspension was centrifuged at 6000 rpm for 10 min to remove the culture medium. The collected mycelia samples were stained with 40 µL of CFW and 40 µL KOH (10%). Then, samples were examined using a confocal laser scanning microscope (CLSM) (Leica TCS SP8 CARS). The fungal culture in PDB without DSEO was used as a control.

2.7. Determination of Cellulase Activity

Briefly, the N. parvum mycelial block was cultured in blank PDB for 2 days. DSEO solutions of different concentrations were added to final concentrations of 0.2, 0.4, and 0.8 µL/mL, and the culture was continued for 3 h. The sample without DSEO treatment acted as a control. The cellulase activity in N. parvum was determined using the 3,5-dinitrosalicylic acid (DNS) colorimetric method [20]. The unit of enzyme activity is defined as follows: the amount of enzyme required to catalyze the production of reducing sugar equivalent to 1 μg of glucose per hour is determined as 1 unit of enzyme activity, expressed in U/mL. The calculation formula of cellulase enzyme activity is as follows:

$$X \text{ (U/mL)} = (W \times 1000 \times n)/(C \times 180 \times t)$$

W: glucose content in the sample; n: dilution multiple of the sample; C: sample amount; 180: glucose molecular weight; t: reaction time.

2.8. Effect of DSEO Coating on the Membrane Integrity

Evans blue staining was used to prove the cell membrane damage caused by the DSEO treatments [21]. The sample processing is the same as in Section 2.6. Mycelia treated with or without DSEO were stained with Evans blue for 5 min. The mycelia were washed with phosphate-buffered saline (PBS) to remove excess dye. Finally, the mycelia were observed under a microscope (Olympus CX23, Beijing, China) to monitor the membrane integrity of *N. parvum*.

2.9. Determination of Ergosterol Content

Hu et al. and Kong et al. described the spectrophotometric determination of ergosterol content [22,23]. A fungal culture is the same as in Section 2.7. The mycelium was collected by filtration, washed 3 times with distilled water, and filter paper was used to absorb the excess water. An amount of 0.5 g of mycelium was weighed and added to 5 ml of 25% alcoholic potassium hydroxide solution, vibrated vigorously for 10 min, and then incubated at 85 °C for 4 h. A mixture of sterile distilled water and n-heptane (1:3) was used to extract sterols. After vortexing for 10 min, the mixture was allowed to stand for about half an hour to collect the n-heptane layer, which was scanned between 230 and 300 nm by ultraviolet spectrophotometer to determine the sterol amount. The calculated formulas of the ergosterol amount are as follows:

$$(\%)/\text{dehydroergosterol} = (A230/E \text{ dehydroergosterol})/\text{net wet weight of mycelia}$$

$$(\%)/\text{ergosterol} = (A282/E \text{ ergosterol})/\text{net wet weight of mycelia} - (\%)/\text{dehydroergosterol}$$

The E value refers to the absorbance of a sample at a specific wavelength through a 1 cm optical path. Among these compounds, E dehydroergosterol = 518, E ergosterol = 90.

2.10. Assessment of Oxidative Stress in N. parvum

A fungal culture is the same as in Section 2.7. The mycelia were collected by filtration and rinsed with distilled water. After the filter paper absorbed excess water, the mycelia were homogenized in an ice bath with different solvents. To determine thiobarbituric acid reactive substances (TBARS), the mycelium was homogenized with 10 times the volume of 10% trichloroacetic acid (TCA) on ice with a glass homogenizer. After centrifuging at 10,000 rpm at 4 °C for 10 min, the supernatant was incubated with 0.67% TCA (1:1) at 95 °C for 30 min, then cooled to room temperature. The reaction product was centrifuged, as described above, and the supernatant was detected at 532 and 600 nm. The TBARS content was expressed as a nmol/mg plot [24]. Superoxide dismutase (SOD) activity and reduced glutathione (GSH) levels were measured using specific kits (Nanjing Jiancheng Bioengineering Institute, Nanjing, China).

2.11. RNA-Sequencing

The mycelia of *N. parvum* cultured for 2 days were exposed to 0.8 μL/mL DSEO, and samples were collected at 1, 2, and 3 h for transcriptome sequencing. The obtained fungal pellets were immediately stored in the liquid nitrogen. All experiments were performed in triplicate, and samples treated without DSEO were used as controls. Total RNA was extracted by TRIzol reagent. The cDNA library construction, RNA-Seq, and the following analysis were performed by Beijing Novogene Bio-Information Technology Co., Ltd. (Beijing, China).

An amount of 1 µg of RNA in each sample was used to prepare subsequent RNA samples. Briefly, the RNA was purified and fragmented with divalent cations, followed by synthesis of first-strand cDNA with random hexamer primers and M-MuLV reverse transcriptase. Second-strand cDNA was synthesized with DNA polymerase I and RNaseH, and the rest of the overhangs were converted into blunt ends by exonuclease/polymerase activity. After adenylation of the 3′ ends of DNA fragments, adaptors with hairpin loop structures were attached for hybridization. The library fragments were purified using the AMPure XP system (Beckman Coulter, Brea, CA, USA) to preferentially select cDNA fragments of 370–420 bp in length. Then, PCR products were purified after a polymerase chain reaction (PCR). The library quality was assessed on the Agilent Bioanalyzer 2100 system, and the library preparations were sequenced on an Illumina Novaseq platform. Then, 150 bp paired-end reads were generated.

The reference genome index (http://ftp.ebi.ac.uk/ensemblgenomes/pub/release-51/fungi/fasta/fungi_ascomycota1_collection/neofusicoccum_parvum_ucrnp2_gca_000385595/dna/, accessed on 5 March 2021) was built using HISAT2 v2.0.5 and paired-end clean reads were compared to the reference genome. Fragments per kilobase million (FPKM) were then calculated for each gene based on the length of the gene and the number of reads localized to that gene. Differential expression gene (DEG) analysis of the two conditions/groups was performed using the DESeq2 R package (1.20.0). A P value of 0.05 and an absolute fold change of 1 were set as the thresholds for significant differential expression.

Gene ontology (GO) enrichment analysis of DEGs was implemented by the clusterProfiler R package, correcting gene length bias. The clusterProfiler R package was used to test for statistical enrichment of DEGs in the Kyoto Encyclopedia of Genes and Genomes (KEGG) pathway, to further clarify the advanced functions and connections of biological systems.

2.12. Verification of the Gene Expression Related to Key Pathways

Sixteen genes associated with metabolic processes, cytoskeleton, genetic information processing, and organelles were selected to confirm the RNA-Seq data by quantitative real-time (qRT)-PCR. Total RNA was extracted from fungal mycelia by the TRIzol method, and the concentration of the purified RNA was determined. Then, the first-strand cDNA was synthesized by reverse transcription, based on an RNA PCR Kit (Vazyme Biotech, Nanjing, Jiangsu, China). The amplification program was as follows: one cycle at 95 °C for 5 min and 40 cycles at 95 °C for 10 s and 60 °C for 30 s. Finally, the 2-$\Delta\Delta$Ct method was used to analyze the gene expression data. The primers used for qRT-PCR are listed in Table A1.

2.13. Antifungal Efficacy of DSEO Fumigation In Vivo

The in vivo antifungal efficacy of DSEO was determined according to previous research [11,25]. Cut off the head of the fresh chestnut (the round one with a hard shell) and keep all chestnuts as cross-sectional of the same size as possible. The trimmed chestnuts were disinfected correctly with 75% ethanol and 2% (v/v) sodium hypochlorite solution and left to dry after rinsing with sterile water. A wound with a depth of 2 mm and width of 2 mm was created with a sterile needle, which was inoculated with 10 µL of *N. parvum* mycelial suspension (1 mg/mL), and then left to air-dry. The inoculated fruits were arranged in moistened sealed containers, with 9 fruits per treatment. Filter paper discs were pasted on the bottom of the container with a volume of 600 mL, and DSEO was added to filter paper discs to obtain container volumes of 0.2, 0.4, and 0.8 µL/mL. The containers were sealed and stored at 28 ± 2 °C. The lesion diameter (Mycelia length) was measured by adsorption onto the activated carbon to visualize the mycelium after 3 days.

2.14. The Effect of DSEO Emulsion on the Storage Period of Chestnut

The DSEO emulsion was formulated by mixing the oil and water phases with emulsifiers. DSEO was used as the oil phase, the emulsifiers were 64% Span 80 and 36% Tween 80, and the aqueous phase was deionized water. The mixture was stirred at a speed of 5000 rpm in a magnetic stirrer, and the oil phase was slowly added to the emulsifier. Then, the aqueous phase was added until the solution was completely dissolved and homogeneous. The DSEO concentration was 15% (v/v), while the emulsifier concentration was 5% (v/v). A blank emulsion without DSEO was prepared using only the emulsifier and the aqueous phase. Positive control was prepared by mixing natamycin (NM), Tween 80, deionized water with 2% (v/v) NM, and 1% of Tween 80.

After the DSEO emulsion was stored at room temperature for 30 days, no delamination was observed. The DSEO emulsion was diluted to the required concentration and sprayed on the chestnut, and then the spoilage of the chestnut was recorded every 3 days for a total of 21 days.

2.15. Statistical Analysis

All data are expressed as the means ± standard deviations (SDs). Each treatment consisted of at least three replicates. The results of the in vitro study were analyzed using a one-way analysis of variance (ANOVA) and Duncan's test. Different letters represent significant differences at $p < 0.05$. Statistical significance was set at * $p < 0.05$, ** $p < 0.01$, *** $p < 0.001$.

3. Results

3.1. Yield Rate and Composition of DSEO

The volatile oil extracted from dill seeds was pale yellow with a characteristic aroma, and the yield rate was 1.83 ± 0.16% (v/w). GC/MS analysis of DSEO identified 10 components representing 99.63% of the total oil. The identified compounds are listed in Table 1. The major components detected in the oil were apiole (39.93%), carvone (23.49%), dihydrocarvone (17.90%), and limonene (16.36%).

Table 1. Chemical composition and content of DSEO.

Peak No.	RI [a]	Compound Name	Content (%)
1	1041	Limonene	16.36
2	1218	Dihydrocarvone	17.90
3	1238	Dihydrocarveol	0.26
4	1246	Carveol	0.17
5	1255	Neodihydrocarveol	0.27
6	1272	Carvone	23.49
7	1549	Myristicin	0.32
8	1576	Elemicin	0.20
9	1652	Apiole	39.93
10	2358	Tetracosanal	0.73
All			99.63

[a] RI is the retention index. It is calculated according to the Van den Dool and Kratz formula using a mixture of C8-C30 n-alkanes.

3.2. In Vitro Antifungal Effect of DSEO

The antifungal effects of different concentrations of DSEO on mycelial growth and fungal biomass of *N. parvum* in vitro are shown in Figure 1. The diameter of mycelia decreased significantly with increased DSEO concentration. The growth of mycelia was completely inhibited at 1 µL/mL DSEO treatment (Figure 1A). DSEO fumigation also has an excellent inhibitory effect on mycelial growth, which can completely inhibit the growth of mycelium at a concentration of 0.2 µL/mL (Figure 1B). Figure 1C shows that, compared to the control group, a significant inhibition rate in *N. parvum* mycelia weight was found at 29.69%, 78.13%, 93.59%, 100.00%, and 100.00%, respectively. These results indicate that the antifungal effects of DSEO are influenced by both the DSEO concentration and the treatment strategy.

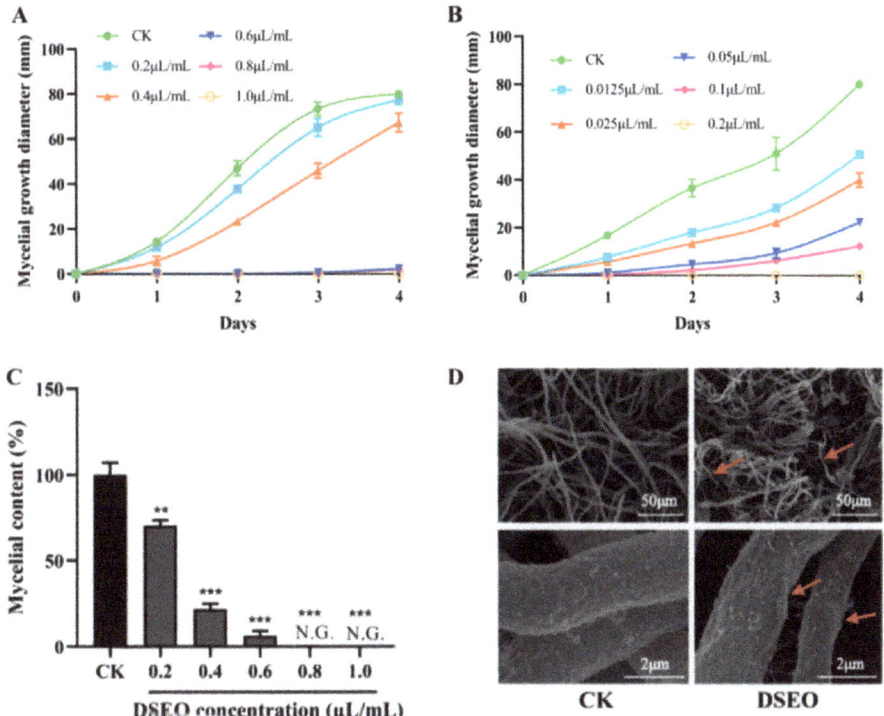

Figure 1. The effect of DSEO on *N. parvum* growth. (**A**) The effect of DSEO on *N. parvum* growth in potato dextrose agar (PDA) medium by direct contact assays; (**B**) The effects of DSEO fumigation on the growth of *N. parvum* in PDA medium by vapor contact assays; (**C**) The effect of DSEO on *N. parvum* growth in potato dextrose broth (PDB) medium, the growth of fungi was shown by the dry weight of mycelium. N.G., No growth. (**D**) Scanning electron microscopy (SEM) images of *N. parvum* with or without DSEO (0.8 μL/mL) treatment. Clear changes were marked with arrows. "CK" represents "control". Statistical significance was set at ** $p < 0.01$, *** $p < 0.001$.

3.3. Effect of DSEO on Mycelial Morphology

The influence of DSEO on the morphology of *N. parvum* was examined using SEM (Figure 1D). The mycelium of the control group had a regular surface morphology, uniform thickness, and smooth cylindrical structure. The mycelia of *N. parvum* treated with DSEO at a concentration of 0.8 μL/mL appeared as multiple folds on the surface, the mycelia were sunken and shriveled, fungal contents leaked, and the thickness of the mycelia was uneven.

3.4. Effect of DSEO on Cell Wall Integrity

Cell wall integrity is critical for fungal growth, an important target for antifungal drugs. The cell wall properties of DSEO-exposed *N. parvum* were examined using the chitin-specific fluorescence dye CFW to analyze the effect of DSEO on the cell wall [26]. As shown in Figure 2A, the mycelia in the control group showed typical blue fluorescence, indicating normal chitin distribution. The mycelia in the DSEO groups showed visibly weaker fluorescence than the control group after 3 h of incubation, indicating that the chitin content was reduced and the cell wall was destroyed. With increasing DSEO concentration, the fluorescence intensity decreased.

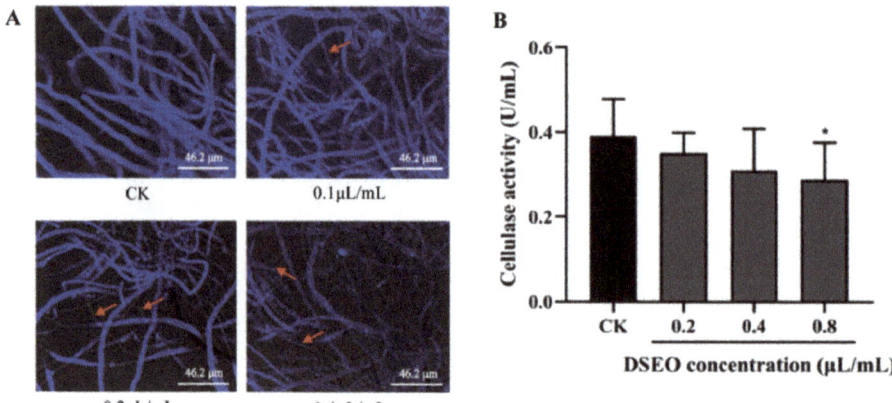

Figure 2. The effects of DSEO on the cell wall of *N. parvum*. (**A**) Merged images under confocal laser scanning microscopy (CLSM) white light and fluorescence after staining with calcium fluorescent white (CFW); Clear changes were marked with arrows. (**B**) The effect of DSEO on the cellulase activity of *N. parvum*. "CK" represents "control". Statistical significance was set at * $p < 0.05$.

To further prove the damage to the fungal cell wall, we measured the activity of cellulase, a cell wall-degrading enzyme [27]. The effect of DSEO on cellulase activity is shown in Figure 2B. Compared with the control group, the cellulase activity did not change significantly when the mycelia were exposed to DSEO at 0.2 µL/mL or 0.4 µL/mL. The cellulase activity decreased significantly when the mycelia were exposed to DSEO at 0.8 µL/mL.

3.5. Effect of DSEO on Plasma Membrane Integrity

The cells were first stained with Evans blue to investigate whether there was a disruption of cell membrane integrity upon exposure to DSEO. As indicated in Figure 3A, when the mycelia were treated with DSEO at a concentration of 0.1 µL/mL, the mycelia were stained blue under a light microscope, suggesting that the cell membranes were compromised after 3 h of treatment with DSEO. With increasing DSEO concentration, the blue color increased significantly, indicating that the plasma membrane was compromised to a greater extent.

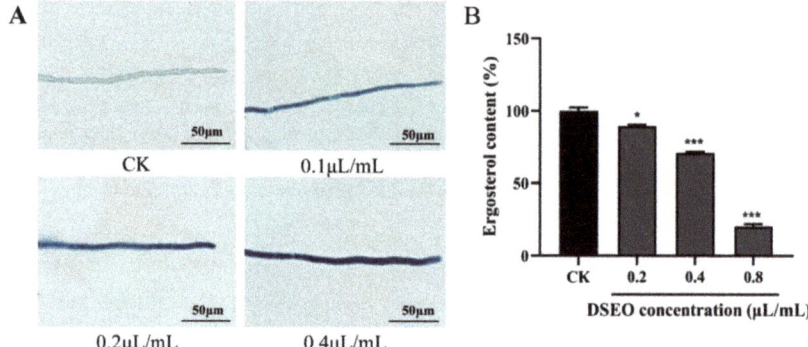

Figure 3. The effects of DSEO on the plasma membrane of *N. parvum*. (**A**) Mycelia were observed under the microscope after staining with Evans blue; (**B**) The effect of DSEO on the content of ergosterol of *N. parvum* after incubation at different concentrations of DSEO for 3 h. "CK" represents "control". Statistical significance was set at * $p < 0.05$, *** $p < 0.001$.

The ergosterol content of *N. parvum* cell membranes was significantly reduced after exposure to DSEO. The inhibition rates of ergosterol biosynthesis in *N. parvum* treated with 0.2 μL/mL, 0.4 μL/mL, and 0.8 μL/mL DSEO were 13.54%, 28.39%, and 71.33%, respectively (Figure 3B).

3.6. Effect of DSEO on the Oxidative Stress Response of N. parvum

As shown in Figure 4A, DSEO induced a pronounced accumulation of TBARS in *N. parvum* at concentrations of 0.2 μL/mL, 0.4 μL/mL, and 0.8 μL/mL, which were 1.50, 2.06, and 2.82 times higher than those in the control group. The results indicated that DSEO could disrupt the fungal cell by oxidative damage.

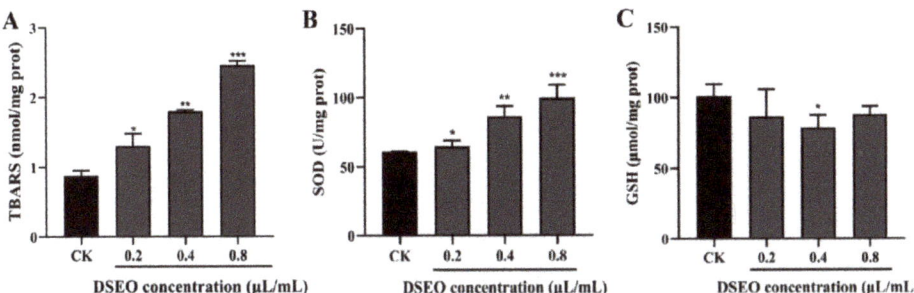

Figure 4. The effects of DSEO on oxidative stress indicators of *N. parvum*, (**A**) MDA levels, (**B**) SOD levels, and (**C**) GSH levels. "CK" represents "control". Statistical significance was set at * $p < 0.05$, ** $p < 0.01$, *** $p < 0.001$.

DSEO exhibited significant enhancement of the cellular antioxidant enzyme SOD, compared to the control. In the control group, the cellular level of SOD was 59.90 units/mg of protein. In contrast, at 0.4 μL/mL and 0.8 μL/mL DSEO treatment, the levels were 86.07 and 98.85 units/mg protein, respectively (Figure 1B).

GSH (γ-glutamylcysteinylglycine) is a tripeptide molecule participating in the nonenzymatic second line of cellular defense and playing an important role in quenching oxyradicals detoxifying xenobiotics. The level of GSH in *N. parvum* decreased after DSEO exposure, and the inhibition rates of DSEO at 0.2 μL/mL and 0.4 μL/mL were 14.21% and 21.99%, respectively. With increasing DSEO concentration, the level of GSH gradually recovered, and the highest concentration of DSEO (0.8 μL/mL) restored the GSH level to 87.19% of the GSH level of the blank group (Figure 4C).

3.7. Transcriptomic Analysis of N. parvum under DSEO Treatment

In this study, 12 samples of *N. parvum* were sequenced by using RNA-seq technology. The principal component analysis (PCA) of variation of the four groups is shown in Figure 5D. Table A2 briefly summarizes the sequencing data information for each sample. A total of 549,876,900 raw reads were preprocessed to obtain 525,036,788 clean reads. High-quality clean data were used to perform subsequent analyses. The Q30 levels were over 92.81%, and the average genome mapping ratio was 94.38%, indicating that the accuracy of the Illumina RNA-Seq data used in the following analysis was reliable.

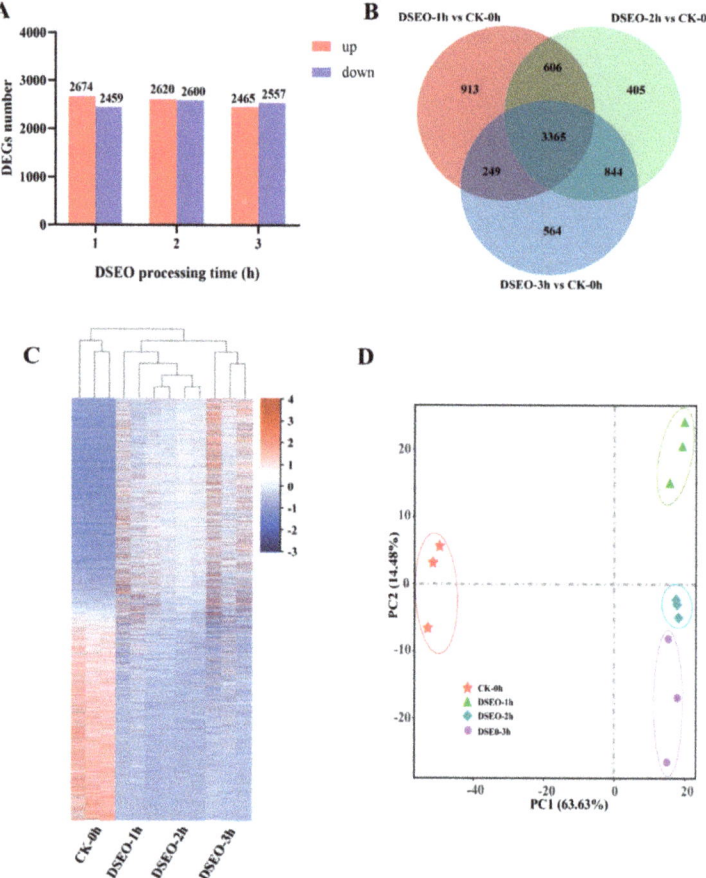

Figure 5. Distribution of DEGs in DSEO-treated and control *N. parvum*. (**A**) Histogram showing the number of DEGs in *N. parvum* after different treatments. (**B**) Venn diagram showing the overlap of DEGs from the 1 h, 2 h, and 3 h datasets. (**C**) Heatmap of DEGs in *N. parvum* before and after DSEO treatment. Each row represents the expression pattern of one gene, and each column corresponds to one sample. (**D**) Principal component analysis (PCA) of transcriptomics data. (n = 3). "CK" represents "control".

Compared with the control group, 5133, 5220, and 5022 genes were found to have changed abundantly after 1 h, 2 h, and 3 h of incubation in the DSEO-treated group (Figure 5A). Based on gene expression analysis, we used Venn diagrams to describe the DEG distribution among several groups (Figure 5B). The overlapping part of the three circles comprised 3365 DEGs. The number of DEGs in DSEO—1 h was the highest, compared with the control group. For a global view of the gene expression profile after exposure to DSEO at different times, a heatmap representing the transcription levels of all DEGs between the four groups was clustered and is shown in Figure 5C; DSEO—1 h and DSEO—2 h clustered into one group, indicating that they have the most similar effects on *N. parvum*. Compared with the control group, the DSEO—1 h group had the most significant number of DEGs, while the DSEO—3 h group had the most significant changes.

All identified DEGs were annotated by gene ontology (GO) analysis and divided into distinct subgroups (Figure 6). GO analysis of DEGs at three treatment timing indicated that the most significant biological process (BP) enrichments were replication, transcription, translation, and lipid, DNA metabolic process. The most significant cellular component

(CC) enrichments occurred in the cytoplasm, intracellular organelle, and cytoskeleton, while molecular function (MF) enrichments occurred in binding and catalytic activity.

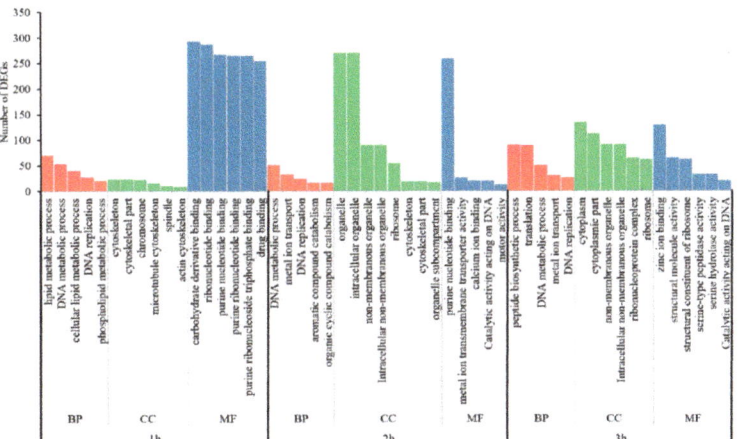

Figure 6. Gene ontology (GO) categorization of DEGs in *N. parvum* after DSEO treatment. GO analysis was performed for three main categories: biological process (BP), cellular component (CC), and molecular function (MF).

Specific biological functions usually result from multiple genes interacting with each other. Based on the Kyoto Encyclopedia of Genes and Genomes (KEGG) pathway database, we aligned the identified DEGs to specific biochemical pathways and outcomes to generate a scatter plot for the top 20 KEGG enrichment results (Figure 7). Pathway classification revealed that most DEGs involved metabolism (such as amino acids, sugars, and lipids) and genetic information processing (such as replication, repair, and recombination). The expression patterns of DEGs were similar at 2 h and 3 h of DSEO treatment. With increasing culture time, the effects of DSEO on the expression of genes involved in the ribosome, secondary metabolite metabolism, and amino acid metabolism were significantly increased.

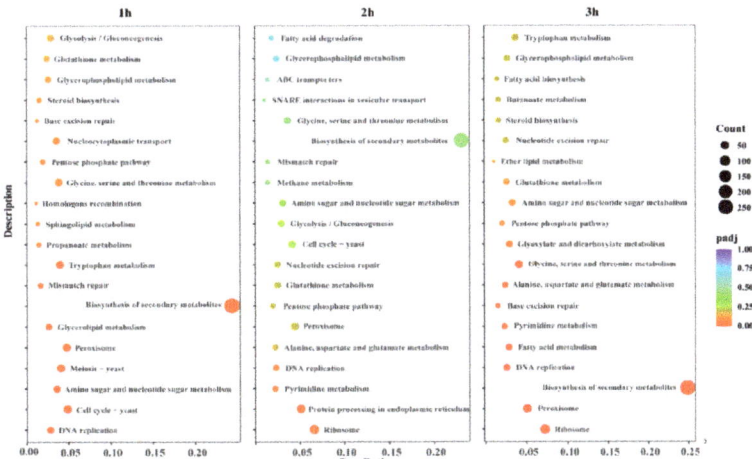

Figure 7. The pathway enrichment statistics of DEGs in *N. parvum* under DSEO stress after 1 h, 2 h, and 3 h of culturing. Gene ratio represents DEGs numbers to all gene numbers annotated in this pathway term. A higher value of the gene ratio means greater intensiveness; padj is a corrected *p*-value ranging from 0 to 1; a lesser *p*-value means greater intensiveness.

3.8. qRT-PCR Validation of Selected DEGs

To verify the reliability of the transcriptome analysis, 16 genes (8 up- and 8 down-regulated) were selected to validate the RNA-seq data by qRT-PCR. As shown in Figure 8, the fold changes in selected DEGs measured by qRT-PCR and RNA-Seq were not precisely consistent. However, they share similar expression profiles, which shows high correlation between RNA-Seq data and transcript abundance detected by qRT-PCR.

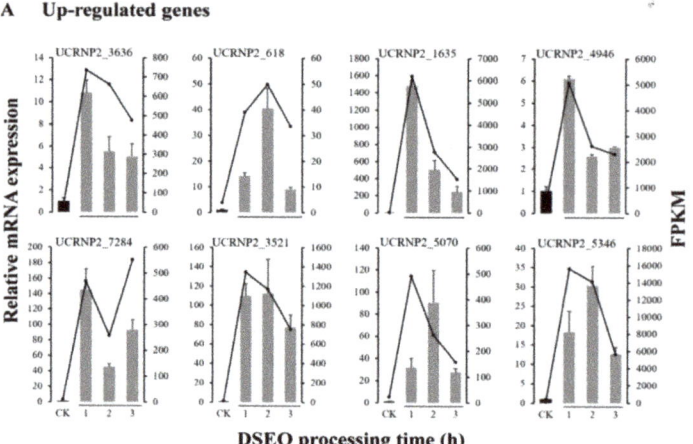

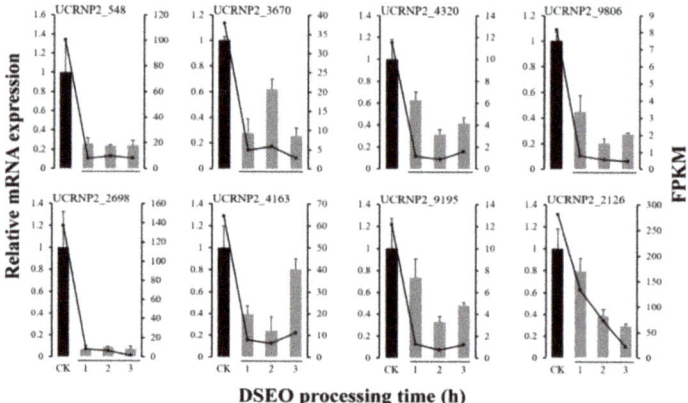

Figure 8. Validation of RNA-Seq data using qRT-PCR data of 16 selected DEGs (8 up- (**A**) and 8 downregulated (**B**)). The central index axis represents relative mRNA expression, and the secondary axis represents the FPKM of genes in transcriptomics. "CK" represents "control".

3.9. Antifungal Efficacy of DSEO Fumigation In Vivo

DSEO has low toxicity and residue advantages, but DSEO has a strong flavor. To avoid affecting the taste of the food, we used the fumigation method to study the antifungal activity of DSEO in vivo. Figure 9 shows the visualized evidence that DSEO could significantly suppress fungal contamination in chestnuts after 3 days of DSEO incubation. With increasing concentrations of DSEO, the protective effect was enhanced. DSEO at a concentration of 0.4 µL/mL significantly inhibited the growth of *N. parvum* in Chinese chestnuts. In comparison, DSEO at a concentration of 0.8 µL/mL almost completely inhibited the growth of *N. parvum* in Chinese chestnuts, ascribed to the gas diffusion capacity of DSEO.

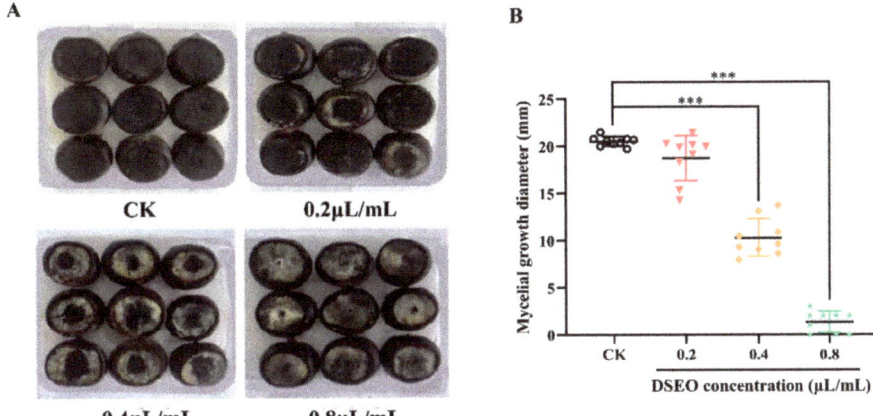

Figure 9. In vivo fumigation antifungal activity of DSEO on the growth of *N. parvum*. (**A**) Representative photos of chestnut samples with or without DSEO treatment during 3 days of incubation and the mycelium length were observed by using the adsorption of activated carbon; (**B**) The effect of DSEO on the growth length of mycelium, n = 9. "CK" represents "control". Statistical significance was set at *** $p < 0.001$.

3.10. DSEO Emulsion Extends the Storage Term of Chestnuts

Although DSEO fumigation can inhibit the growth of pathogens of rot disease in a short time, long-term storage of chestnuts is still a problem and has not yet been solved due to the high volatility of DSEO. To achieve the purpose of long-term storage of chestnut, a DSEO emulsion was employed to prolong the action time of essential oil and improve its bioavailability. The decay rate of chestnuts was expressed by chestnut number and chestnut weight. As shown in Figure 10, both DSEO emulsion concentrations reduced the chestnut decay rate. In the early stage of chestnut storage, the anticorruption effect of the 7.5 μL/mL DSEO emulsion was higher than the anticorruption effect of the 1.5 μL/mL DSEO emulsion, and this significance gradually disappeared with the extension of storage time. Moreover, the DSEO emulsion's effect on prolonging the chestnut storage life is much better than the effect of NM. Therefore, we can choose the concentration of the DSEO emulsion according to the storage time of the Chinese chestnut.

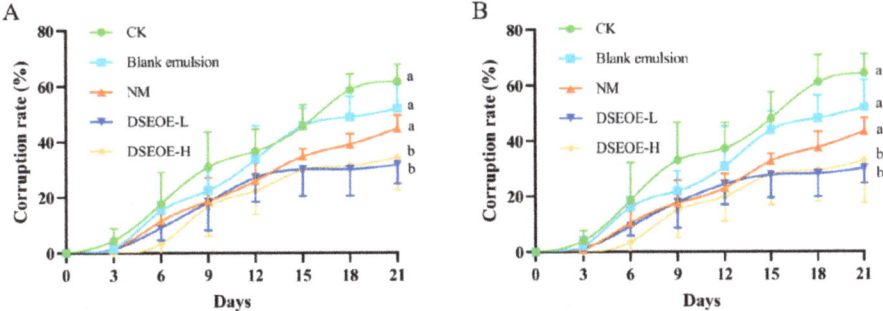

Figure 10. The effect of DSEO emulsion on the storage period of chestnut. (**A**) The number of spoiled chestnuts calculates the spoilage rate of chestnuts; (**B**) The weight of spoiled chestnuts calculates the spoilage rate of chestnuts. NME, 200 μg/mL NM emulsion; DSEOE-L, 1.5 μL/mL DSEO emulsion; DSEOE-H, 7.5 μL/mL DSEO emulsion. Different letters (a and b) indicate a significant difference ($p < 0.05$). "CK" represents "control".

4. Discussion

There has been a growing need for effective and eco-friendly agents to control fungal contamination in food, given the health threat from synthetic chemical fungicides. Essential oil, a natural product from aromatic medicinal plants, was reported to possess antibacterial, antifungal, antioxidant, and anti-inflammatory activities. DSEO is a botanical fungicide in food with broad-spectrum antifungal activity inhibiting pathogens' growth [28,29]. The content of DSEO components varies with geographical conditions, varieties, cultivation patterns, harvest time, extraction methods, etc. [29]. Previous research demonstrated that the main components of DSEO were carvone (41.5%), limonene (32%), and apiole (16.79%), three major compounds [29]. Another study showed that DSEO was primarily composed of carvone (40.36%), limonene (19.31%), and apiole (17.50%) [30]. The main components of our DSEO are apiole (39.93%), carvone (23.49%), dihydrocarvone (17.90%), and limonene (16.36%). Thus, apiole, carvone, and limonene are common major components in the DSEO.

Our study found that the antifungal effects of DSEO against *N. parvum* in vitro could be influenced by the treatment strategy, wherein the growth inhibitory effects of DSEO in gas diffusion were higher than those of liquid culture or that of solid diffusion-induced growth inhibition. In this regard, we investigated the in vivo inhibitory activity of DSEO against *N. parvum* by the fumigation method. Our results showed that DSEO fumigation had a strong protective effect on chestnut storage, which also met the original sensory requirements of foods benefiting from the volatility of DSEO.

There are currently no available reports on the antifungal mechanism of the action of DSEO on *N. parvum*. Our SEM results showed that DSEO treatment resulted in the collapse of the mycelial structure and apparent deformation, indicating loss of cytoplasm and damage to organelles. The fungal cell wall is vital in maintaining cells' inherent morphology and integrity, supporting normal cell metabolism, ion exchange, and osmotic pressure [26]. The experimental data obtained in this work revealed that DSEO treatment significantly reduced the chitin content of the fungal cell wall. This result is consistent with previous studies showing that DSEO exerts antifungal activity by disrupting the cell wall [6,21,31].

In the process of phytopathogenic fungi infecting the host, the pathogenic fungi secrete cellulase to degrade the cell wall of the host plant, which is conducive to the invasion and spread of the pathogenic fungi [20]. Under DSEO treatment, the content of extracellular cellulase in *N. parvum* decreased, and the ability to infect the host was also reduced accordingly. Evans blue staining was employed to investigate whether DSEO affects plasma membrane integrity. Our results showed that the permeability of the cell membrane was compromised by DSEO treatment. Ergosterol is an important component of fungal cell membrane, which can maintain the fluidity and bioregulation of cell membrane [32]. This is also illustrated by the inhibition of ergosterol synthesis by DSEO. Multiple studies have reported that botanical fungicides exert antifungal effects by lowering the membrane ergosterol content [31]. Two major inhibitory mechanisms were explained; one is to disrupt the biosynthesis of ergosterol, and the other is to damage the cell membrane of eukaryotic cells by linking sterols, causing membrane perforation and rupture [19]. Therefore, we reasoned that DSEO might disrupt the cell membrane of eukaryotic cells by linking sterols, affecting its integrity, and, consequently, resulting in macromolecules leakage, thus, we assume that the cell membrane is a target of DSEO against *N. parvum*. Interestingly, similar results have been observed in antifungal studies of other essential oils [16]. Based on their lipophilic character, Eos can penetrate the cell wall through passive diffusion and further destroy the cell membrane [16]. The release of intracellular components leads to cell shrinkage, physiological dysregulation, and even cell death. Blue staining was observed in fungal cells treated with DSEO, confirming cell death [33].

This paper investigated the effect of different concentrations of DSEO on the antioxidant system of fungi. TBARS mainly refers to malondialdehyde (MDA), and MDA is an indicator of cell injury that is generated after exposure to reactive oxygen species (ROS) and is one of the most important biomarkers of lipid peroxidation [34]. The results showed that TBARS levels were significantly increased after DSEO treatment, indicating that DSEO could damage fungal cells through oxidative damage. In addition, antioxidant systems, such as defense-related antioxidant enzymes, might be activated [35]. Our study found that SOD activity in fungi was elevated to dose-dependent after exposure to DSEO, indicating that DSEO activated the antioxidant enzymes of *N. parvum*. The elevation in SOD activity may be an adaptive response that counteracts some of the negative effects of elevated TBARS in DSEO-exposed *N. parvum* [36]. In addition, a decrease in GSH levels was recorded for *N. parvum* exposed to DSEO. Similar supportive observations have been reported earlier in some studies. The reduction in GSH may cause redox imbalance and impair cell function [36,37].

To better understand the interaction between DSEO and *N. parvum*, RNA-seq methods were used to detect DSEO-induced transcriptome changes in *N. parvum*. Transcriptomic analysis revealed that DSEO affects a wide range of cytoskeleton-related genes. Actin forms filaments ("F- actin" or microfilaments) that are the skeleton fibers in eukaryotic cells, which participate in muscle contraction, deformation, and cytoplasmic separation [38]. The cytoskeleton participates in the formation and division of the cell wall and is a part of the cell wall. At the same time, it is also necessary to maintain cell morphology, material exchange, transport, and cell division. This may explain the cell wall changes observed after the DSEO treatment of *N. parvum*. In earlier reports, the cell membrane was identified as an important target for DSEO to exert its antifungal activity [31,39]. After DSEO treatment, the expression levels of DEGs in metabolic pathways related to cell membrane homeostases, such as phospholipid metabolism and lipid metabolism, were significantly changed. At the same time, with increasing DSEO treatment time, the DEGs regulating the cytoplasm and organelles were downregulated considerably, which supported the previous conclusion that DSEO destroyed the cell wall and cell membrane structure and caused the fungal mycelium to collapse and shrink [21,28]. Most fungi's DEGs related to the ribosome and amino acid metabolism were downregulated. Ribosomes are the site of protein synthesis, and amino acid metabolism mainly synthesizes proteins unique to the body [40]. Therefore, the inhibition of protein biosynthesis may partially explain the antifungal activity of DSEO against fungal pathogens.

5. Conclusions

This study investigated the in vitro antifungal effect of DSEO on *N. parvum*. It explored its mode of action while evaluating its potential as a natural preservative in chestnut storage in vivo. Our results showed that DSEO exhibited solid antifungal activity via gas diffusion and was well applied to food storage. In addition, DSEO can inhibit fungal growth and activity by disrupting the cell wall (cytoskeleton) and cell membrane integrity and inducing intracellular redox imbalance and protein biosynthesis. Most importantly, its high antifungal activity against chestnut samples during storage demonstrates that DSEO may be a promising economic antifungal agent in practice.

Author Contributions: Data curation, T.-T.L., L.-J.G., G.Z., W.-R.D. and Y.C.; Formal analysis, T.-T.L.; Funding acquisition, Y.-X.C.; Methodology, Q.C.; Project administration, Y.-X.C.; Resources, H.Z.; Software, T.-T.L.; Supervision, Q.C. and Y.-X.C.; Writing—original draft, T.-T.L.; Writing—review and editing, H.Z. and Q.C. All authors have read and agreed to the published version of the manuscript.

Funding: This work was supported by the National Natural Science Foundation of China (82104536), Hubei Provincial Natural Science Foundation of China (2020CFB197), Collaborative Grant-in-Aid of the HBUT National "111" Center for Cellular Regulation and Molecular Pharmaceutics (XBTK-2020003, XBTK-2022009), Open Grant from Xinjiang Production & Construction Corps Key Laboratory of Protection and Utilization of Biological Resources in Tarim Basin (BRZD2002), and Doctoral Start-up Foundation of Hubei University of Technology (BSQD2020034, BSQD2020040).

Acknowledgments: We thank Zhi-Jie Liu from the Hubei University of Technology for the GC-MS experiment.

Conflicts of Interest: The authors declared that there are no conflict of interest.

Appendix A

Table A1. Primers used for qRT-PCR.

Gene Symbol	Gene Description	Forward Primer	Reverse Primer
UCRNP2_4935 (actin)	Internal reference gene	GGATGTGCAGGTCATCACAC	GACCACCGAGAAGAGCAAAG
UCRNP2_3636	phospholipid metabolic process	GATGCGGGAGTAGCCTGAC	CCCCAACCAAATCGGTAAGC
UCRNP2_7284	lipid metabolic process	AACTGGGATTAGCGGAAGCC	GACCTGATCAGCGAGCCTAC
UCRNP2_618	DNA metabolic process	GGAAGAGGGGCTGCTTTCTT	GTCCGGATTTGGTGGTCCAT
UCRNP2_3521	DNA binding	GGGGTCAATGTGGTCGAAGT	CCGAGAGGCCCTTTCAAACT
UCRNP2_1635	purine ribonucleotide binding	TCGACCGTCTCCTCTTCCTT	ACCGTTGCGTTCGAAGTAGT
UCRNP2_5070	cell cycle-yeast; cytoskeleton	ATGGCTACACCTCATCCACC	CATGAAGGTCGCCTCATTGTC
UCRNP2_4946	ribosome	AGAAGCGCAAGTCAGCTCAT	CAGAGTCACCTGAGAAGGCG
UCRNP2_5346	hypothetical protein	TCCTCCATCCGCTCTTCGTA	GGCGTTTCCGAGAACAACAC
UCRNP2_548	Amino acid and sugar metabolism	TGGTGGTTTGGAACAGCTTC	ATTGACGCCGGCACATCA
UCRNP2_2698	Fatty acid metabolism	GGTCTTGTCCTGGACTTCGG	CTTCGGCTCTAAGCTCCTCG
UCRNP2_3670	cytoskeleton (actin cytoskeleton)	TGGGGGATTACCAGATTGCG	TCCCACAGGTACACGCTAGA
UCRNP2_4163	cytoskeleton (cytoskeleton part)	GGATATCGCAGGGATGGCAA	GGTACGTCAGGCACTGTGAA
UCRNP2_4320	cytoskeleton	AAAACTTGCCGAGGAGAGGG	ATCGTTATCTCTGCCTGCCG
UCRNP2_9195	organelle	ATGGACTTGGACGCCAACAT	CGACCCAATCGGTCAAGCTA
UCRNP2_9806	non-membrane-bounded organelle	GTGGCTGACGAGGAGGAAAA	GTACTCCTCGAAGCCGATGG
UCRNP2_2126	structural constituent of ribosome	GACCAAGTTCAAGGTCCGCT	CCCATGAGTGTCGTGAGAGG

Table A2. Summary of RNA-seq reads in CK-0h and treatments (DSEO—1 h, DSEO—2 h, DSEO—3 h) groups of *N. parvum*.

Sample	Base Number	Clean Reads	Q30 (%)	GC Content (%)	Mapped Reads	Mapped Ratio (%)
CK-0h-1	46714618	45152836	95.04	59.85	42847419	94.89
CK-0h-2	47740962	45556208	92.81	59.4	42973482	94.33
CK-0h-3	46182706	44575174	94.94	59.8	42355868	95.02
DSE0-1h-1	46699708	45752188	92.55	60.11	42753528	93.45
DSE0-1h-2	46274846	42891208	93.24	60.22	40365528	94.11
DSE0-1h-3	41420784	38231250	93.62	60.41	36136015	94.52
DSE0-2h-1	44093534	41476198	93.43	59.82	38946635	93.90
DSE0-2h-2	44771022	41943828	93.31	59.79	39277988	93.64
DSE0-2h-3	45907572	43027896	93.46	60.39	40713819	94.62
DSE0-3h-1	42078438	39608058	93.36	60.42	37505962	94.69
DSE0-3h-2	48817802	47684086	95.13	59.82	44991433	94.35
DSE0-3h-3	46233786	45365574	95.03	59.51	43080355	94.96

References

1. Li, Q.; Shi, X.; Zhao, Q.; Cui, Y.; Ouyang, J.; Xu, F. Effect of cooking methods on nutritional quality and volatile compounds of Chinese chestnut (*Castanea mollissima* Blume). *Food Chem.* **2016**, *201*, 80–86. [CrossRef] [PubMed]
2. LaBonte, N.R.; Zhao, P.; Woeste, K. Signatures of selection in the genomes of Chinese chestnut (*Castanea mollissima* Blume): The roots of nut tree domestication. *Front. Plant Sci.* **2018**, *9*, 810. [CrossRef] [PubMed]
3. Shuttleworth, L.A.; Liew, E.C.Y.; Guest, D.I. Survey of the incidence of chestnut rot in south-eastern Australia. *Australas. Plant Path.* **2013**, *42*, 63–72. [CrossRef]
4. Seddaiu, S.; Mello, A.; Sechi, C.; Cerboneschi, A.; Linaldeddu, B.T. First Report of Neofusicoccum parvum associated with chestnut nut rot in Italy. *Plant Dis.* **2021**, *105*, 3743. [CrossRef] [PubMed]

5. Waqas, M.; Guarnaccia, V.; Spadaro, D. First report of nut rot caused by *Neofusicoccum parvum* on hazelnut (*Corylus avellana*) in Italy. *Plant Dis.* **2021**, *106*, 1987. [CrossRef] [PubMed]
6. Swanson, H.A.; Svenning, J.C.; Saxena, A. History as grounds for interdisciplinarity: Promoting sustainable woodlands via an integrative ecological and socio-cultural perspective. *One Earth* **2021**, *4*, 226237. [CrossRef]
7. Chen, Y.; Zeng, H.; Tian, J.; Ban, X.; Ma, B.; Wang, Y. Antifungal mechanism of essential oil from *Anethum graveolens* seeds against Candida albicans. *J. Med. Microbiol.* **2013**, *62*, 1175–1183. [CrossRef] [PubMed]
8. Jana, S.; Shekhawat, G.S. *Anethum graveolens*: An Indian traditional medicinal herb and spice. *Pharmacogn. Rev.* **2010**, *4*, 179–184. [CrossRef]
9. Teneva, D.; Denkova, Z.; Denkova-Kostova, R.; Goranov, B.; Kostov, G.; Slavchev, A.; Degraeve, P. Biological preservation of mayonnaise with Lactobacillus plantarum LBRZ12, dill, and basil essential oils. *Food Chem.* **2021**, *344*, 128707. [CrossRef]
10. Kaur, V.; Kaur, R.; Bhardwaj, U.; Kaur, H. Antifungal potential of dill (*Anethum graveolens* L.) seed essential oil, its extracts and components against phytopathogenic maize fungi. *J. Essent. Oil Bear. Plants* **2021**, *24*, 1333–1348. [CrossRef]
11. Boukaew, S.; Prasertsan, P.; Sattayasamitsathit, S. Evaluation of antifungal activity of essential oils against aflatoxigenic *Aspergillus flavus* and their allelopathic activity from fumigation to protect maize seeds during storage. *Ind. Crops Prod.* **2017**, *97*, 558–566. [CrossRef]
12. Kumar, N.; Khurana, S.M.; Pandey, V.N. Application of clove and dill oils as an alternative of salphos for chickpea food seed storage. *Sci. Rep.* **2021**, *11*, 10390. [CrossRef]
13. Yang, K.; Liu, A.; Hu, A.; Li, J.; Zen, Z.; Liu, Y.; Li, C. Preparation and characterization of cinnamon essential oil nanocapsules and comparison of volatile components and antibacterial ability of cinnamon essential oil before and after encapsulation. *Food Control* **2021**, *123*, 107783. [CrossRef]
14. Xu, D.; Wei, M.; Peng, S.; Mo, H.; Huang, L.; Yao, L.; Hu, L. Cuminaldehyde in cumin essential oils prevents the growth and aflatoxin B1 biosynthesis of *Aspergillus flavus* in peanuts. *Food Control* **2021**, *125*, 107985. [CrossRef]
15. Ju, J.; Xie, Y.; Yu, H.; Guo, Y.; Cheng, Y.; Zhang, R.; Yao, W. Major components in *Lilac* and *Litsea cubeba* essential oils kill *Penicillium roqueforti* through mitochondrial apoptosis pathway. *Ind. Crops Prod.* **2020**, *149*, 112349. [CrossRef]
16. Sun, Q.; Li, J.; Sun, Y.; Chen, Q.; Zhang, L.; Le, T. The antifungal effects of cinnamaldehyde against *Aspergillus niger* and its application in bread preservation. *Food Chem.* **2020**, *317*, 126405. [CrossRef]
17. Jiang, N.; Wang, L.; Jiang, D.; Wang, M.; Liu, H.; Yu, H.; Yao, W. Transcriptomic analysis of inhibition by eugenol of ochratoxin A biosynthesis and growth of *Aspergillus carbonarius*. *Food Control* **2020**, *135*, 108788. [CrossRef]
18. Long, Y.; Huang, W.; Wang, Q.; Yang, G. Green synthesis of garlic oil nanoemulsion using ultrasonication technique and its mechanism of antifungal action against *Penicillium italicum*. *Ultrason. Sonochem.* **2020**, *64*, 104970. [CrossRef]
19. OuYang, Q.; Okwong, R.O.; Chen, Y.; Tao, N. Synergistic activity of cinnamaldehyde and citronellal against green mold in citrus fruit. *Postharvest Biol. Technol.* **2020**, *162*, 111095. [CrossRef]
20. Khaledi, N.; Taheri, P.; Tarighi, S. Antifungal activity of various essential oils against *R. hizoctonia solani* and *Macrophomina phaseolina* as major bean pathogens. *J. Appl. Microbiol.* **2015**, *118*, 704–717. [CrossRef]
21. Oliveira, R.C.; Carvajal-Moreno, M.; Mercado-Ruaro, P.; Rojo-Callejas, F.; Correa, B. Essential oils trigger an antifungal and anti-aflatoxigenic effect on *Aspergillus flavus* via the induction of apoptosis-like cell death and gene regulation. *Food Control* **2020**, *110*, 107038. [CrossRef]
22. Hu, Y.; Zhang, J.; Kong, W.; Zhao, G.; Yang, M. Mechanisms of antifungal and anti-aflatoxigenic properties of essential oil derived from turmeric (*Curcuma longa* L.) on *Aspergillus flavus*. *Food Chem.* **2017**, *220*, 1–8. [CrossRef]
23. Kong, J.; Zhang, Y.; Ju, J.; Xie, Y.; Guo, Y.; Cheng, Y.; Yao, W. Antifungal effects of thymol and salicylic acid on cell membrane and mitochondria of *Rhizopus stolonifer* and their application in postharvest preservation of tomatoes. *Food Chem.* **2019**, *285*, 380–388. [CrossRef]
24. Yan, J.; Wu, H.; Shi, F.; Wang, H.; Chen, K.; Feng, J.; Jia, W. Antifungal activity screening for mint and thyme essential oils against *Rhizopus stolonifer* and their application in postharvest preservation of strawberry and peach fruits. *J. Appl. Microbiol.* **2021**, *130*, 1993–2007. [CrossRef]
25. Zhao, Y.; Yang, Y.H.; Ye, M.; Wang, K.B.; Fan, L.M.; Su, F.W. Chemical composition and antifungal activity of essential oil from *Origanum vulgare* against *Botrytis cinerea*. *Food Chem.* **2021**, *365*, 130506. [CrossRef]
26. Xu, T.; Cao, L.; Zeng, J.; Franco, C.M.; Yang, Y.; Hu, X.; Zhu, Y. The antifungal action mode of the rice endophyte Streptomyces hygroscopicus OsiSh-2 as a potential biocontrol agent against the rice blast pathogen. *Pestic. Biochem. Phys.* **2019**, *160*, 58–69. [CrossRef]
27. Xu, T.; Li, Y.; Zeng, X.; Yang, X.; Yang, Y.; Yuan, S.; Zhu, Y. Isolation and evaluation of endophytic Streptomyces endus OsiSh-2 with potential application for biocontrol of rice blast disease. *J. Sci. Food Agric.* **2017**, *97*, 1149–1157. [CrossRef]
28. Tian, J.; Ban, X.; Zeng, H.; He, J.; Chen, Y.; Wang, Y. The mechanism of antifungal action of essential oil from dill (*Anethum graveolens* L.) on *Aspergillus flavus*. *PLoS ONE* **2012**, *7*, e30147. [CrossRef]
29. Ma, B.; Ban, X.; Huang, B.; He, J.; Tian, J.; Zeng, H.; Wang, Y. Interference and mechanism of dill seed essential oil and contribution of carvone and limonene in preventing Sclerotinia rot of rapeseed. *PLoS ONE* **2015**, *10*, 131733. [CrossRef]
30. Li, H.; Zhou, W.; Hu, Y.; Mo, H.; Wang, J.; Hu, L. GC-MS analysis of essential oil from *Anethum graveolens* L (dill) seeds extracted by supercritical carbon dioxide. *Trop. J. Pharm. Res.* **2019**, *18*, 1291–1296. [CrossRef]

31. Wei, J.; Bi, Y.; Xue, H.; Wang, Y.; Zong, Y.; Prusky, D. Antifungal activity of cinnamaldehyde against *Fusarium sambucinum* involves inhibition of ergosterol biosynthesis. *J. Appl. Microbiol.* **2020**, *129*, 256–265. [CrossRef] [PubMed]
32. OuYang, Q.; Tao, N.; Jing, G. Transcriptional profiling analysis of *Penicillium digitatum*, the causal agent of citrus green mold, unravels an inhibited ergosterol biosynthesis pathway in response to citral. *BMC Genom.* **2016**, *17*, 599. [CrossRef] [PubMed]
33. Donadu, M.G.; Peralta-Ruiz, Y.; Usai, D.; Maggio, F.; Molina-Hernandez, J.B.; Rizzo, D.; Chaves-Lopez, C. Colombian essential oil of *Ruta graveolens* against nosocomial antifungal resistant *Candida* strains. *J. Fungi* **2021**, *7*, 383. [CrossRef] [PubMed]
34. Dan, W.; Gao, J.; Li, L.; Xu, Y.; Wang, J.; Dai, J. Cellular and non-target metabolomics approaches to understand the antifungal activity of methylaervine against *Fusarium solani*. *Bioorg Med. Chem. Lett.* **2021**, *43*, 128068. [CrossRef] [PubMed]
35. Hua, C.; Kai, K.; Bi, W.; Shi, W.; Liu, Y.; Zhang, D. Curcumin induces oxidative stress in *Botrytis cinerea*, resulting in a reduction in gray mold decay in kiwifruit. *J. Agric. Food Chem.* **2019**, *67*, 7968–7976. [CrossRef] [PubMed]
36. Kiran, S.; Prakash, B. Toxicity and biochemical efficacy of chemically characterized Rosmarinus officinalis essential oil against *Sitophilus oryzae* and *Oryzaephilus surinamensis*. *Ind. Crops Prod.* **2015**, *74*, 817–823. [CrossRef]
37. Kumar, A.; Kujur, A.; Yadav, A.; Pratap, S.; Prakash, B. Optimization and mechanistic investigations on antifungal and aflatoxin B1 inhibitory potential of nanoencapsulated plant-based bioactive compounds. *Ind. Crops Prod.* **2019**, *131*, 213–223. [CrossRef]
38. Pollard, T.D.; Cooper, J.A. Actin, a central player in cell shape and movement. *Science* **2009**, *326*, 1208–1212. [CrossRef]
39. Chen, Y.; Zeng, H.; Tian, J.; Ban, X.; Ma, B.; Wang, Y. Dill (*Anethum graveolens* L.) seed essential oil induces *Candida albicans* apoptosis in a metacaspase-dependent manner. *Fungal Biol.* **2014**, *118*, 394–401. [CrossRef]
40. Batista-Silva, W.; Heinemann, B.; Rugen, N.; Nunes-Nesi, A.; Araújo, W.L.; Braun, H.P.; Hildebrandt, T.M. The role of amino acid metabolism during abiotic stress release. *Plant Cell Environ.* **2019**, *42*, 1630–1644. [CrossRef]

Article

Optimization of Steam Distillation Process and Chemical Constituents of Volatile Oil from *Angelicae sinensis* Radix

Na Wan [1], Jing Lan [2], Zhenfeng Wu [1], Xinying Chen [2], Qin Zheng [1,*] and Xingchu Gong [1,2,3,*]

[1] Key Laboratory of Modern Preparation of TCM, Ministry of Education, Jiangxi University of Chinese Medicine, No. 1688 Meiling Avenue, Xinjian District, Nanchang 330004, China; wanna988@163.com (N.W.); zfwu527@163.com (Z.W.)

[2] Pharmaceutical Informatics Institute, College of Pharmaceutical Science, Zhejiang University, Hangzhou 310058, China; 22119063@zju.edu.cn (J.L.); 3180102503@zju.edu.cn (X.C.)

[3] Jinhua Institute of Zhejiang University, Jinhua 321016, China

* Correspondence: 20060903@jxutcm.edu.cn (Q.Z.); gongxingchu@zju.edu.cn (X.G.)

Citation: Wan, N.; Lan, J.; Wu, Z.; Chen, X.; Zheng, Q.; Gong, X. Optimization of Steam Distillation Process and Chemical Constituents of Volatile Oil from *Angelicae sinensis* Radix. *Separations* **2022**, *9*, 137. https://doi.org/10.3390/separations9060137

Academic Editors: Paraskevas D. Tzanavaras and Stefania Garzoli

Received: 22 April 2022
Accepted: 24 May 2022
Published: 30 May 2022

Publisher's Note: MDPI stays neutral with regard to jurisdictional claims in published maps and institutional affiliations.

Copyright: © 2022 by the authors. Licensee MDPI, Basel, Switzerland. This article is an open access article distributed under the terms and conditions of the Creative Commons Attribution (CC BY) license (https://creativecommons.org/licenses/by/4.0/).

Abstract: In this study, the steam distillation process of volatile oil from *Angelicae sinensis* Radix was optimized according to the concept of quality-by-design. A homemade glass volatile oil extractor was used to achieve better cooling of the volatile oil. First, the soaking time, distillation time, and liquid–material ratio were identified as potential critical process parameters by consulting the literature. Then, the three parameters were investigated by single factor experiments. The volatile oil yield increased with the extension in the distillation time, and first increased and then decreased with the increase in soaking time and liquid–material ratio. The results confirmed that soaking time, distillation time, and liquid–material ratio were all critical process parameters. The kinetics models of volatile oil distillation from *Angelicae sinensis* Radix were established. The diffusion model of spherical particle was found to be the best model and indicated that the major resistance of mass transfer was the diffusion of volatile oil from the inside to the surface of the medicinal herb. Furthermore, the Box–Behnken experimental design was used to study the relationship between the three parameters and volatile oil yield. A second-order polynomial model was established, with R2 exceeding 0.99. The design space of the volatile oil yield was calculated by a probability-based method. In the verification experiments, the average volatile oil yield reached 0.711%. The results showed that the model was accurate and the design space was reliable. In this study, 21 chemical constituents of volatile oil from *Angelicae sinensis* Radix were identified by gas chromatograph-mass spectrometer(GC-MS), accounting for 99.4% of the total volatile oil. It was found that the content of Z-ligustilide was the highest, accounting for 85.4%.

Keywords: Radix *Angelicae sinensis*; volatile oil; Box–Behnken design; steam distillation; extraction kinetics; design space

1. Introduction

Angelicae sinensis Radix is the dry root of *Angelica sinensis* (Oliv.) Diels [1]. It is widely used as a herbal medicine and has the effects of clearing heat and promoting dieresis, invigorating qi, and blood, etc. Volatile oil is an important component of *Angelicae sinensis* Radix [2,3]. The content of neutral oil in the volatile oil of *Angelicae sinensis* Radix is the highest. The volatile oil of *Angelicae sinensis* Radix mainly contains Z-ligustilide and other components [4,5]. Volatile oil is also considered as the main effective component of *Angelicae sinensis* Radix, which has the effects of treating hypertension [6], analgesic and anti-inflammatory [7], and anti-tumor [8]. The preparation process of the *Angelicae sinensis* Radix volatile oil exists in the production of many Chinese medicines such as Danggui Tiaojing granules, Ruhe Sanjie tablets, Compound Herba Leonuri capsules, and Yangxueyin oral liquid, which are all included in the 2020 edition of the Chinese Pharmacopoeia (1st Part).

At present, the reported preparation methods of angelica volatile oil include steam distillation, supercritical CO_2 extraction, organic solvent extraction, microwave assisted extraction, ultrasound assisted extraction, and so on [9]. Microwave assisted extraction has the advantages of rapidity [10], but a high microwave power may lead to a decrease in the yield of essential oil because of the loss of active compounds [9]. Ultrasound assisted extraction has the advantages of less solvent consumption and better efficiency [11], but there are difficulties in scale-up production. In industry, steam distillation is most widely used to obtain angelica volatile oil [4]. Steam distillation has the advantages of simple equipment, low cost, and a safe solvent [12–14]. In the Chinese Pharmacopoeia, there are more than 30 preparations that use the volatile oil of *Angelicae sinensis* Radix, where more than 80% used the steam distillation process. It is also the method used to determine the volatile oil content of drugs in the Chinese Pharmacopoeia.

However, there are some practical problems in the distillation of volatile oil from *Angelicae sinensis* Radix such as low yield, easy emulsification, and long-time consumption, which bring great challenges to the manufacturing processes and the quality control of Chinese medicines containing *Angelicae sinensis* Radix. Therefore, it is necessary to optimize the distillation process of volatile oil from *Angelicae sinensis* Radix.

In recent years, the quality-by-design concept has been widely applied to optimize the pharmaceutical processes of Chinese medicines [15]. Its implementation steps include: defining the critical process parameters (CPPs), establishing quantitative models of CPPs and pharmaceutical process evaluation indices, establishing the design space, determining the quality control strategies, and continuous improvement, etc. [16]. Before building mathematical models, response surface methodology (RSM) is usually used to collect data with small number of experiments [10]. The Box–Behnken design is often used in many studies [17] as it suggests how to select points from a three-level factorial arrangement, which allows for the efficient estimation of the first- and second-order coefficients of the mathematical model [18].

In this work, steam distillation was used to prepare the volatile oil from *Angelicae sinensis* Radix. A volatile oil extractor with enhanced cooling was used to collect more volatile oil. The CPPs of the distillation process were determined, and the distillation kinetic models were established. According to the results, the main mass transfer resistance of the volatile oil distillation was determined. Then, taking the volatile oil yield as the index, the Box–Behnken experimental design was used to study the effects of the distillation time, soaking time, and liquid–material ratio [10]. The design space of the distillation process of volatile oil from *Angelicae sinensis* Radix was obtained with a probability-based method, which was verified by experiments. Finally, the chemical constituents in the volatile oil of *Angelicae sinensis* Radix were determined by GC-MS, and the results were compared with those reported in the literature.

2. Materials and Methods

2.1. Materials, Reagents, and Instruments

Angelicae sinensis Radix was obtained from the Jiangxi Zhangshu Tianqitang Chinese Herbal Pieces Co., Ltd. (batch number: 2009007, origin: Gansu). All id the above medicinals were identified by Professor Ge Fei from the Identification Teaching and Research Section of the Jiangxi University of Chinese Medicine. The moisture content of Angelica sinensis was $10.0 \pm 0.5\%$. Anhydrous sodium sulfate (batch number: 180408) and anhydrous ethanol (batch number: 200707) were purchased from Xilong Scientific Co., Ltd. (Shantou, China). Distilled water was self-made in the laboratory.

2.2. Distillation Method

Gansu Province is a genuine producing area of *Angelicae sinensis* Radix. Therefore, the *Angelicae sinensis* Radix from Gansu Province was used in this work. The steam distillation was used to prepare the volatile oil from *Angelicae sinensis* Radix. Figure 1 shows the essential oil extractor. With the homemade essential oil extractor, better cooling could be achieved. Compared to that of using a conventional extractor, more essential oil can be extracted.

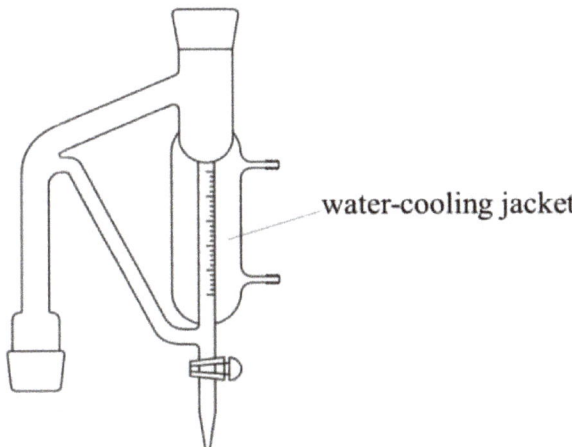

Figure 1. Homemade essential oil extractor.

A proper amount of *Angelicae sinensis* Radix was taken and then crushed with a crusher (FD200T, Shanghai Traditional Chinese Medicine Machinery Factory, Shanghai, China). A total of 100 g of *Angelicae sinensis* Radix powder was taken and added into a 2000 mL flask. Water and several glass beads were added into the flask. After that, the flask was shaken to wet the herbal powders and was connected with a condenser tube. Before the experiment, water was added into the volatile oil extractor. An electric heating jacket (KDM type, Shandong Zhencheng Hualu Electric Equipment Co., Ltd., Qingdao, China) was used to heat the flask. After boiling, the electric heating jacket was adjusted to keep boiling. The heating was stopped after a certain time of distillation. After more than 1 h, the height of the volatile oil was measured with a vernier caliper (0–150 mm, 3V type, Guilin Ganglu Digital Measurement and Control Co., Ltd., Guilin, China). The volume of volatile oil was calculated, then the volatile oil was collected. The obtained volatile oil was dehydrated and dried with anhydrous sodium sulfate overnight to obtain an oil-like product until the amount of volatile oil no longer increased. Thee *Angelicae sinensis* Radix volatile oil was sealed in brown reagent bottles and stored in a refrigerator at 4 °C. The volatile oil yield was calculated using Equation (1).

Volatile oil yield (%) = [volume of oil (mL)/weight of decoction pieces (g)] × 100%   (1)

2.3. Optimization of Distillation Process Parameters

After the literature review, the process parameters of distilling the volatile oil from herbal medicines by steam distillation were found and are shown in an Ishikawa diagram (Figure 2).

According to the pre-experiments, it was considered that the potential CPPs affecting the volatile oil yield by steam distillation were distillation time (A), soaking time (B), and the liquid–material ratio (C). In this study, the above three factors were investigated by single factor experiments. During the investigation, a total of 100 g of *Angelicae sinensis* Radix powder was taken, soaked for a certain time, and then heated to distill the volatile oil. After reaching the distillation time, the heating was stopped. After the volatile oil was cooled, the volume of the volatile oil was accurately read, and the volatile oil yield was calculated. When the distillation time was studied, the liquid–material ratio was 10:1 mL·g^{-1} and the soaking time was 2 h. When the soaking time was studied, the ratio of the liquid to solid was 10:1 mL·g^{-1} and the distillation time was 4 h. When the liquid–material ratio was studied, the distillation time was 4 h and the soaking time was 2 h.

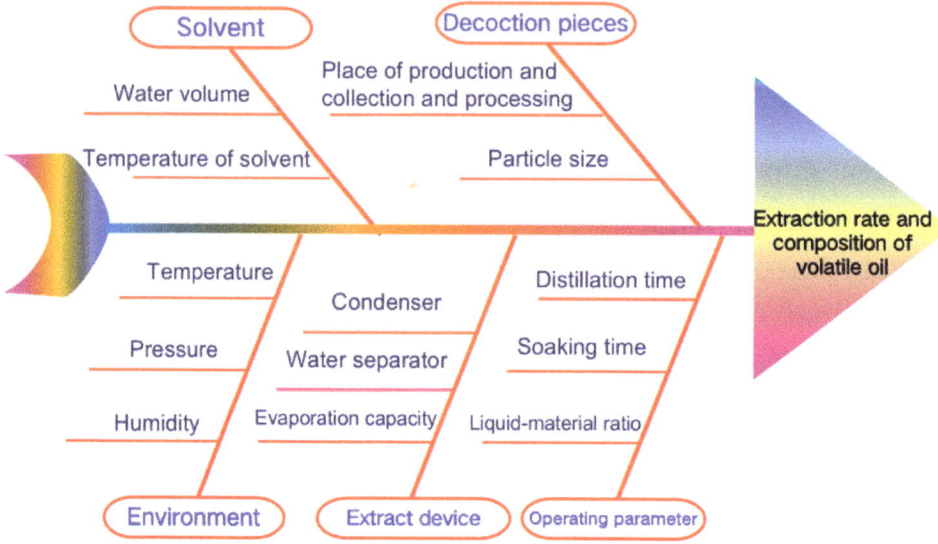

Figure 2. The Ishikawa diagram of the steam distillation process for volatile oil distillation.

Next, the Box–Behnken design was used to optimize the distillation process of the volatile oil from *Angelicae sinensis* Radix with the distillation time, soaking time, and liquid–material ratio as the factors. Compared with some other commonly used response surface designs, the number of experiments in the Box–Behnken design is relatively smaller when there are three 3-level factors. For example, there are 29, 17, and 13 runs for the full factorial design, central composite design, and the Box–Behnken design when there are three repetitions of the center point, respectively. The volatile oil yield was considered as the evaluation index. The factor levels are shown in Tables 1 and 2.

Table 1. Factors and levels of the Box–Behnken design.

Factor	Level		
	Low (−1)	Medium (0)	High (1)
A: Distillation time (h)	6	8	10
B: Soaking time (h)	2	3	4
C: Liquid–material ratio (mL·g^{-1})	6:1	10:1	14:1

Table 2. The Box–Behnken designed experiments and results.

Serial Number	A (Distillation Time/h)	B (Soaking Time/h)	C [Liquid-Material Ratio (mL·g^{-1})]	Y (Volatile Oil Yield/%)
1	−1	−1	0	0.58
2	1	−1	0	0.63
3	−1	1	0	0.59
4	1	1	0	0.65
5	−1	0	−1	0.55
6	1	0	−1	0.61
7	−1	0	1	0.52
8	1	0	1	0.56
9	0	−1	−1	0.61

Table 2. Cont.

Serial Number	A (Distillation Time/h)	B (Soaking Time/h)	C [Liquid-Material Ratio (mL·g^{-1})]	Y (Volatile Oil Yield/%)
10	0	1	−1	0.65
11	0	−1	1	0.56
12	0	1	1	0.57
13	0 (8 h)	0 (3 h)	0 (10:1)	0.70
14	0 (8 h)	0 (3 h)	0 (10:1)	0.71
15	0 (8 h)	0 (3 h)	0 (10:1)	0.71
16	0 (8 h)	0 (3 h)	0 (10:1)	0.72
17	0 (8 h)	0 (3 h)	0 (10:1)	0.71

2.4. Analysis of Chemical Constituents of Volatile Oil from Angelicae sinensis Radix

A total of 100 µL of volatile oil was measured accurately, diluted to 10 mL with ethanol, filtered with a 0.22 µm microporous filter membrane, and loaded into a sample bottle. The chemical constituents of the volatile oil samples were determined by gas chromatography-mass spectrometry (Agilent 7890A/5975C, Agilent, Santa Clara, CA, USA).

The gas chromatographic conditions were as follows [19]: HP-5MS capillary quartz column (30 m × 250 µm × 0.25 µm); injection volume: 1.0 µL; the inlet temperature: 230 °C; no diversion; heating procedure: 80 °C, keeping for 0 min; heating at 3 °C min^{-1} to 167 °C for 2.5 min, at 2 °C min^{-1} to 180 °C for 0 min, at 30 °C min^{-1} to 280 °C for 1 min; the transmission line temperature: 300 °C; the carrier gas was high purity helium, the flow rate was 1 mL min^{-1}, and the injection volume was 1 µL. Mass spectrometry conditions: Ion source temperature: 230 °C, ionization source was EI, electron energy: 70 eV; the quadrupole temperature: 150 °C, and the scanning quality range was 30–550 amu.

The identification of volatile compounds was performed by computer matching their mass spectra with those stored in a digital library of mass spectral data (NIST 14). The identification results were tentatively identified by the EI-MS spectrum and further experiments are planned to confirm their identification by authentic standards.

2.5. Data Processing Method

2.5.1. Kinetic Models

Volatile oil yield was fitted by Equations (2)–(4), as shown below.
First-order kinetic model:

$$C = C_{eq}\left(1 - e^{-kt}\right) \quad (2)$$

where C_{eq} is the concentration of solution at equilibrium; k is the total distillation rate constant; and t is time.

Peleg's model:

$$C = \frac{t}{(k_1 + k_2 t)} \quad (3)$$

where k_1 is the mass transfer rate constant and k_2 is the concentration of the solution at equilibrium.

The diffusion model of spherical particle:

$$C = C_{eq}\left[1 - \frac{6}{\pi^2}\sum_{n=1}^{\infty}\frac{1}{n^2}\exp\left(-\frac{Dn^2\pi^2 t}{R^2}\right)\right] \quad (4)$$

where D is the solute diffusion coefficient in the solvent and R is the particle radius. In the calculation, only the first three terms ($n = 3$) were taken for the sum of the infinite order.

In order to evaluate the fitting effect of each model, R^2 can be calculated according to Equation (5).

$$R^2 = 1 - \frac{\sum(C_{act} - C_{fit})^2}{\sum(C_{act} - \overline{C_{act}})^2} \tag{5}$$

where C_{act} is the experimental value; $\overline{C_{act}}$ is the average value of the measured value, and C_{fit} is the model fitting value. The software MATLAB R2019b (American Math Works Company) was used to analyze and process the data. Except for R^2, other indices of RMSE, MSE, SSE, and MAE were also calculated according to Equations (6)–(9).

$$RMSE = \sqrt{\frac{\sum(C_{act} - C_{fit})^2}{n}} \tag{6}$$

$$MSE = \frac{\sum(C_{act} - C_{fit})^2}{n} \tag{7}$$

$$SSE = \sum(C_{act} - C_{fit})^2 \tag{8}$$

$$AE = \frac{\sum|C_{act} - C_{fit}|}{n} \tag{9}$$

The software Microsoft Office Excel (American Microsoft Company) was used to analyze and process the data.

2.5.2. Data Processing of Volatile Oil Distillation Rate Obtained from Experimental Design

The software "Design-Expert 8.0.6" was used to analyze the experimental data collected from the Box–Behnken designed experiments. The volatile oil yield was taken as the evaluation index (Y), and a second-order polynomial equation fitting was carried out. Equation (10) was adopted as the mathematical model between the evaluation index and the three process parameters.

$$Y = b_0 + b_1 A + b_2 B + b_3 C + b_4 AB + b_5 AC + b_6 BC + b_7 A^2 + b_8 B^2 + b_9 C^2 \tag{10}$$

where b_{1-9} is the partial regression coefficient and b_0 is the intercept.

2.5.3. Optimization of Distillation Parameters of Volatile Oil by the Monte Carlo Method

The design space was calculated by the Monte Carlo method with a parameter optimization software compiled by MATLAB R2018b (Math Works Company of America) [20]. According to previous work, the design space calculated with this method is more reliable [20]. Three factors affecting the volatile oil yield were simulated randomly. The combination of parameters with the probability of no less than 0.80 of attaining the preset volatile oil yield value was taken as the design space. In the calculation, the step lengths of distillation time, soaking time, and liquid—material ratio were set to 0.04, 0.02, and 0.08, respectively. The simulations were conducted 2000 times.

3. Results and Discussion

3.1. Critical Process Parameters of Volatile Oil Distillation

The distillation time was changed to 2, 4, 6, 8, and 10 h, respectively, and the volatile oil yield is shown in Figure 3. The volatile oil yield increased continuously within the first 8 h of distillation time. When the distillation time reached 8 h, the volatile oil yield reached 0.68%, and the yield tended to be stable when the distillation time was prolonged.

The soaking time was changed to 0, 1, 2, 3, and 4 h, the yield of volatile oil from *Angelicae sinensis* Radix is shown in Figure 4. When the soaking time was 3 h, a maximum oil yield of 0.67% was obtained. After soaking time reached 3 h, the oil yield decreased with the extension in the soaking time. The reason may be that the aqueous extract became viscous when soaking for too long, which was not conducive to the distillation of the volatile oil.

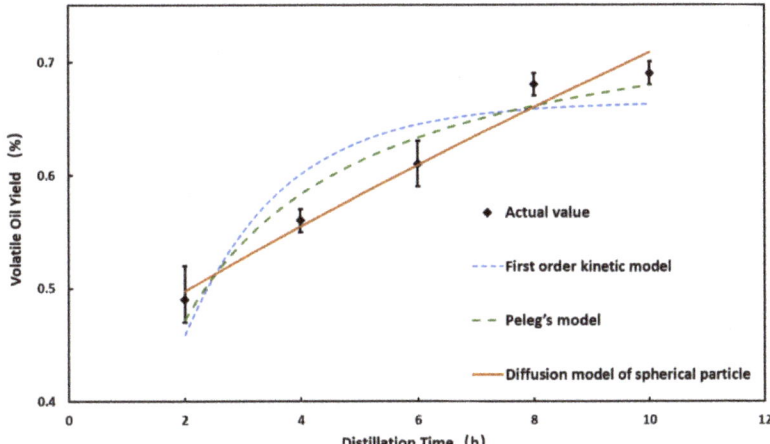

Figure 3. The effect of distillation time on the volatile oil yield ($n = 3$).

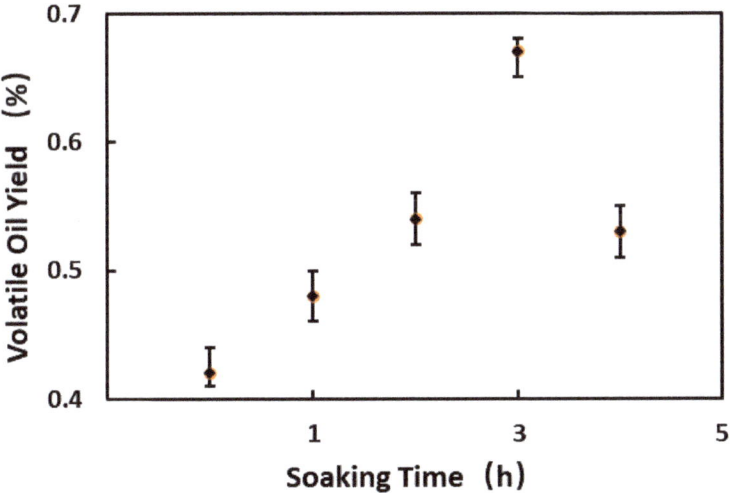

Figure 4. The effect of soaking time on the volatile oil yield ($n = 3$).

The liquid–material ratio was changed to 6:1, 8:1, 10:1, 12:1, and 14:1 (mL·g^{-1}), respectively, and the volatile oil yield is shown in Figure 5. When the liquid–material ratio was not higher than 10:1, the volatile oil yield increased as the liquid–material ratio increased. The reason may lie in the more uniform heating when the liquid–material ratio increased, which was beneficial to the diffusion and dissolution of the volatile oil. In Figure 4, when the liquid–material ratio was greater than 10:1, the volatile oil yield showed an obvious downward trend, which may be due to the loss caused by the dissolution of the volatile oil in water.

The above results indicate that the distillation time, soaking time, and liquid–material ratio all had a great influence on the yield of the volatile oil from *Angelicae sinensis* Radix and were all indeed the CPPs.

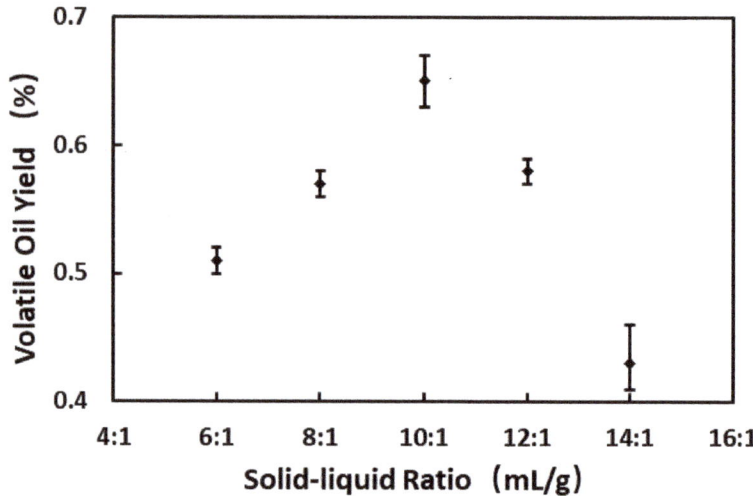

Figure 5. The effect of solid–liquid ratio on the volatile oil yield ($n = 3$).

3.2. Study on Kinetics of Volatile Oil Distillation

Three models were used to fit the kinetic data of the volatile oil distillation, and the fitting results are shown in Figure 2. The first-order kinetic model fitting results C_{eq} and k were 0.6644% and 0.5861 h^{-1}, respectively, and R^2 was 0.8198. The Peleg's model fitted k_1 and k_2 were 1.6156 h·%$^{-1}$ and 1.3105 %$^{-1}$, respectively, and the R^2 was 0.9338. The diffusion model of spherical particle fitted C_{eq} and $\frac{D}{R^2}$ were 2.5317% and 0.0007 h^{-1}, respectively, and the R^2 was 0.9710. Among the three models, the diffusion model of spherical particle fitted R^2 was the highest, which could explain more than 97% of the variance in experimental data.

Table 3 shows the errors between the predicted values and the actual values of different models. The prediction values of the diffusion model were closest to the actual values.

Table 3. The error results of the different functions.

Model	RMSE (10^{-2} mL/g)	SSE (10^{-4} mL^2/g^2)	MSE (10^{-4} mL^2/g^2)	MAE (10^{-2} mL/g)
First-order kinetic model	0.032	0.0051	0.0010	0.031
Peleg's model	0.019	0.0019	0.00037	0.019
Diffusion model of spherical particle	0.013	0.00082	0.00016	0.010

The diffusion model of spherical particle assumed that the major resistance of mass transfer in the distillation process of volatile oil was the diffusion of volatile oil from the inside to the surface of the medicinal materials. The mass transfer rate of volatile oil from the surface of the medicinal materials to the extraction solution and the mass transfer rate in the air inside the extractor were all much faster. These results indicate that a rapid distillation of volatile oil can probably be realized by lowering the average particle size of *Angelicae sinensis* Radix.

3.3. Optimization of Distillation Parameters of Volatile Oil

3.3.1. Data Processing and Model Fitting

By modeling the oil yield data, the multivariate binomial regression model obtained was as follows: $Y = 0.710 + 0.026A + 0.010B - 0.026C + 2.50 \times 10^{-3}AB - 5.00 \times 10^{-3}AC - 7.50 \times 10^{-3}BC - 0.068A^2 - 0.030B^2 - 0.082C^2$. The model determination coefficient

R^2 was 0.9917. The variance analysis of each item in the model is shown in Table 4. The F value of the model was 93.13, and $p < 0.0001$, which showed that the model was extremely significant. The model R^2 exceeded 0.99, which showed that the model could well explain the data variation in the experiment. A ($p < 0.0001$) and C ($p < 0.0001$) were extremely significant, and B ($p < 0.05$) was significant. Among the quadratic terms, A^2 ($p < 0.0001$), B^2 ($p = 0.0003$), and C^2 ($p < 0.0001$) were all extremely significant terms, which showed that the influence of the three parameters on the oil yield was nonlinear. The results were consistent with the previous single factor experimental results. The p values of AB, BC, and AC were all greater than 0.10, which showed that the interaction between the three factors was small. Figure 6 shows the Pareto chart. It can be concluded that all the linear terms and quadratic terms were significant.

Table 4. The results of the analysis of variance.

Variance Source	Sum of Square	Degree of Freedom	Mean Square	F Value	p Value
Model	0.069	9	7.65×10^{-3}	93.13	<0.0001
A	5.513×10^{-3}	1	5.513×10^{-3}	67.11	<0.0001
B	8×10^{-4}	1	8×10^{-4}	9.74	0.0168
C	5.512×10^{-3}	1	5.512×10^{-3}	67.11	<0.0001
AB	2.5×10^{-5}	1	2.5×10^{-5}	0.3	0.5983
AC	1×10^{-4}	1	1×10^{-4}	1.22	0.3064
BC	2.25×10^{-4}	1	2.25×10^{-4}	2.74	0.1419
A22	0.019	1	0.019	233.55	<0.0001
B22	3.789×10^{-3}	1	3.789×10^{-3}	46.13	0.0003
C22	0.029	1	0.029	348.88	<0.0001
Residual error	5.75×10^{-4}	7	8.214×10^{-5}		
Misfit term	3.75×10^{-4}	3	1.25×10^{-4}	2.5	0.1985
Pure error	2×10^{-4}	4	5×10^{-5}		
Total error	0.069	16			

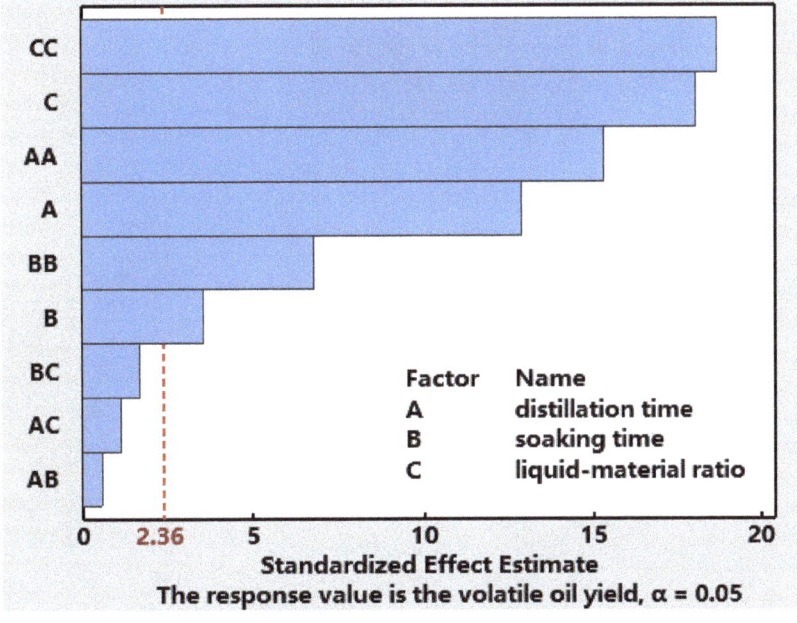

Figure 6. The Pareto chart of the parameters.

3.3.2. Response Surface Diagram and Contour Diagram

The response surface diagram and contour diagram of the distillation process of the volatile oil from *Angelicae sinensis* Radix are shown in Figure 7. The figure reflects the effects of the distillation time, soaking time, and liquid–material ratio on the oil yield. When the liquid–material ratio is fixed, the oil yield increases first and then decreases gradually with the increase in distillation time. When the distillation time is too long, some volatile components in the volatile oil may volatilize and be lost. When the soaking time is fixed, the volatile oil yield increases first and then decreases gently with the increase in the distillation time, and increases first and then decreases with the increase in the liquid–material ratio. When the distillation time is fixed, the oil yield increases first and then stabilizes with the increase in soaking time, and increases first and then decreases with the increase in the liquid–material ratio.

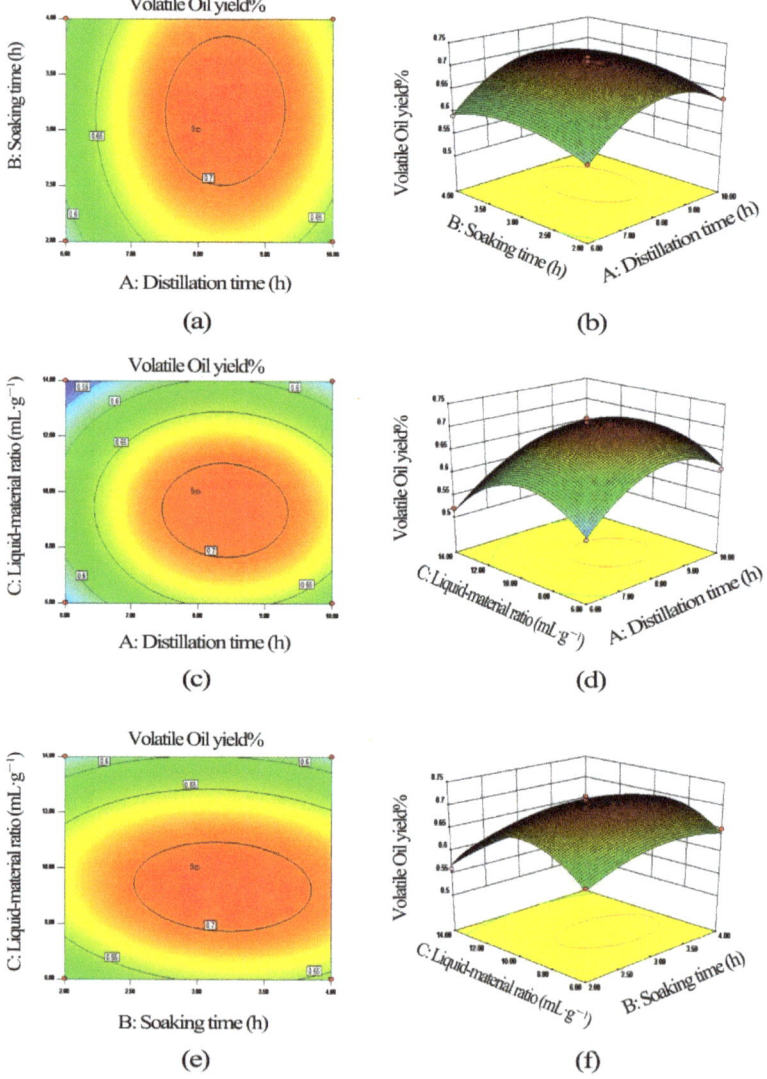

Figure 7. The 3D response surface and 2D contour map for the distillation process of volatile oil.

3.3.3. Design Space Calculation and Verification

The acceptable lower limit of volatile oil yield was set at 0.65%, and the lowest probability of target attainment was set at 0.80. The design space calculated by the probability-based method is shown in Figure 7. Figure 7a is a three-dimensional design space diagram, and Figure 8b is a two-dimensional space diagram after fixing the liquid–material ratio at 9.3:1. It can be seen from the figure that the design space was close to the shape of a football.

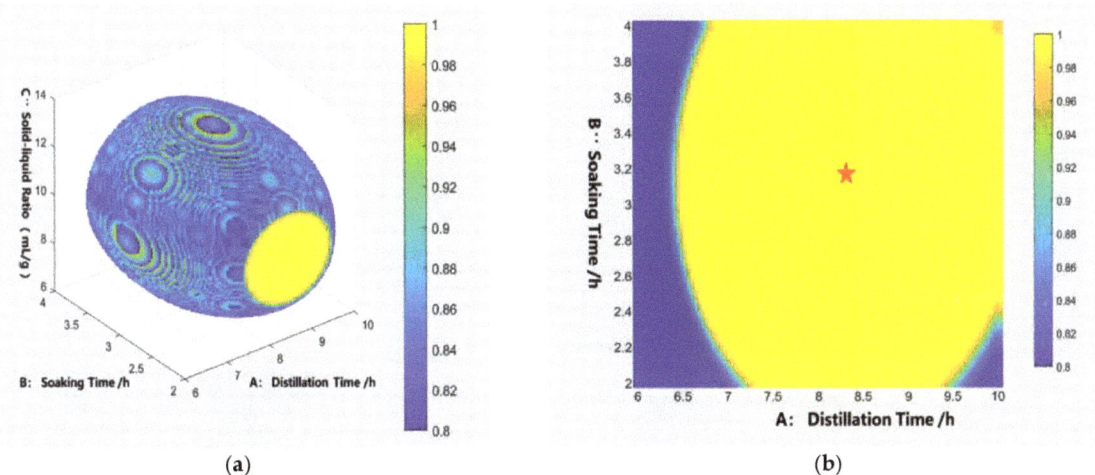

(a) (b)

Figure 8. The design space. (**a**) Panorama of the design space. (**b**) The design space when the liquid–material ratio = 9.3:1 (mL·g^{-1}). The red star in the figure represents the condition of the verification experiments, and the color bar indicates the probability of the target attainment.

The design space was verified with a selected condition of 8.4 h of distillation time, 3.2 h of soaking time, and 9.3:1 (mL·g^{-1}) of liquid–material ratio. Under this condition, the probability for target attainment was 1.0 and the predicted volatile oil yield was 0.715%. Three parallel validation tests were carried out, and the oil yield of *Angelicae sinensis* Radix volatile oil was 0.711%, 0.709%, and 0.712%, respectively. The average oil yield was 0.711% and the relative standard deviation was 0.21%. The density of the collected volatile oil was 1.0075 g/cm^3 at room temperature. The mathematical model was developed according to the results of the Box–Behnken designed experiments. The measured values were close to the predicted values, indicating that the prediction of the model was accurate. The volatile oil yield was higher than the preset standard, which shows that the design space was reliable.

3.4. Qualitative Analysis of Chemical Constituents of Volatile Oil

The GC-MS analysis of the volatile oil from *Angelicae sinensis* Radix was carried out, and the total ion flow diagram is shown in Figure 9. The results of the mass spectrometry are shown in Table 5. Twenty chemical constituents were identified, accounting for 99.426% of the total volatile oil. Among them, Z-ligustilide had the highest relative content, and its relative content reached 85.385%. Its structural formula can be found in SciFinder [21]. Other components with relatively high contents were α-pinene, **β-ocimene**, 2-methyl−1,3-benzoxazole, **butylidenephthalide**, and **E-ligustilide**.

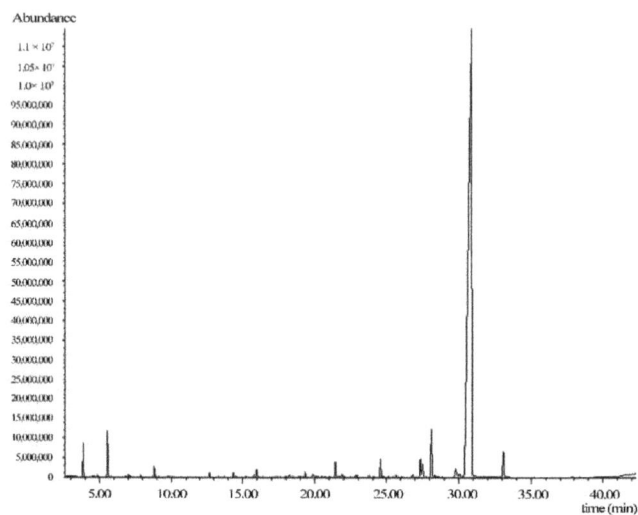

Figure 9. The total ion chromatogram of the angelica volatile oil.

Table 5. The tentative identification results of the chemical composition of angelica volatile oil.

Serial Number	T/min	Compound	Chemical Formula	Relative Content/%	m/z Value of the Fragment with Maximum Abundance	Abundance
1	3.85	α-Pinene 2,6,6-trimethylbicyclo[3.1.1]hept-2-ene	$C_{10}H_{16}$	1.21	93.1	239,680
2	5.55	β-Ocimene 3,7-dimethylocta−1,3,6-triene	$C_{10}H_{16}$	1.67	93.1	298,432
3	8.84	6-butylcyclohepta−1,4-diene	$C_{11}H_{18}$	0.53	79.1	61,936
4	12.68	Diamyl Ketone undecan-6-one	$C_{11}H_{22}O$	0.21	71.1	22,096
5	14.33	4-ethenyl-2-methoxyphenol	$C_9H_{10}O_2$	0.30	150.1	32,720
6	15.95	Duraldehyde 2,4,5-trimethylbenzaldehyde	$C_{10}H_{12}O$	0.46	147.1	34,272
7	18.24	β-Cedrene (1S,2R,5S,7S)-2,6,6-trimethyl-8-methylidenetricyclo[5.3.1.01,5]undecane	$C_{15}H_{24}$	0.27	161.2	7463
8	19.36	4-(1,2-dimethylcyclopent-2-en−1-yl)butan-2-one	$C_{11}H_{18}O$	0.25	93.1	12,845
9	19.88	Sesquichamene 2,4a,8,8-tetramethyl−1,1a,4,5,6,7-hexahydrocyclopropa[j]naphthalene	$C_{15}H_{24}$	0.02	93.1	4696
10	21.45	(1R,2R)−1-ethenyl−1-methyl−4-propan-2-ylidene-2-prop−1-en-2-ylcyclohexane	$C_{15}H_{24}$	0.75	121.1	50,288
11	21.90	β-Bisabolene (4S)−1-methyl−4-(6-methylhepta−1,5-dien-2-yl)cyclohexene	$C_{15}H_{24}$	0.17	69.1	9638
12	23.77	Alloaromadendrene 1,1,7-trimethyl−4-methylidene-2,3,4a,5,6,7,7a,7b-octahydro−1aH-cyclopropa[e]azulene	$C_{15}H_{24}$	0.11	121.1	4412

Table 5. Cont.

Serial Number	T/min	Compound	Chemical Formula	Relative Content/%	m/z Value of the Fragment with Maximum Abundance	Abundance
13	24.57	Spathulenol (7S)−1,1,7-trimethyl−4-methylidene−1a,2,3,4a,5,6,7a,7b-octahydrocyclopropa[h]azulen-7-ol	$C_{15}H_{24}O$	0.96	91.1	22,384
14	27.36	2-methyl−1,3-benzoxazole	C_8H_7NO	1.06	133.0	240,832
15	27.52	1-oxido−4-[2-(1-oxidopyridin−1-ium−4-yl)ethyl]pyridin−1-ium	$C_{12}H_{12}N_2O_2$	0.71	108.1	70,000
16	27.75	Cyclopentadiene 2,5,5-trimethylcyclopenta−1,3-diene	C_8H_{12}	0.09	159.0	373,632
17	28.12	Butylidenephthalide (3E)-3-butylidene-2-benzofuran−1-one	$C_{12}H_{12}O_2$	2.79	159.0	56,168
18	30.06	Senkyunolide A (3S)-3-butyl−4,5-dihydro-3H-2-benzofuran−1-one	$C_{12}H_{16}O_2$	0.36	107.1	27,088
19	30.88	Z-ligustilide (3Z)-3-butylidene−4,5-dihydro-2-benzofuran−1-one	$C_{12}H_{14}O_2$	85.4	161.1	1,526,784
20	32.34	E-ligustilide (3E)-3-butylidene−4,5-dihydro-2-benzofuran−1-one	$C_{12}H_{14}O_2$	2.04	161.1	86,504

4. Discussion

Li Tao et al. [22] found that the content of volatile oil from wild *Angelicae sinensis* Radix (1.14%) was more than twice as high than that from the artificially cultivated *Angelicae sinensis* Radix (0.4%). Yan Hui et al. [23] found that longer sunshine was not conducive to the increase in volatile components that mainly consisted of ligustilide. Lin Haiming [24] found that the content of the volatile oil and alcohol-soluble extract decreased gradually with the increase in the drying temperature. Ji Peng et al. [25] determined and analyzed the chemical components of volatile oil in raw *Angelicae sinensis* Radix and different processed products of *Angelicae sinensis* Radix by GC-MS. It was found that different processing methods would affect the total amount of volatile oil in *Angelicae sinensis* Radix and the contents of ligustilide and butenyl phthalolactone in the volatile oil [25]. Li Runhong et al. [26] studied the difference in the volatile oil composition and content in *Angelicae sinensis* Radix from different producing areas. The results showed that the highest volatile oil content of wild *Angelicae sinensis* Radix in Linzhi (Tibet Province) was 2-hydrobutenyl phthalide (70.184%), followed by n-butenyl phthalide (9.288%) [26]. In the volatile oil of wild *Angelicae sinensis* Radix in Pingwu (Sichuan Province), the content of 2-hydrobutenyl phthalide (92.551%) was the highest, followed by butenyl phthalide (3.037%) [26]. These results were different compared with those reported in this study and most of the literature, in which ligustilide was found to be the most abundant constituent. Generally speaking, the composition and content of volatile oil in *Angelicae sinensis* Radix are affected by the growth environment, harvest time, drying method, processing method, extraction technology, etc. [14,27]. Therefore, the composition and content of volatile oil in *Angelicae sinensis* Radix may have significant differences.

The results of some published works [28] have shown that the yield of volatile oil obtained from Angelica sinensis by steam distillation was low (about 0.3–0.5%). In industry, the yield is even lower. Sometimes, only the volatile oil aqueous solution can be obtained. In this work, an improved steam distillation with enhanced cooling was used. Compared with that of using a traditional volatile oil extractor, the collected volatile oil could be increased by 30–58%. The collected volatile oil in Angelica sinensis was different to that in the literature [22,29,30], which may be attributed to germplasm resources, harvesting time, planting altitude, processing technology, and so on. However, the efficiency of steam distillation was not high. More volatile oil may be obtained when using other techniques such as supercritical fluid extraction [31].

5. Conclusions

In this work, an improved volatile oil extractor with better cooling was applied to collect the volatile oil from *Angelicae sinensis* Radix by steam distillation. The distillation process was optimized according to the concept of quality-by-design. The soaking time, distillation time, and liquid–material ratio were determined as CPPs. It was observed that the volatile oil yield increased with the increase in the distillation time. Furthermore, three models were used to fit the distillation kinetics data of the volatile oil. The fitting effect of the diffusion model of spherical particle was best, and the R^2 exceeded 0.97. This indicates that a more rapid distillation process can probably be realized when smaller *Angelicae sinensis* Radix can be used. The volatile oil yield increased first and then decreased with the increase in the soaking time and liquid–material ratio. Then, the distillation time, soaking time, and liquid–material ratio were investigated with the Box–Behnken designed experiments. The R^2 of the established second-order polynomial model exceeded 0.99. Then, the design space for distilling the volatile oil from *Angelicae sinensis* Radix was obtained with a probability-based method. The design space was verified at a condition shown as follows: Distillation time: 8.41 h, soaking time: 3.20 h, liquid–material ratio: 9.3:1 (mL·g^{-1}). The probability of the target attainment was 1 at this condition. The actual volatile oil yield was 0.711%, which was close to the predicted value of 0.715% and higher than the preset standard, indicating that the model could predict accurately and the design space was reliable. The volatile oil of *Angelicae sinensis* Radix was also analyzed by GC-MS, and 21 chemical constituents were found, accounting for 99.426% of the total volatile oil. The content of Z-ligustilide in the volatile oil was the highest, and the relative content reached 85.38%.

Author Contributions: Conceptualization, X.G.; Investigation, N.W., J.L. and Z.W.; data curation, N.W., J.L., X.C. and Q.Z.; writing—original draft preparation, N.W., X.C., J.L. and X.G.; writing—review and editing, Q.Z. and X.G.; supervision, Q.Z. and X.G.; funding acquisition, X.G. and Z.W. All authors have read and agreed to the published version of the manuscript.

Funding: This work was supported by the National Natural Science Foundation of China (82060720); the Jiangxi University of Traditional Chinese Medicine 1050 Youth Talent Project; the Open Fund Project of Key Laboratory of Modern Preparation of TCM; the Ministry of Education; the Jiangxi University of Chinese Medicine (zdsys-202110); and the Chinese Medicine Innovation Team and Talent Support Program of The National Administration of Traditional Chinese Medicine (ZYYCXTD-D-202002).

Institutional Review Board Statement: Not applicable.

Informed Consent Statement: Not applicable.

Data Availability Statement: All data generated or analyzed during this study are included in this published article.

Acknowledgments: The authors are grateful to Wanying Wang for her help during the manuscript preparation.

Conflicts of Interest: The authors declare no conflict of interest.

References

1. Chen, Z.; Wu, G.; Sun, M.; Du, L.; Ren, Y. Research progress of ligustilide in *Angelica sinensis*. *J. Gansu Univ. Chin. Med.* **2018**, *35*, 102–105.
2. Zhao, X.; Wang, D.; Zhao, D.; Ji, H.; Pei, Y.; Bai, J. Isolation and identification of the chemical constituents from roots of *Angelica sinenses* (Oliv.) Diels. *J. Shenyang Pharm. Univ.* **2013**, *30*, 182–185.
3. Wu, H.; Hua, Y.; Guo, Y.; Wei, Y. Extraction and Analysis of Volatile Oil in Different Parts of Radix *Angelicae Sinensis* from Min County in Gansu. *Nat. Prod. Res. Dev.* **2012**, *24*, 1225–1229.

4. Sun, M.; Ma, Q.; Liu, F.; Ren, Y. New Progress in the Study of Volatile Oil from *Angelica sinensis*. *Abstr. World Curr. Med. Inf.* **2019**, *19*, 56–58.
5. Cui, F.; Feng, L.; Hu, J. Factors affecting stability of z-ligustilide in the volatile oil of radix angelicae sinensis and ligusticum chuanxiong and its stability prediction. *Drug Dev Ind Pharm.* **2006**, *32*, 747–755. [CrossRef] [PubMed]
6. Wang, Y.; Zhang, X.; Zou, J.; Jia, Y.; Wang, C.; Shi, Y.; Guo, D. Network analysis of the pharmacological mechanism of *Angelica sinensis* volatile oil in the treatment of hypertension. *Nat. Prod. Res. Dev.* **2021**, *33*, 657–666.
7. Liu, X.; Huang, X.; Zhang, X.; Wang, L.; Zhu, L.; Xu, J.; Chen, Q.; Yang, M.; Wang, F. Q-Marker prediction of volatile oil of *Angelicae Sinensis* Radix based on GC-MS analysis combined with network pharmacology. *Chin. Tradit. Herb. Drugs* **2021**, *52*, 2696–2706.
8. Zhu, L.; Luo, J.; Song, R.; Wang, L.; Zhang, A.; Zang, K. Effect and mechanism of volatile oil of angelica on apoptosis and autophagy in human colorectal cancer SW480 cells. *Chin. J. Clin. Pharmacol.* **2021**, *37*, 3253–3256.
9. Kusuma, H.S.; Mahfud, M. Chemical composition of essential oil of Indonesia sandalwood extracted by microwave-assisted hydrodistillation. *AIP Conf. Proc.* **2016**, *1755*, 50001.
10. Kusuma, H.S.; Mahfud, M. Box-Behnken design for investigation of microwave-assisted extraction of patchouli oil. *AIP Conf. Proc.* **2015**, *1699*, 50014.
11. Priyadarshi, S.; Kashyap, P.; Gadhave, R.K.; Jindal, N. Effect of ultrasound-assisted hydrodistillation on extraction kinetics, chemical composition, and antimicrobial activity of *Citrus jambhiri* peel essential oil. *J. Food Process Eng.* **2021**, *44*, e13904. [CrossRef]
12. Lan, Z.; Wang, L.; Li, Q.; Wang, S.; Meng, J. Analysis of volatile oil components of different species of *Curcumae rhizoma* based on GC-MS and chemometrics. *China J. Chin. Mater. Med.* **2021**, *46*, 3614–3624.
13. Zou, J.; Zhang, X.; Shi, Y.; Guo, D.; Cheng, J.; Cui, C.; Tai, J.; Liang, Y.; Wang, Y.; Wang, M. Kinetic study of extraction of volatile components from turmeric by steam distillation. *China J. Tradit. Chin. Med. Pharm.* **2020**, *35*, 1175–1180.
14. Wu, Z.; Xie, L.; Li, Y.; Wang, Y.; Wang, X.; Wan, N.; Huang, X.; Zhang, X.; Yang, M. A novel application of the vacuum distillation technology in extracting *Origanum vulgare* L. essential oils. *Ind. Crop. Prod.* **2019**, *139*, 111516. [CrossRef]
15. Gong, X.; Chen, T.; Qu, H. Research advances in secondary development of Chinese patentmedicines based on quality by design concept. *China J. Chin. Mater. Med.* **2017**, *42*, 1031–1036.
16. Tai, Y.; Qu, H.; Gong, X. Design Space Calculation and Continuous Improvement Considering a Noise Parameter: A Case Study of Ethanol Precipitation Process Optimization for Carthami Flos Extract. *Separations* **2021**, *8*, 74. [CrossRef]
17. El-Shamy, A.M.; El-Hadek, M.A.; Nassef, A.E.; El-Bindary, R.A. Box-Behnken design to enhance the corrosion resistance of high strength steel alloy in 3.5 wt.% NaCl solution. *Moroc. J. Chem.* **2020**, *8*, 788–800.
18. Kusuma, H.S.; Sudrajat, R.G.M.; Febrilliant, D.; Susanto, D.F.; Gala, S.; Mahfud, M. Response Surface Methodology (RSM) Modeling of Microwave-Assisted Extraction of Natural Dye from Swietenia mahagony: A comparation Between Box-Behnken and Central Composite Design Method. *AIP Conf. Proc.* **2015**, *1699*, 50009.
19. Zhao, M.; Yang, S.; Sun, Y.; Li, F.; Zhang, S.; Jiao, J. Gas Chromatography-mass Spectrometry Analysis of Volatile Oil of *Angelica sinensis* Root Dry Chemical Composition. *Chem. World* **2018**, *59*, 231–234.
20. Shao, J.; Qu, H.; Gong, X. Comparison of two algorithms for development of design space-overlapping method and probability-based method. *China J. Chin. Mater. Med.* **2018**, *43*, 2074–2080.
21. Zhang, Y.; Zhang, Y.; Han, Y.; Tian, Y.; Wu, P.; Xin, A.; Wei, X.; Shi, Y.; Zhang, Z.; Su, G.; et al. Pharmacokinetics, tissue distribution, and safety evaluation of a ligustilide derivative (LIGc). *J. Pharm. Biomed. Anal.* **2020**, *182*, 113140. [CrossRef]
22. Li, T.; He, X. Analysis of essential oil from the roots of *Angelica sinensis* by GC-MS. *West China J. Pharm. Sci.* **2015**, *30*, 249–250.
23. Yan, H.; Zhang, X.-B.; Zhu, S.-D.; Qian, D.-W.; Guo, L.-P.; Huang, L.-Q.; Duan, J.-A. Production regionalization study of Chinese angelica based on MaxEnt model. *China J. Chin. Mater. Med.* **2016**, *41*, 3139–3147.
24. Liu, P.; Chen, J.; Zhou, B.; Xu, Y.; Qian, D.W.; Duan, J.A. Analysis of variation of coumarin and volatile compounds in Angelica Dahuricae radix in different drying methods and conditions. *Zhongguo Zhong Yao Za Zhi* **2014**, *39*, 2653–2659.
25. Ji, P.; Hua, Y.; Xue, W.; Wu, H.; Guo, Y.; Wei, Y. Extraction and Composition Analysis of Essential Oil from Raw Radix *Angelicae Sinensis* and Its Different Processed Products. *Nat. Prod. R D.* **2012**, *24*, 1230–1234, 1238.
26. Li, R. Analysis of Volatile Oil Ingredients in Wild *Angelica sinensis* from Different Origin. *Chin. J. Ethnomed. Ethnopharm.* **2019**, *28*, 41–44.
27. Li, Z.; Cui, J.; Fu, Z.; Mu, J. Supercritical CO_2 Fluid Extraction and Chemical Composition Analysis of Volatile Oil from *Atractylodis macrocephalae rhizoma*. *Food Drug* **2019**, *21*, 269–273.
28. Wang, T.; Cheng, Z. Research on Angelica Naphtha Components form Different Regions. *Pharm. Biotechnol.* **2013**, *20*, 535–537.
29. Wang, X.; Zhao, Z.; Zhang, Y.; Shi, X.; Cheng, F.; Guo, M.; Wen, X. Comparative analysis of volatile compounds in *Angelica Sinensis* by different acquisition methods with GC-MS technique. *J. Tradit. Chin. Vet. Med.* **2018**, *37*, 61–65.
30. Li, D.; Ma, X.; Song, P.; Zhao, J.; Ding, Y. Analysis of Three Methods for Extracting Volatile Constituent from DangGui by Combination of Gas Chromatography and Mass Spectrometry. *West. J. Tradit. Chin. Med.* **2013**, *26*, 15–18.
31. Yue, H.; Jia, H.; Wang, J. Study on the Extraction Technology of the Volatile Oil in *Angelica sinensis* (Oliv.) Diels Based on Supercritical CO_2 Extraction. *J. Anhui Agric. Sci.* **2010**, *1*, 47–50.

Article

Enantiomeric Separation and Molecular Modelling of Bioactive 4-Aryl-3,4-dihydropyrimidin-2(1*H*)-one Ester Derivatives on Teicoplanin-Based Chiral Stationary Phase

Isabella Bolognino [1,*,†], Antonio Carrieri [1], Rosa Purgatorio [1], Marco Catto [1], Rocco Caliandro [2], Benedetta Carrozzini [2], Benny Danilo Belviso [2], Maria Majellaro [3], Eddy Sotelo [3], Saverio Cellamare [1] and Cosimo Damiano Altomare [1,*]

[1] Department of Pharmacy-Pharmaceutical Sciences, University of Bari Aldo Moro, Via Orabona 4, 70125 Bari, Italy; antonio.carrieri@uniba.it (A.C.); rosa.purgatorio@uniba.it (R.P.); marco.catto@uniba.it (M.C.); saveriocellamare56@gmail.com (S.C.)

[2] Institute of Crystallography, CNR, Via Giovanni Amendola, 122/O, 70126 Bari, Italy; rocco.caliandro@ic.cnr.it (R.C.); benedetta.carrozzini@ic.cnr.it (B.C.); danilo.belviso@ic.cnr.it (B.D.B.)

[3] Singular Research Center in Biological Chemistry and Molecular Materials (CIQUS), University of Santiago de Compostela, 15782 Santiago de Compostela, Spain; ma.majellaro@usc.es (M.M.); e.sotelo@usc.es (E.S.)

* Correspondence: isabella.bolognino@unibg.it (I.B.); cosimodamiano.altomare@uniba.it (C.D.A.); Tel.: +39-080-5442781 (C.D.A.)

† Current address: Department of Engineering and Applied Sciences, University of Bergamo, Viale G. Marconi 5, 24044 Dalmine, Italy.

Abstract: The enantiomeric separation of 15 racemic 4-aryl-3,4-dihydropyrimidin-2(1*H*)-one (DHP) alkoxycarbonyl esters, some of which proved to be highly active as A_{2B} adenosine receptor antagonists, was carried out by HPLC on Chirobiotic™ TAG, a chiral stationary phase (CSP) bearing teicoplanin aglycone (TAG) as the chiral selector. The racemic compounds were separated under polar organic (PO) conditions. Preliminarily, the same selectands were investigated on three different Pirkle-type CSPs in normal-phase (NP) conditions. A baseline separation was successfully obtained on TAG-based CSPs for the majority of compounds, some of which achieved high enantioselectivity ratios ($\alpha > 2$) in contrast with the smaller α values (1–1.5) and the lack of baseline resolution observed with the Pirkle-type CSPs. In particular, the racemic tetrazole-fused DHP ester derivatives, namely compounds **8** and **9**, were separated on TAG-based HPLC columns with noteworthy α values (8.8 and 6.0, respectively), demonstrating the potential of the method for preparative purposes. A competition experiment, carried out with a racemic analyte (**6**) by adding *N*-acetyl-D-alanine (NADA) to the mobile phase, suggested that H-bonding interactions involved in the recognition of the natural dipeptide ligand D-Ala-D-Ala into the TAG cleft should be critical for enantioselective recognition of 4-aryl DHPs by TAG. The X-ray crystal structure of TAG was elucidated at a 0.77 Å resolution, whereas the calculation of molecular descriptors of size, polar, and H-bond interactions, were complemented with molecular docking and molecular dynamics calculations, shedding light on repulsive (steric effects) and attractive (H-bond—polar and apolar) interactions between 4-aryl DHP selectands and TAG chiral selectors.

Keywords: 4-aryl-3,4-dihydropyrimidin-2(1*H*)-ones; teicoplanin aglycone; chiral HPLC; enantioselectivity; molecular docking; molecular dynamics

1. Introduction

The development and optimisation of methods to obtain enantiomers with high optical purity remain an important goal to be achieved in the early stages of drug discovery, considering that single enantiomers and diastereoisomers could often have distinct profiles in pharmacodynamics, pharmacokinetics, metabolism, and toxicity [1]. Within the variety of methods used, enantiomers' separation by high-performance liquid chromatography

(HPLC) has great impact in pharmaceutical research. A literature survey indicates that chiral stationary phases (CSPs) in enantioselective HPLC is the method of choice as it allows the overcoming of limitations such as, to name a few, the demand on the enantiomeric purity of the derivatizing agent when opting for off-line diastereomeric formation, chiral selector consumption, and detection interference when adding a chiral auxiliary to the mobile phase.

In this study, we report on the enantiomeric separation of a novel class of chiral bioactive 3,4-dihydropyrimidin-2(1H)-one (DHP) alkoxycarbonyl esters (Figure 1), mostly bearing an aryl group at C4, discovered by some of us as being potent and selective A_{2B} adenosine receptor (A_{2B}AR) antagonists. The investigated DHP derivatives were assembled in excellent yields by modification of the versatile and highly robust Biginelli reaction [2,3], a three-component transformation involving the catalysed condensation of a 1,3-dinucleophile (urea/thiourea or cyclic derivatives), and an aldehyde and β-ketoester [4]. A diverse series of DHP derivatives were explored [5–11] and optimal substituents in different positions of the heterocyclic core were identified as follows: (i) 2- or 3-furyl as well as 2- or 3-thienyl moiety at C4, (ii) ethyl or isopropyl ester group at C5, (iii) a methyl group at C6, and (iv) NH at position 1. Further diversification of the original 4-aryl-DHP scaffold [5,12–15] produced new ligands with improved affinity and selectivity profiles. Among others DHP derivatives, compounds **6**, **7**, **12**, and **13** were disclosed as highly potent A_{2B}AR selective antagonists [5,6,9], and the observed structure-activity relationships (SARs) were fully supported by molecular docking calculations.

Regarding the enantioselective binding of racemic DHPs to A_{2B}AR, it was demonstrated that the receptor affinity is almost exclusively due to the (*S*)-enantiomer [6,8–11]. The pharmacological evaluation of chiral DHP ligands investigated herein, **6**, **7**, and **13**, resolved into their enantiomers by chiral HPLC with polysaccharide-based CSPs in NP mode, demonstrated highly enantioselective recognition at the A_{2B}AR binding site, with the eutomers (*S*) showing K_i values in the low nanomolar range (6.30, 15.1, and 11.1 nM for **6**, **7**, and **13**, respectively) and the distomers (*R*) achieving less than 25% inhibition in all cases [6,11]. The enantioselective binding modes were investigated by molecular docking calculations, which suggested key H-bonding interactions between the DHP ligands and A_{2B}AR [6,11].

Fifteen representative racemic 4-aryl DHP derivatives (Figure 1), synthesised in the framework of the above-mentioned long-lasting research program, were selected from a large molecular library of the Sotelo group, considering the structural diversity and spread of pharmacological potency as selective A_{2B}AR antagonists. In previous studies, a good semi-preparative enantiomeric resolution of a number of racemic DHP derivatives was achieved using chiral HPLC with the polysaccharide-based (cellulose or amylose backbone) CSPs, whereas in this study we report on the investigation of teicoplanin-based CSPs, namely Chirobiotic™ TAG, which bears teicoplanin aglycone (TAG) as the chiral selector [16] in polar organic mode (POM) [17] as a suitable chiral HPLC method for the enantiomeric resolution of the racemic bioactive DHPs **1–15**.

TAG was preferred over the native glycosylated teicoplanin (TE) as chiral selector, based on the achievements of several studies [18]. Among racemic compounds resolved with teicoplanin-based stationary phases, for 5-methyl-5-phenylhydantoin and α-methyl-α-phenylsuccinimide the enantioselectivity factor (α) values determined on TAG-based CSPs were greater than those achieved with the native glycosylated TE chiral selector. Aryl sulfoxides proved particularly useful in studies aimed at understanding the multifactorial enantiorecognition mechanism on CSPs containing macrocyclic glycopeptide antibiotic as chiral selectors [19,20], and data showed a superior enantiodiscrimination capacity of TAG compared to TE under polar organic conditions [21].

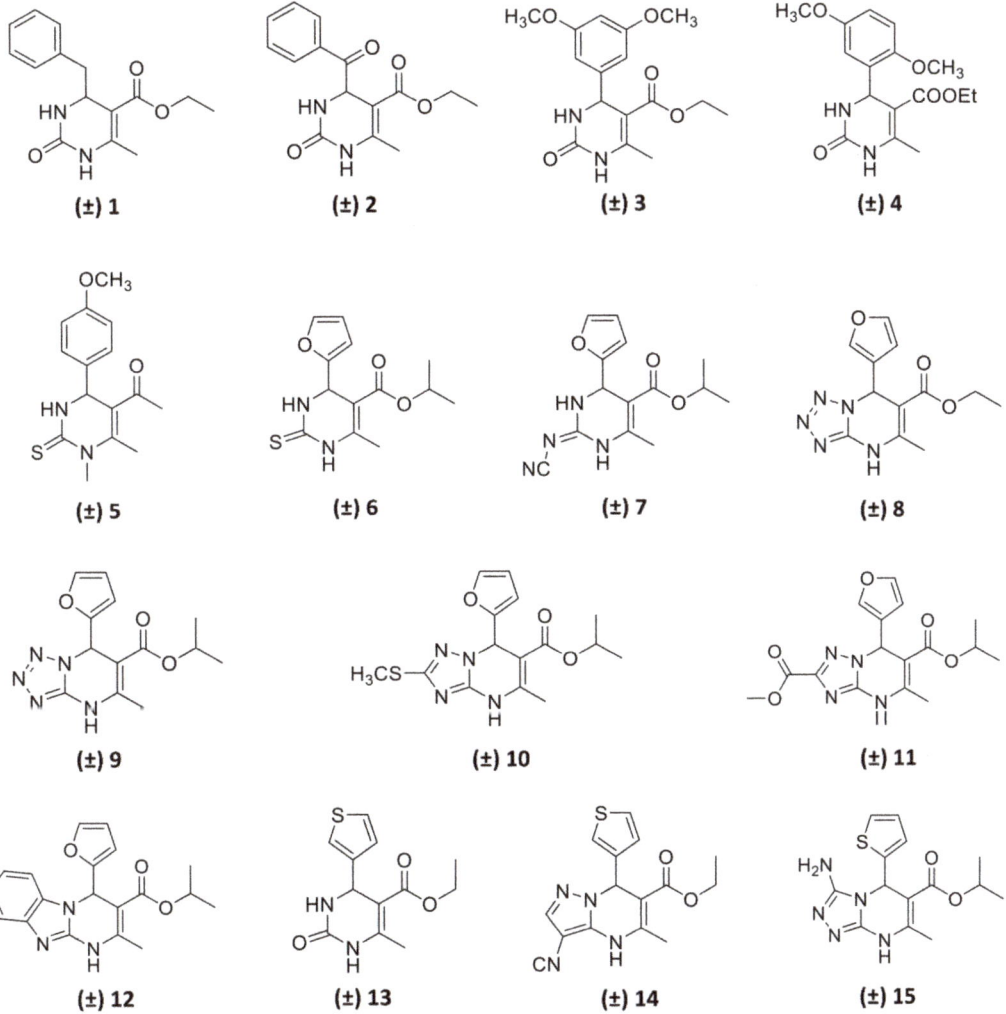

Figure 1. Structures of the examined chiral 4-aryl-substituted 3,4-dihydropyrimidin-2(1*H*)-one (DHP) ester derivatives. The biological activity of compounds **8–11** and **14–15** will be published elsewhere.

The enantioselective separations of the anticancer drug ifosfamide and its metabolites were successfully achieved on TAG under PO conditions, and a computational model suggested the network of HB interactions and cation-π interactions between the phosphoramide moiety and the aromatic components of the aglycon basket, which are fully consistent with the observed enantioselectivity [22].

Preliminary to the study of the chiral separation of DHP selectands on Chirobiotic™ TAG, three diverse Pirkle-type CSPs (Figure 2) were investigated in isocratic normal-phase (NP) conditions, with the aim of evaluating the effects on chiral recognition of π-π interactions, H-bonds, and repulsive steric factors. We use a computational protocol, similar to that previously adopted by us for another class of chiral selectands and diverse CSPs [23], to gain insights into the structure–enantioselective retention relationships and physicochemical interactions primarily responsible for the chiral recognition of the DHP selectands by TAG chiral selector, whose crystal structure has been solved herein by X-ray diffraction.

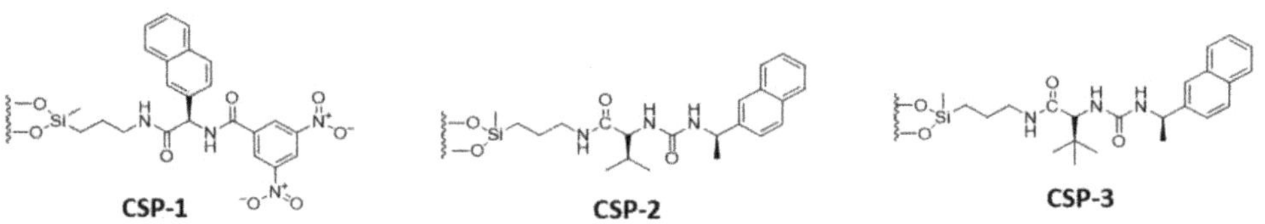

CSP-1

CSP-2

CSP-3

Pirkle-type chiral selector in CSPs 1–3

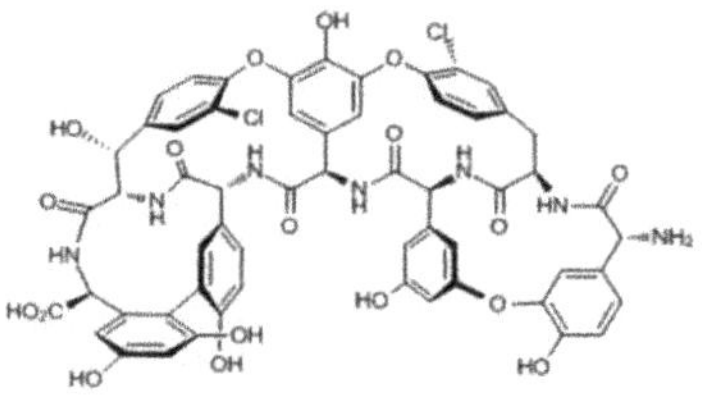

Teicoplanin aglycone chiral selector

Figure 2. Chiral selectors of the CSPs investigated in this study.

2. Materials and Methods

2.1. Chemicals

All the analytical reagents and HPLC-grade solvents were purchased from Sigma-Aldrich (Milan, Italy). Ultrapure water was purified by a Milli-Q system.

Most of the racemic 4-substituted 3,4-dihydropyrimidin-2(1H)-one ester derivatives **1–15** were already synthesised by the Sotelo group in the Singular Research Center in Biological Chemistry and Molecular Materials (CIQUS), University of Santiago de Compostela (Spain), with Biginelli-based procedures using different 1,3-dinucleophiles, aldehydes, and β-ketoesters [5–11]. The UV spectra of racemic compounds **6**, **11** and **12**, taken as representative of the different chromophores examined in this study are shown in Figure S3 (Supplementary Materials).

Compounds **5**, **8**, **9**, **10**, **11**, **14**, and **15** were prepared according to the following general procedure. Analytical, physicochemical, and biological data of these compounds will be reported elsewhere.

Compounds **5**, **10**, and **11**: A mixture of the appropriate 1,3-dinucleophile (N-methyl thiourea, 3-(methylthio)-3H-1,2,4-triazol-5-amine or methyl 5-amino-3H-1,2,4-triazole-3-carboxylate; 1.5 eq), the corresponding aldehyde (4-methoxy benzaldehyde, 2-furaldehyde or 3-furaldehyde; 1 eq), the desired β-ketoester (pentane-2,4-dione, or isopropyl 3-oxobutanoate; 1 eq), and chloroacetic acid (0.1 eq) in THF (3 mL) was stirred at 90 °C for 12 h. The reaction was monitored by TLC and upon completion, the solvent was evaporated and the residue was purified by flash chromatography on silica gel (Hex:AcOEt).

Compounds **8** and **9**: A mixture of 1H-tetrazol-5-amine (1.5 eq), the corresponding aldehyde (2-furaldehyde or 3-furaldehyde; 1 eq), β-ketoester (ethyl 3-oxobutanoate or isopropyl 3-oxobutanoate; 1 eq), and L-proline (0.1 eq) in DMF (3 mL) was stirred at 90 °C for 12 h. The reaction was monitored by TLC, and upon completion, the solvent was evaporated and the residue was purified by flash chromatography on silica gel (Hex:AcOEt).

Compounds **14** and **15**: A mixture of the appropriate heterocyclic 1,3-dinucleophile (5-amino-1H-pyrazole-4-carbonitrile or 4H-1,2,4-triazole-3,5-diamine; 1.5 eq), the corresponding aldehyde (2-thiophenecarboxaldehyde or 3-thiophenecarboxaldehyde; 1 eq), β-ketoester (ethyl 3-oxobutanoate or isopropyl 3-oxobutanoate; 1 eq), and chloroacetic acid (0.1 eq) in DMF (3 mL) was stirred at 90 °C for 15 h. The reaction was monitored by TLC, and upon completion, the solvent was evaporated and the residue was purified by flash chromatography on silica gel (Hex:AcOEt).

2.2. Chromatography

The analytical HPLC measurements were performed on a Waters 1525 HPLC System equipped with a Waters 2487 variable-wavelength UV-Vis detector and a Waters 717 plus autosampler. The chromatographic data were acquired using the Waters Breeze Software (Waters, Milan, Italy). Chromatographic separations were performed on the teicoplanin-based column Chirobiotic™ TAG (250 × 4.6 mm id, 5 µm particle size) from ASTEC (Whippany, NJ, USA). Analyses were also carried out by using three commercial Pirkle-type CHIREX® columns (50 × 4.6 mm id, 5 µm particle size), namely (R)-1-naphtylglycine and 3,5-dinitrobenzoic acid (CSP-1), (S)-Valine and (R)-1-(α-naphthyl)ethylamine (CSP-2), and (S)-tert-Leucine and (R)-1-(α-naphthyl)ethylamine (CSP-3), from Phenomenex (Castel Maggiore, Bologna, Italy). Stock solutions were prepared by dissolving the analytes in MeOH (1 mg/mL). The injection volume was 5 µL and all the chromatographic analyses were performed at room temperature in triplicate. The dead time (t_0) was measured as the retention time of methanol-d_4. The column flow rate was 0.7 mL/min, and the chromatograms were recorded by UV detection at wavelengths of 254 nm and 280 nm. The solvents for mobile phases were filtered through a polytetrafluoroethylene (PTFE) or Nylon-66 membrane (0.45 µm) before use. All the DHP analytes (Figure 1) were evaluated with different mobile phases under normal-phase (NP) and polar organic modes (POM).

In the competition experiments, N-acetyl-D-alanine (NADA) was added directly in the mobile phase. Three different concentrations of competitor were tested: 0.1–0.5% and 1%.

2.3. X-ray Crystallography

X-ray diffraction data were collected at the I03 Beamline of the Diamond Light Source, UK (wavelength 0.7293 Å, temperature 100 K, detector DECTRIS PILATUS3 S 6M). The XDS program [24] was used to perform data reduction, while POINTLESS and AIMLESS programs [25] were used for space group identification, scaling, and merging of diffraction data.

TAG crystallised in the space group C2, with one molecule in the asymmetric unit (85 non-H atoms); unit cell parameters a = 35.72 Å, b = 13.11 Å, c = 21.71 Å, and β = 123.34. The Matthews coefficient of such crystals (Vm) is 1.76 Å^3/Da and the solvent content is equal to 30%. The structure solution was achieved by the Modern Direct Methods (MDM) ab initio phasing procedure implemented in the package SIR2014 [26]. It was preliminarily refined by the SHELXL full-matrix least-squares technique based on F^2 [27], by using reflections with I ≥ 2σ(I) (7457/9030). Non-hydrogen atoms were refined with anisotropic thermal parameters. As a final step, DMSO and water molecules were manually added by using the graphics program COOT [28] and iteratively refined by using REFMAC5 [29]. Atomic coordinates and structure factor amplitudes of TAG were deposited in the Protein Data Bank with PDB ID code 6TOV. Data collection and structure refinement parameters are summarised in Table S4 and the crystal structure is shown in Figure S1 (Supplementary Materials).

2.4. Molecular Docking and Molecular Dynamics Calculations

The molecular skeleton of DHP compounds were built within the MAESTRO software package [30] with standard values of bond lengths and valence angles, and then passed to OPENBABEL [31] for 10,000 steps of steepest descent minimisation using the Universal Force Field. Six representative DHP derivatives among those listed in Figure 1 were chosen, considering three resolved (**6**, **8**, and **9**) and three not-resolved (**12**, **13**, and **15**) compounds, each in its two enantiomers. Docking calculations were performed on each of them considering the resolved TAG crystal structure, herein determined as the target. The *molcharge* complement of QUACPAC [32] was used to achieve Marsili–Gasteiger charges for both chiral selector and selectands. The affinity maps were calculated on a 0.375 Å spaced 85 × 85 × 85 Å^3 cubic box centred on the TAG crystal structure, and the accessibility of the binding site was explored throughout 1000 runs of the Lamarckian Genetic Algorithm (LGA) implemented in AUTODOCK 4.2.6 [33] using the GPU-OpenCL algorithm version [34], the population size and the number of energy evaluation figures were set to 300 and 10,000,000, respectively. All the achieved hits were clustered according to a RMSD threshold <2.0, and afterwards the best energy and most populated group was selected as input for further molecular dynamics (MD) calculations for monitoring the time-dependence of the interactions of the derivatives with the chiral selector using DESMOND [35].

The TAG/DHP complexes were assembled with the system builder tool implemented in Maestro [36] into an orthorhombic box filled with methanol as the explicit solvent. All simulations were performed on a NVDIA Quadro M4000 GPU at a constant temperature (300 K) and pressure (1 bar) for a total of 1 µs using the default settings and relaxation protocol of DESMOND, with energy and time steps for trajectory recording intervals of 0.5 ps, and to mimic the effect of the anchoring stationary phase, a harmonic constraint of 100 kcal/mol was applied on the TA Csp^3 carbon. From the achieved trajectories, a total number of 2000 frames were afterwards sampled for the data set calculations.

MD trajectories have been analysed by Principal Component Analysis (PCA) implemented in the program RootProf [37] and by scripts of the VMD program [38].

3. Results and Discussion

3.1. Chiral Separation with Pirkle-Type Chiral Stationary Phases (CSPs)

Pirkle-type CSPs are designed for achieving enantiomeric separations exploiting π-π stacking interactions between electron-rich and electron-deficient aromatic systems as the primary attractive interaction forces. Besides the aromatic ones, polar and H-bonding interactions, as well as steric repulsive factors, can play a role in modulating the enantioselective recognition of racemic selectands by chiral selectors in Pirkle-type HPLC columns [39].

The structures of the Pirkle-type CSPs, which differ for the π-acceptor/donor group, and α-amino acid linked through a n-propyl bridge to the silica matrix, are shown in Figure 3. Preliminarily, for some representative racemic compounds the mobile-phase composition varied from Hex-EtOH (95:5, v/v) to more polar mixtures, such as MeOH (100%, v/v) or MeOH/H$_2$O (95:5, v/v) mixtures, which resulted in significant decreases of the chromatographic selectivity. Therefore, the mobile phase was fixed as Hex-EtOH (95:5, v/v) and all the fifteen racemic DHPs were analysed at room temperature and flow rates of 0.7 mL/min. The chromatographic data, k_1 and α values, are summarised in Table 1.

Table 1. Chromatographic data: retention factors of first eluted enantiomer (k_1) and separation factor (α) for the chiral DHP analytes on Pirkle-type CSPs in fixed experimental conditions [a].

Analyte	CSP 1		CSP 2		CSP 3	
	k_1	α	k_1	α	k_1	α
1	5.07	1	2.51	1.15	2.38	1.17
2	23.0	1.09	11.0	1.11	10.4	1.09
3	19.9	1.23	7.70	1	6.67	1.06
4	15.7	1.21	5.61	1	4.86	1.06
5	23.5	1	20.3	1	22.6	1
6	8.22	1.09	5.06	1.22	4.24	1.27
7	11.5	1	11.2	1.05	6.13	1.11
8	6.97	1.08	5.34	1.06	3.85	1.12
9	6.92	1.08	6.17	1.07	4.34	1.09
10	2.00	1	0.93	1.43	1.06	1.43
11	11.0	1.40	4.37	1.24	4.27	1.38
12	2.25	1.22	0.98	1.39	0.99	1.51
13	11.5	1.12	7.08	1.08	6.62	1.13
14	5.82	1.17	2.19	1	1.62	1
15	17.03	1	5.66	1.1	4.59	1.18

[a] Mobile phase: n-Hex-EtOH (95:5, v/v); flow rate: 0.7 mL·min^{-1}; room temperature; UV detection: 254 nm.

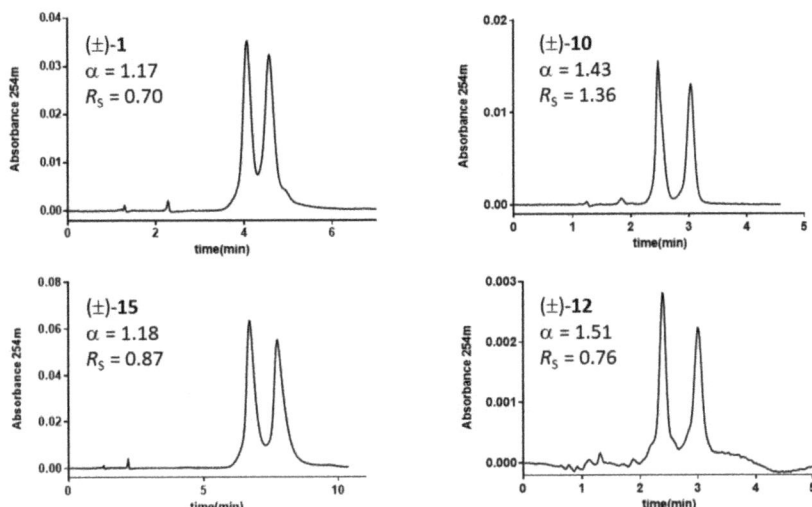

Figure 3. Chromatograms of DHP analytes **1**, **10**, **12**, and **15** on Pirkle-type 3020-(S)-*tert*-leucine and (R)-1-(α-naphthyl)ethylamine (**CSP-3**) columns. Mobile phase: Hex-EtOH (95:5, v/v); flow rate: 0.7 mL/min; room temperature; UV detection: 254 nm.

The Pirkle-type CPSs, in the given experimental conditions, separated the majority of the racemic DHPs selectands, but in no case a baseline resolution was obtained. The α values achieved with CSP1, bearing the π-acidic 3,5-dinitro benzoyl (3,5-DNB) moiety and napht-2-yl as amino acid sidechain, did not correlate at all with α values determined on CSP2 (r^2 = 0.006) and CSP3 (r^2 = 0.061), which in turn, given their more similar chemical structures, returned linearly related α values (r^2 = 0.923), with a slope close to unity (1.09) and an intercept close to zero (−0.05). CSP3, bearing napht-2-yl as the only aromatic binding site and *t*-butyl as the amino acid side-chain, provided slightly better enantiomeric separation of the examined DHP chiral derivatives, as (i) only two racemic compounds were not separated (in contrast to four and five non-resolved racemic selectands with CSP2 and CSP1, respectively), and (ii) the spread of the α ratios (Δα = 0.45) was slightly larger compared with CSP2 (Δα = 0.38) and CSP1 (Δα = 0.32).

Due to the methylation of N1, the racemic compound **5** is not separated on any of the three Pirkle-type columns, highlighting the critical role of the H-bond achieved by N1-H as HB donor. Compounds **10**, **11**, and **12**, which are among the bulkiest molecules in the investigated series, achieved better enantioselective separation with CSP2 and CSP3, bearing branched amino acid sidechains like *i*-Pr and *t*-Bu, respectively. HPLC traces on CSP3 for some diverse DHP derivatives, along with enantioselectivity α and resolution R_S values, are shown in Figure 3.

Overall, the chiral HPLC data from the three-Pirkle-type CSPs in NP mode suggest that through replacing the π-acidic 3,5-DNB aromatic binding site in CSP1 and gradually introducing a site of steric repulsive interaction, namely the sidechains *i*-Pr (CSP2) and *t*-Bu (CSP3), the retention times tend to diminish, and separation factors (α) tend to increase. Within the limits of the chemical space explored, CSP3 appears a more suitable and versatile Pirkle-type column for chiral DHP derivatives. However, in no case was a baseline separation achieved, but there should be room for improvement aimed at preparative applications.

3.2. Chiral Separation with Teicoplanin Aglycone Selector

Macrocyclic glycopeptide antibiotics have been successfully used as versatile and selective tools for enantioseparation, evaluation of enantiomeric purity, and the pharmacokinetic studies of a great variety of chiral molecules. Multiple stereogenic centres, different functional groups, and primarily inclusion cavities are responsible for the good outcomes of these CSPs [40].

In this study the separation of the enantiomeric DHP mixtures was performed with a Chirobiotic™ TAG column under a polar organic mode where MeOH, EtOH, and MeOH-ACN (50:50, v/v) were used as mobile phases. Different proportions of EtOH-ACN and MeOH-H$_2$O were preliminarily evaluated for a few racemic DHP-based selectands (Table S1, Supplementary Materials).

The TAG chiral selector consists of four fused rings forming a semi-rigid 'basket' that contains seven aromatic rings, two of which bear chloro-substituents and five bear phenol groups (Figure 2). The HPLC results obtained on TAG-based column eluting with the above mobile phases are summarised in Table 2, whereas the chromatograms of racemic DHP analogs are shown in Figure 4.

Table 2. Chromatographic data: retention factors of the first eluted enantiomer (k_1) and separation factor (α), for the chiral DHP analytes on a Chirobiotic™ TAG column with three organic polar mobile phases [a].

Analyte	A		B		C	
	k_1	α	k_1	α	k_1	α
1	0.72	4.24	2.13	4.00	0.66	2.65
2	0.84	3.88	2.46	2.75	0.64	2.55
3	0.86	2.96	2.40	2.42	0.61	2.90
4	0.89	1.61	2.17	1.58	0.61	1.46
5	0.73	1	1.9	1	0.63	1
6	0.22	2.41	0.50	2.12	0.14	1.64
7	0.32	2.00	0.56	2.28	0.31	1.32
8	0.47	8.76	0.85	9.68	0.26	4.77
9	0.46	6.04	0.94	5.52	0.21	3.67
10	0.27	1	0.37	1	0.24	1
11	0.46	1.50	0.94	1.45	0.18	1.11
12	0.35	1	0.52	1	0.30	1
13	0.79	1	2.32	1	0.66	1
14	0.28	1	0.41	1	0.15	1
15	0.40	1	0.75	1	0.38	1

[a] Mobile phases: A = MeOH, B = EtOH and C = MeOH-ACN (50:50, v/v); flow rate: 0.7 mL·min^{-1}; room temperature; UV detection: 254 nm.

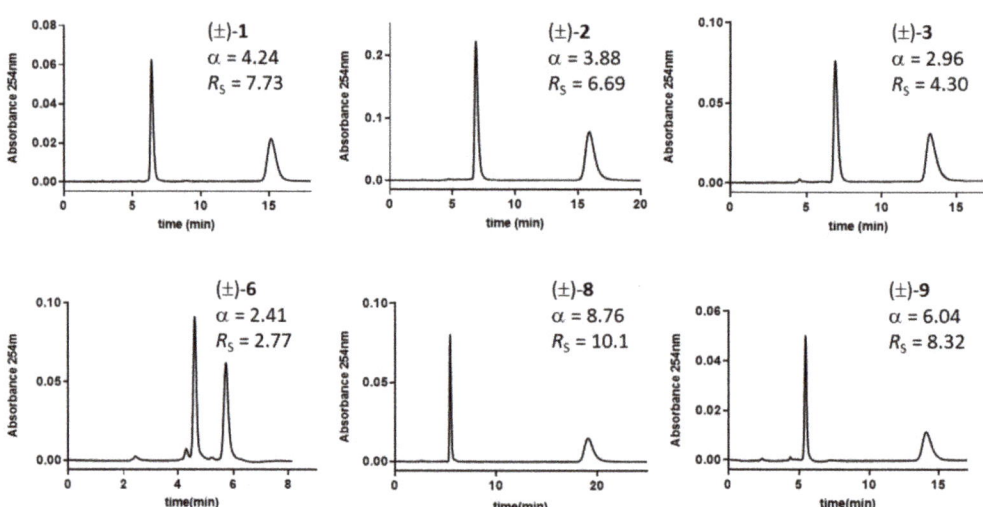

Figure 4. Chromatograms of DHP analytes on Chirobiotic™ TAG column. Mobile phase: MeOH 100%, v/v; flow rate: 0.7 mL·min^{-1}; room temperature; UV detection: 254 nm.

The majority of racemic DHP mixtures (60%) achieved on TAG-based CSP baseline enantiomer separations showed in some cases (e.g., *rac*-**8** and *rac*-**9**) excellent enantioselectivity ratios ($\alpha > 6$) and *Rs* values from 2 to 10. With *rac*-**6**, the previously resolved (*R*)-enantiomer was the most retained one (Figure S2, Supplementary Materials. The decrease of polarity of the mobile phase from MeOH 100%, *v/v*, to EtOH 100%, *v/v*, and the mixture MeOH/ACN (50:50, *v/v*) does decrease α values. The three-mobile-phase-furnished enantioseparation factors were highly intercorrelated ($r^2 > 0.88$). However, better separations were obtained with MeOH (A) and EtOH (B). The α values, which spanned a range of about 7.8 and 8.7 units with MeOH and EtOH, respectively, was linearly interrelated ($r^2 = 0.966$), with a slope close to unity (1.03), and an intercept equal to -0.17. The low toxicity of the solvents examined, compared to those generally employed in NPM, is an advantage to be considered.

The molecules under investigation are not congeneric; therefore, structure–enantioselectivity relationships (QSERs) are not always clearly explainable. Looking at the α values determined using MeOH as eluent (and EtOH as well), the best enantiomers resolution was achieved with the racemic analytes **8** and **9**, most likely because of the fused 1*H*-tetrazole ring bearing three further H-bond acceptor sites. In contrast, the loss of a critical H-bond donor group, due to the methylation of N1, may explain the lack of enantioselectivity of the TAG chiral selector toward *rac*-**5**. Repulsive steric effects experienced by the selectands into the TAG chiral basket(s) may cause the lack of enantiomeric resolution of DHP derivatives **10**, **12**, **14**, and **15** ($\alpha = 1$) and the poor enantioseparation ($\alpha = 1.50$) of **11**.

To account for steric effects, H-bonding, and polar interactions, the following molecular descriptors were calculated and collected in Table S2 (Supplementary Materials) for the whole set of DHP analytes: MV (molar volume), MSA (molecular surface area), PSA (polar surface area), HBA (count of H-bond acceptors), and HBD (count of H-bond donors). The squared correlation matrix of the enantioseparation factor and calculated descriptors of *rac*-**1–15** (Table S3) shows that, excluding the expected correlations of MV with MSA ($r^2 = 0.723$) and PSA with HBA count ($r^2 = 0.858$), no other noteworthy correlation exists among the molecular descriptors and between single descriptors and α value. Nevertheless, there must be a detrimental effect of bulkiness, as assessed by MV and MSA, on the enantioselectivity of TAG. As matter of fact, poorly or non-separated compounds ($\alpha < 1.5$), i.e., **10**, **11**, **12**, **14**, and **15**, have MSA values volumes falling within the area of 473 ± 13 Å^2, whereas better separated 4-substituted DHPs ($\alpha > 2$; **1–4** and **6–9**) have MSAs equalling 417 ± 38 Å^2. A similar trend can be inferred from the calculation of the mean MV $\pm$ SD. In contrast, no difference in the mean value of PSA, accounting for 'attractive' intermolecular forces, for the two clusters of compounds, which is 87 ± 15 Å^2 and 85 ± 13 Å^2, were found for the two groups of selectands achieving $\alpha < 2$ and $\alpha > 2$, respectively. This trend suggests that DHP derivatives with sizes larger than a threshold value (reasonably, MSA > 460 Å^2) could not access the TAG cleft, where highly directional polar and H-bond interactions may favour enantiomer separations.

Two remarkable exceptions to the above rules are represented by *rac*-**5** (MSA = 443 Å^2) and *rac*-**13** (MSA = 354 Å^2), which, despite their bulkiness under the threshold, did not achieve any enantiomer separation. While the lack of an N1-H HBD group could explain the lack of enantiomer separation with **5**, the absence of the enantioselective binding of **13** into the TAG chiral selector cleft was not so obvious. Compound **13**, with its MSA clearly below the threshold value of 460 Å^2, should enter the chiral cleft of TAG. It cannot be excluded that the S atom of the thiophen-3-yl group at the chiral C4, compared with the O atom in the furanyl moiety in other DHPs examined herein, serves as a poorer H-bond acceptor, due to unsuitable bond length and angles [41].

Teicoplanin-related antibiotics exert their activities because they stereospecifically bind to the precursor peptidoglycan peptide terminus *N*-acetyl-D-Ala-D-Ala during bacterial cell wall biosynthesis. *N*-acetyl-D-alanine (NADA) is capable of binding specifically to the pocket of TAG, and therefore was used as a competing agent in displacement studies. The displacement concept was applied to investigate the enantioselective and non-selective binding of test analytes on the Chirobiotic™ TAG column [42].

Interestingly, the competition experiments performed on *rac*-6 showed the retention factor of the more retained enantiomer (k'_2), and α value decrease when the concentration of NADA added to the mobile phase increases, thus suggesting that the enantiorecognition of the examined analytes takes place mainly through HBs and polar interactions in the same site where the terminal D-Ala of the 'natural' ligand binds. As shown in Figure 5, the experiment translates graphically with the tendency of the chromatographic peaks to coalescence.

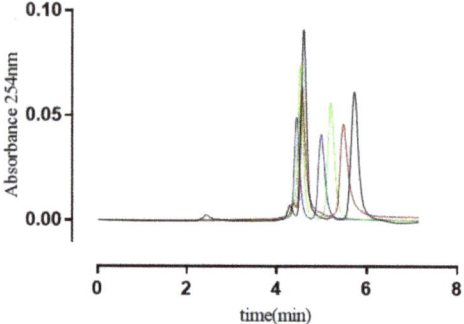

Figure 5. Superposition of chromatograms of (±)-**6** on TAG-based CSP using MeOH 100%, v/v as the mobile phase containing increasing concentrations of *N*-acetyl-D-Ala (NADA). Retention factors of the first and second eluted enantiomers (k'_1 and k'_2), and separation ratios (α) are summarised in the table; flow rate: 0.7 mL·min^{-1}; room temperature; UV detection: 254 nm.

3.3. Molecular Dockings of Selectands

For better understanding the intermolecular forces mainly responsible for the enantioselective recognition of 4-aryl DHPs by the TAG chiral selector, molecular docking and molecular dynamics calculations were performed with four racemic compounds (i.e., **6**, **8**, **9**, and **12**) chosen, exploring the range of α values (MeOH) from 8.76 (**8**) to 1 (**12**). The high-resolution (0.77 Å) X-ray crystal structure of TAG (Figure 6) was solved, deposited in the Protein Data Bank (PDB ID code: 6TOV) and used in molecular docking calculations with molecular models of the investigated 4-aryl DHP-containing selectands.

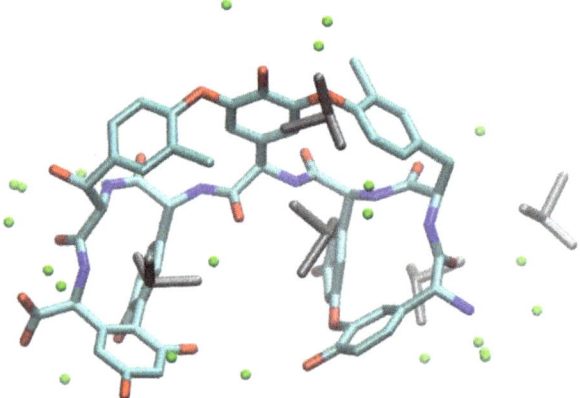

Figure 6. X-ray crystal structure of teicoplanin aglycone (TAG) (oxygen, red; nitrogen, blue; carbon, cyan) with DMSO (grey) and water molecules (green spheres).

The computational study allowed us to check for any kind of shape- and atom-type complementarities between the selectands and their anchoring points on the CS molecular surface, and at the same time to achieve plausible and low energy conformations for TAG–DHP complexes that might further constitute input for MD runs. In more detail, the selectands **6**, **8**, **9**, and **12** were initially docked into the basket-like cavity of TAG, and the relative energies of binding, together with a breakdown for the different contributions arising from the key attractive and/or repulsive interactions, were evaluated (Table 3). Starting from these data, timeline monitoring of geometrical and chemical features of TAG-selectands' interactions were indeed deduced from the analysis of MD trajectories carried out with principal component analysis (PCA).

Table 3. Free energy of binding (FEB) and relative contributions [a] estimated from docking calculations.

cmpd	Chirality	FEB	EFF [a]	VdW + HB + ELE + DES [a]	VdW + HB [a]	ELE [a]	DES [a]
6	R	−6.80	−0.296	−7.08	−8.87	0.02	1.78
	S	−7.55	−0.328	−7.60	−9.82	0.04	2.19
8	R	−6.74	−0.337	−7.13	−9.03	−0.12	2.02
	S	−6.68	−0.334	−7.07	−8.96	−0.05	1.94
9	R	−7.39	−0.352	−7.40	−9.29	−0.05	1.93
	S	−7.36	−0.350	−7.51	−9.43	−0.14	2.06
12	R	−7.56	−0.302	−7.38	−9.46	−0.25	2.33
	S	−7.16	−0.286	−7.10	−9.28	−0.26	2.44

[a] EFF, ligand efficiency (FEB/number of heavy atoms); VdW, Van der Walls; HB, hydrogen bond; ELE, electrostatic; DES, desolvation.

As it may be inferred from the docking poses (Figure 7), the binding mode is quite similar for structurally related selectands achieving comparable α values (i.e., **8** and **9**), whereas compounds with a diverse scaffold (i.e., **6** and **12**) substantially fit the TAG surface in a different manner. Indeed, the binding modes of the most retained 4-furanyl DHPs, such as **8** and **9**, are enforced by highly directional H-bonds comprising the nitrogen of the tetrazole ring, with the backbone peptide bond close to the charged N-terminal of TAG; moreover, the furan ring attached to the chiral centre is oriented, depending upon the configuration, towards diverse areas, namely the chlorine atom or the exterior moiety of the TAG basket exposed to solvent. It might be proposed that (R)-**8** and (S)-**9** should correspond to the most retained enantiomers according to a unique kind of π-π stacking involving the halogen atom and the five-membered heterocyclic rings. In contrast, (S)-**8** a®(R)-**9** should be more easily eluted by the mobile phase.

Regarding the two other selectands, the above considerations cannot be applied for both the enantiomers in **12**, since it is rather clear from the dockings that it cannot achieve any polar and persistent anchoring interactions due to the steric hindrance (exceeding volume) of the benzimidazole ring, while thiopyrimidinone **6** shows a diverse interaction pattern. The (R) configuration obliges the ligand to orient the thiopyrimidinone ring outside the TAG basket, while in the (S) configuration it is more deeply embedded in a narrow gorge, and consequently the furan ring stacks with the phenol ring close to the C-terminus of TAG. According to this interpretation, the elution order from the least to the most retained DHP enantiomers (i.e., (R)-**12**, (S)-**12**, (R)-**6**, (S)-**6**, (S)-**9**, (R)-**8**, (R)-**9**, and (S)-**8**) it is not consistent with the HPLC data. Deeper insights were instead gained from molecular dynamics calculations carried out on the complexes from the docking outputs of both the enantiomers of each virtually investigated DHP selectand. Regardless of the evidence emerging from the PCA hereafter reported, timeline plots of the radius of gyration highlighted some intriguing insights and divergences (Figure S4, Supplementary Materials). Both enantiomers of **8** and **9** are constantly patched on TAG, and this also applies to (R)-**6** in the first 200 ns. On the contrary, (S)-**6**, (R)-**12**, and (S)-**12** are substantially steered out from the basket-like TAG cavity. This may, at least in part, explain the differences in the experimentally determined separation factors of 4-aryl DHPs on TAG CSP.

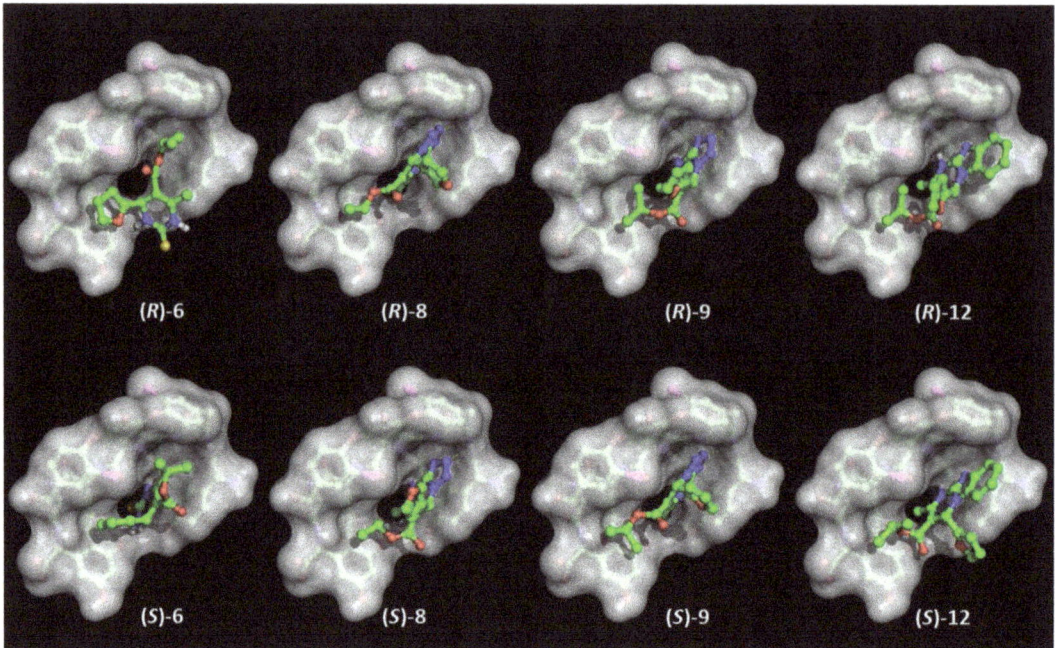

Figure 7. Binding poses of the enantiomers of compounds **6**, **8**, **9**, and **12** to X-ray structure of TAG rendered with transparent solid surface. Molecular models are colored by atom type: C (green), O (red), N (blue), S (yellow); H atoms are not shown.

3.4. Structural Interpretation of the Chiral Separation

The structural determinants of the chiral separation of DHP derivatives by TAG have been further investigated. The persistence of the docking pose, monitored by considering the distance between the barycentre of TAG and DHP selectands, did not appear related to the experimental data (Figure 8). In fact, both enantiomers of **8** and **9** persist on the initial docking position throughout the simulation with an average distance of about 5 Å; both enantiomers of **6**, **12**, and **13** move away from the initial pose, having an average distance of about 15 Å, while the enantiomers of **15** behave differently.

Conformational flexibility, which is instead strongly influenced by the interaction with TAG, was analysed in this study by the following procedure: (i) the molecules are aligned along the trajectory and amended of an initial equilibration time of 200 ns; (ii) the root mean square fluctuation (RMSF) of the DHP atoms is calculated; (iii) RMSF profiles of different compounds are scaled so that they have same average value and standard deviation to account for the different complexities of the compounds (Figure S5 and Table S5 in Supplementary Materials); (iv) scaled fluctuation profiles are compared by PCA. The PCA results are illustrated in Figure 9 (score plots), which shows the distribution of representative points of each compound in the score plots of the first two principal components.

Both enantiomers of compounds **6** and **13** are separated along the first principal component (PC1), explaining about 66% of the total data variance (Figure 9a). These differences are due to the way the molecules move during the simulation, regardless of their relative position with TAG. This could be due to the presence of sulphur (a poorer HBA), or to the specific interactions they have with TAG. To make the set of structures homogeneous, the comparative analysis of RMSF profiles has been repeated without these compounds (Figure 9b). In this case, PC1, which accounts for 61.2% of the total variance, clearly separates the (*R*)-enantiomers of **8** and **9**, while the same does not occur for the enantiomers of **12** and **15**, in full agreement with the experimentally observed chiral separation. Therefore, the fluctuation of the atomic positions of the common DHP nucleus, as influenced by the presence of TAG, regardless of whether the selectand remains bound to it or not, can explain the observed chiral separation. It is worth noting that the chiral separation of **8** and **9** was already present in Figure 9a but hidden by the presence of heterogeneous compounds **6** and **13**. Individual RMSF profiles, grouped according to clusters shown in Figure 9, are reported in Supplementary Materials (Figure S4).

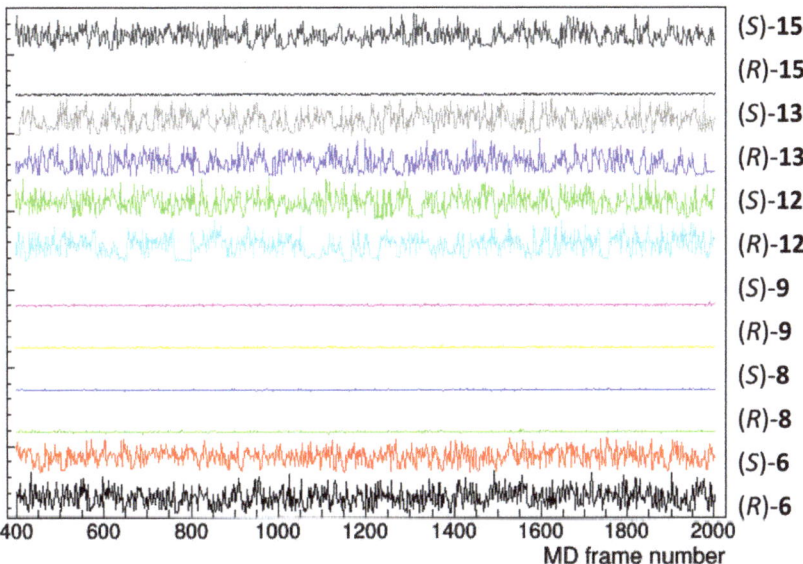

Figure 8. Distance between TAG and the compound barycentre as a function of the frame number of the MD trajectory (1 frame = 0.5 ns). Initial 400 frames, corresponding to 200 ns, have been not considered to account for system equilibration.

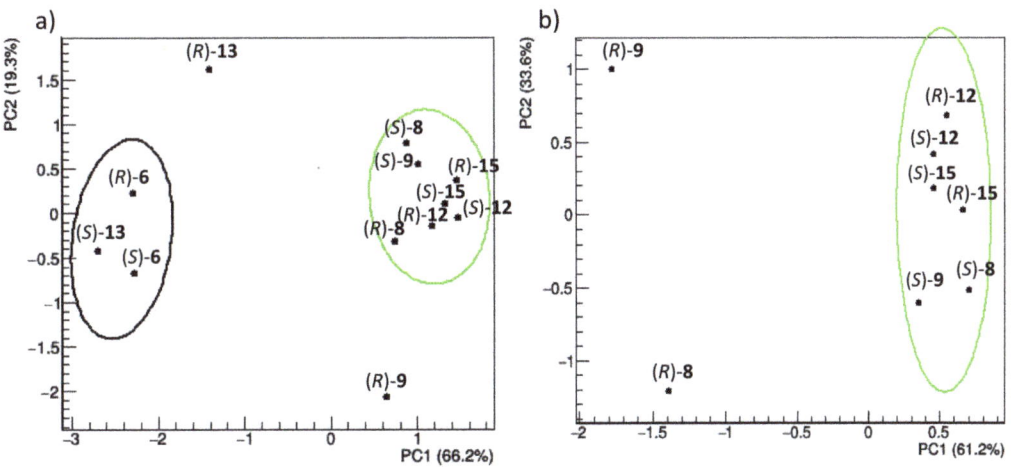

Figure 9. Score plots of the first two principal components obtained by PCA applied to RMSF profiles of the (R) and (S) enantiomers of racemic **6**, **8**, **9**, **12**, **13**, and **15** (**a**) and **8**, **9**, **12**, and **15** (**b**). Representative points of each enantiomer are grouped according to a hierarchic clustering algorithm; 85% confidence level ellipses are shown. The percentage of the data variance explained by each principal component is reported in parentheses on the axes.

To possibly gain further insight into the molecular features favouring the chiral separation, we compared the solvent accessibility of the two enantiomers in the whole conformer population sampled by MD calculations. The ratio between the solvent-accessible surface areas (SASA) of the DHP atoms of the (R) and (S) enantiomers is shown in Figure 10, which indicates, albeit not a statistically significant trend, the largest deviation from unity for compounds **8** and **9**.

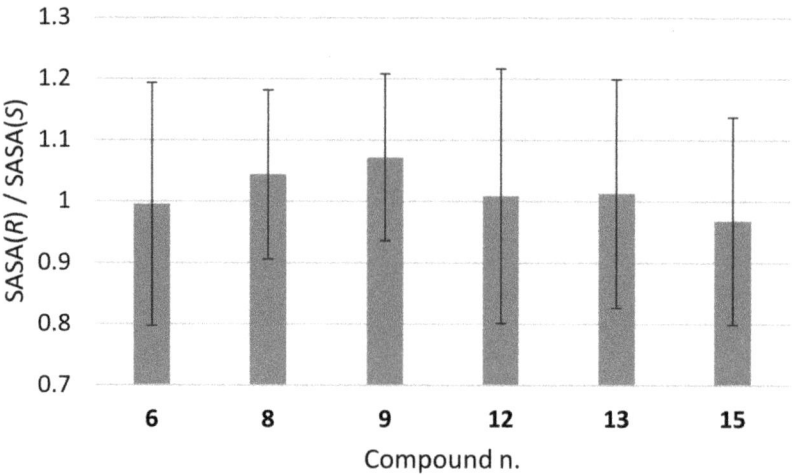

Figure 10. Ratio between average values of solvent accessible surface area (SASA) of (R) and (S) enantiomers, as calculated from MD simulations.

4. Conclusions

Chirobiotic™ TAG, a CSP bearing teicoplanin aglycone as the chiral selector, and 100%, v/v MeOH or EtOH as the mobile phase in PO conditions, proved to be a suitable method for the efficient enantiomer resolution of several 4-aryl-3,4-dihydropyrimidin-2(1H)-one (DHP) alkoxycarbonyl esters, pharmacologically active as A_{2B} adenosine receptor antagonists. Compared to three herein examined Pirkle-type CSPs, eluted in normal mode, the TAG-based CSP provided a baseline separation and high enantioselectivity factors ($\alpha > 2$) for the majority (60%) of the fifteen diverse racemic DHP derivatives investigated, demonstrating in those cases high potential for preparative purposes.

Quantitative information analysis of calculated molecular descriptors of size and polar/H-bonding interactions shed light on the main repulsive (steric effects) and attractive (H-bonds—polar and apolar) intermolecular interactions governing the binding of DHP selectands and the TAG chiral selector. Within the limits of the explored physicochemical space, it appears that only compounds with MV < 260 Å^3 and MSA < 460 Å^2 may enter the chiral cleft/cavities of TAG and form enantioselective H-bonds, both polar and apolar (e.g., π-stacking) interactions, which stabilize the selectand/selector complexes. Bulkier compounds (MV > 280 Å^3, MSA > 470 Å^2) should be less likely to form complexes with TAG that are stable over time. PSA or a count of H-bonding atoms, taken individually, are unable to distinguish between separated and non-separated DHP selectands.

Useful insights into the mechanism of chiral separation on TAG-based CSP were obtained by molecular dynamics calculations on selector-selectand, as obtained by docking calculation. Multivariate analysis (PCA and clustering) of MD simulation data revealed that the movements of the DHP selectands in the simulation box are influenced by the presence of TAG, even if they are not closely bound to it. Structural determinants, such as positional fluctuation and solvent accessibility to 4-aryl DHP selectands, were found to be related to the observed chiral separation.

Supplementary Materials: The following supporting information can be downloaded at https://www.mdpi.com/article/10.3390/separations9010007/s1, Figures S1–S5 and Tables S1–S5 showing graphics/plots or collecting data related to HPLC chiral separation, X-ray crystal structure of teicoplanin aglycone, molecular descriptors, molecular dynamics trajectories, etc.

Author Contributions: Conceptualisation, I.B., A.C., R.C., S.C. and C.D.A.; methodology, I.B., A.C., B.C., B.D.B., R.C. and S.C.; software, A.C., R.C. and C.D.A.; investigation, I.B., R.P., A.C., B.C., B.D.B., R.C. and S.C.; data curation, I.B., R.P., M.C., M.M., E.S., S.C. and C.D.A.; writing—original draft preparation, I.B., M.M., A.C., R.C. and S.C.; writing—review and editing, M.C., R.C., S.C. and C.D.A.; supervision, S.C. and C.D.A.; funding acquisition, R.P., A.C., M.C. and C.D.A. All authors have read and agreed to the published version of the manuscript.

Funding: This research was financially supported by the Italian Ministry of Universities and Research (Progetti di Rilevante Interesse Nazionale-Call 2017, PRIN 2017, Grant 201744BN5T_004).

Institutional Review Board Statement: Not applicable.

Informed Consent Statement: Not applicable.

Data Availability Statement: Not applicable.

Acknowledgments: The authors gratefully acknowledge the instrumental support of the University of Bari Aldo Moro, Department of Pharmacy–Pharmaceutical Sciences, and the Institute of Crystallography, Italian National Council of Research (C.N.R.), Bari. The authors gratefully acknowledge the Diamond Light Source for the provision of beamtime (proposal number MX15832).

Conflicts of Interest: The authors declare no conflict of interest.

Abbreviations

DHP, 3,4-dihydropyrimidin-2(1H)-one; CSP, chiral stationary phase; NARP, non-aqueous reversed phase; NP, normal phase; NADA, N acetyl d alanine; TAG, teicoplanin aglycone; TE, teicoplanin; MeOH, methanol; EtOH, ethanol; IPA, isopropanol; ACN, acetonitrile; Hex, hexane.

References

1. Calcaterra, A.; D'Acquarica, I. The market of chiral drugs: Chiral switches versus de novo enantiomerically pure compounds. *J. Pharm. Biomed. Anal.* **2018**, *14*, 7323–7340. [CrossRef]
2. Kappe, C.O. Biologically active dihydropyrimidones of the Biginelli-type: A literature survey. *Eur. J. Med. Chem.* **2000**, *35*, 1043–1052. [CrossRef]
3. Kappe, C.O. Recent Advances in the Biginelli dihydropyrimidine synthesis. New tricks from an old dog. *Acc. Chem. Res.* **2000**, *33*, 879–888. [CrossRef] [PubMed]
4. Zhu, J.; Bienayme, H. Asymmetric isocyanide-based MCRs. In *Multicomponent Reactions*, 1st ed.; Wiley-VCH: Weinheim, Germany, 2005.
5. Crespo, A.; El Maatougui, A.; Biagini, P.; Azuaje, J.; Coelho, A.; Brea, J.; Loza, M.I.; Cadavid, M.I.; García-Mera, X.; Gutiérrez-de-Terán, H.; et al. Discovery of 3,4-dihydropyrimidin-2(1H)-ones as a novel class of potent and selective A2B adenosine receptor antagonists. *ACS Med. Chem. Lett.* **2013**, *4*, 1031–1036. [CrossRef] [PubMed]
6. Carbajales, C.; Azuaje, J.; Oliveira, A.; Loza, M.I.; Brea, J.; Cadavid, M.I.; Masaguer, C.F.; García-Mera, X.; Gutiérrez de Terán, H.; Sotelo, E. Enantiospecific recognition at the A2B adenosine receptor by alkyl 2-cyanoimino-4-substituted-6-methyl-1,2,3,4-tetrahydropyrimidine-5-carboxylates. *J. Med. Chem.* **2017**, *60*, 3372–3382. [CrossRef]
7. Crespo, A.; El Maatougui, A.; Azuaje, J.; Escalante, L.; Majellaro, M.; Loza, M.I.; Brea, J.; Cadavid, M.I.; Gutiérrez de Terán, H.; Sotelo, E. Exploring the influence of the substituent at position 4 in a series of 3,4-dihydropyrimidin-2(1H)-one A2B adenosine receptor antagonists. *Chem. Heterocycl. Compd.* **2017**, *53*, 316–322. [CrossRef]
8. Mallo-Abreu, A.; Majellaro, M.; Jespers, W.; Azuaje, J.; Caamaño, O.; García-Mera, X.; Brea, J.M.; Loza, M.I.; Gutiérrez-De-Terán, H.; Sotelo, E. Trifluorinated pyrimidine-based A2B antagonists: Optimization and evidence of stereospecific recognition. *J. Med. Chem.* **2019**, *62*, 9315–9330. [CrossRef] [PubMed]
9. El Maatougui, A.; Azuaje, J.; González-Gómez, M.; Miguez, G.; Crespo, A.; Carbajales, C.; Escalante, L.; García-Mera, X.; Gutiérrez de Terán, H.; Sotelo, E. Discovery of potent and highly selective A2B adenosine receptor antagonist chemotypes. *J. Med. Chem.* **2016**, *59*, 1967–1983. [CrossRef]
10. Mallo-Abreu, A.; Prieto-Díaz, R.; Jespers, W.; Azuaje, J.; Majellaro, M.; Velando, C.; García-Mera, X.; Caamaño, O.; Brea, J.; Loza, M.I.; et al. Nitrogen-Walk Approach to Explore Bioisosteric Replacements in a Series of Potent A2B Adenosine Receptor Antagonists. *J. Med. Chem.* **2020**, *63*, 7721–7739. [CrossRef]

11. Majellaro, M.; Jespers, W.; Crespo, A.; Núñez, M.J.; Novio, S.; Azuaje, J.; Prieto-Díaz, R.; Gioé, C.; Alispahic, B.; Brea, J.; et al. 3,4-Dihydropyrimidin-2(1*H*)-ones as Antagonists of the Human A2B Adenosine Receptor: Optimization, Structure–Activity Relationship Studies, and Enantiospecific Recognition. *J. Med. Chem.* **2021**, *64*, 458–480. [CrossRef]
12. Kappe, C.O.; Stadler, A. The Biginelli dihydropyrimidinone synthesis. *Org. React.* **2004**, *63*, 1–116. [CrossRef]
13. Bhosale, R.S.; Wang, T.; Zubaidha, P.K. An efficient, high yield protocol for the one-pot synthesis of dihydropyrimidin-2(1*H*)-ones catalyzed by iodine. *Tetrahedron Lett.* **2004**, *45*, 9111–9113. [CrossRef]
14. Reddy, K.R.; Reddy, C.V.; Mahesh, M.; Raju, P.V.K.; Narayana Reddy, V.V. New environmentally friendly solvent free synthesis of dihydropyrimidinones catalysed by *N*-butyl-*N*,*N*-dimethyl-α-phenylethylammonium bromide. *Tetrahedron Lett.* **2003**, *44*, 8173–8175. [CrossRef]
15. Balalaie, S.; Soleiman-Beigia, M.; Rominerb, F. Novel one-pot synthesis of new derivatives of dihydropyrimidinones and unusual polysubstituted imidazolin-2-ones: X-ray crystallographic structure. *J. Iran. Chem. Soc.* **2005**, *2*, 319–329. [CrossRef]
16. Berthod, A.; Chen, X.; Kullman, J.P.; Armstrong, D.W.; Gasparrini, F.; D'Acquarica, I.; Villani, C.; Carotti, A. Role of the Carbohydrate Moieties in Chiral Recognition on Teicoplanin-Based LC Stationary Phases. *Anal. Chem.* **2000**, *72*, 1767–1780. [CrossRef] [PubMed]
17. Jandera, P. Comparison of various modes and phase systems for analytical HPLC. In *Separations Methods in Drug Synthesis and Purification*; Valkò, K., Ed.; Elsevier: Amsterdam, The Netherlands, 2000; pp. 1–71.
18. Péter, A.; Arki, A.; Tourwé, D.; Forró, E.; Fülöp, F.; Armstrong, D.W. Comparison of the separation efficiencies of chirobiotic T and TAG columns in the separation of unusual amino acids. *J. Chromatogr. A* **2004**, *1031*, 159–170. [CrossRef] [PubMed]
19. Berthod, A.; Xiao, T.L.; Liu, Y.; McCulla, R.D.; Jenks, W.S.; Armstrong, D.W. Separation of chiral sulfoxides by liquid chromatography using macrocyclic glycopeptide chiral stationary phases. *J. Chromatogr. A* **2002**, *955*, 53–69. [CrossRef]
20. Altomare, C.; Carotti, A.; Cellamare, S.; Fanelli, F.; Gasparrini, F.; Villani, C.; Carrupt, P.; Testa, B. Enantiomeric resolution of sulfoxides on a DACH-CNB chiral stationary phase: A quantitative structure-enantioselective retention relationship (QSERR) study. *Chirality* **1993**, *5*, 527–537. [CrossRef]
21. Meričko, D.; Lehotay, J.; Armstrong, D.W. Effect of temperature on retention and enantiomeric separation of chiral sulfoxides using teicoplanin aglycone chiral stationary phase. *J. Liq. Chromatogr. Relat. Technol.* **2006**, *29*, 623–638. [CrossRef]
22. Ravichandrana, S.; Collins, J.R.; Singh, N.; Wainer, I.W. A molecular model of the enantioselective liquid chromatographic separation of (R,S)-ifosfamide and its N-dechloroethylated metabolites on a teicoplanin aglycon chiral stationary phase. *J. Chromatogr. A* **2012**, *1269*, 218–225. [CrossRef]
23. Pisani, L.; Rullo, M.; Catto, M.; de Candia, M.; Carrieri, A.; Cellamare, S.; Altomare, C. Structure-property relationship study of the HPLC enantioselective retention of neuroprotective 7-[(1-alkylpiperidin-3-yl) methoxy]coumarin derivatives on an amylose-based chiral stationary phase. *J. Sep. Sci.* **2018**, *41*, 1376–1384. [CrossRef]
24. Kabsch, W. XDS. *Acta Cryst.* **2010**, *D66*, 125–132. [CrossRef]
25. Winn, M.D.; Ballard, C.C.; Cowtan, K.D.; Dodson, E.J.; Emsley, P.; Evans, P.R.; Keegan, R.M.; Krissinel, E.B.; Leslie, A.G.; McCoy, A.; et al. Overview of the CCP4 suite and current developments. *Acta Cryst.* **2011**, *D67*, 235–242.
26. Burla, M.C.; Caliandro, R.; Carrozzini, B.; Cascarano, G.L.; Cuocci, C.; Giacovazzo, C.; Mallamo, M.; Mazzone, A.; Polidori, G. Crystal structure determination and refinement via SIR2014. *J. Appl. Cryst.* **2015**, *48*, 306–309. [CrossRef]
27. Sheldrick, G.M. Crystal structure refinement with SHELXL. *Acta Crystallogr.* **2015**, *C71*, 3–8.
28. Emsley, P.; Cowtan, K. Coot: Model-building tools for molecular graphics. *Acta Cryst.* **2004**, *D60*, 2126–2132. [CrossRef]
29. Murshudov, G.N.; Vagin, A.A.; Dodson, E.J. Refinement of macromolecular structures by the maximum-likelihood method. *Acta Cryst.* **1997**, *D53*, 240–255. [CrossRef] [PubMed]
30. *Schrödinger Release 2020-4, Maestro*, Schrödinger, LLC: New York, NY, USA, 2020.
31. O'Boyle, N.M.; Banck, M.; James, C.A.; Morley, C.; Vandermeersch, T.; Hutchison, G.R. Open Babel: An open chemical toolbox. *J. Cheminf.* **2021**, *3*, 33. [CrossRef] [PubMed]
32. *QUACPAC 2.1.0.4*, OpenEye Scientific Software: Santa Fe, MX, USA. Available online: http://www.eyesopen.com (accessed on 21 December 2021).
33. Morris, G.M.; Goodsell, D.S.; Halliday, R.S.; Huey, R.; Hart, W.E.; Belew, R.K.; Olson, A.J. Automated docking using a Lamarckian genetic algorithm and an empirical binding free energy function. *J. Comput. Chem.* **1998**, *19*, 1639–1662. [CrossRef]
34. Santos-Martins, D.; Solis-Vasquez, L.; Tillack, A.F.; Sanner, M.F.; Koch, A.; Forli, S. Accelerating AutoDock4 with GPUs and Gradient-Based Local Search. *J. Chem. Theory Comput.* **2021**, *17*, 1060–1073. [CrossRef] [PubMed]
35. Bowers, K.J.; Chow, E.; Xu, H.; Dror, R.O.; Eastwood, M.P.; Gregersen, B.A.; Klepeis, J.L.; Kolossvary, I.; Moraes, M.A.; Sacerdoti, F.D.; et al. Scalable Algorithms for Molecular Dynamics Simulations on Commodity Clusters. In Proceedings of the ACM/IEEE Conference on Supercomputing (SC06), Tampa, FL, USA, 11–17 November 2006.
36. *Maestro-Desmond Interoperability Tools*, Schrödinger: New York, NY, USA, 2020.
37. Caliandro, R.; Belviso, B.D. RootProf: Software for multivariate analysis of unidimensional profiles. *J. Appl. Cryst.* **2014**, *47*, 1087–1096. [CrossRef]
38. Humphrey, W.; Dalke, A.; Schulte, K. VMD: Visual molecular dynamics. *J. Mol. Graph.* **1996**, *14*, 33–38. [CrossRef]
39. Forjan, D.M.; Gazic, I.; Vinkovic, V. Role of the weak interactions in enantiorecognition of racemic dihydropyrimidinones by novel Brush-type chiral stationary phases. *Chirality* **2000**, *19*, 446–452. [CrossRef] [PubMed]

40. Fernandes, C.; Tiritan, M.E.; Cass, Q.; Kairys, V.; Fernandes, M.X.; Pinto, M. Enantioseparation and chiral recognition mechanism of new chiral derivatives of xanthones on macrocyclic antibiotic stationary phases. *J. Chromatogr. A* **2012**, *1241*, 60–68. [CrossRef] [PubMed]
41. Zhou, P.; Tian, F.; Lv, F.; Shang, Z. Geometric characteristics of hydrogen bonds involving sulfur atoms in proteins. *Proteins* **2009**, *76*, 151–163. [CrossRef] [PubMed]
42. Slama, I.; Ravelet, C.; Villet, A.; Ravel, A.; Grosset, C.; Peyrin, E. Displacement study on a vancomycin-based stationary phase using N-acetyl-D-Alanine as a Competing Agent. *J. Chromatogr. Sci.* **2002**, *40*, 83–86. [CrossRef] [PubMed]

Article

Quality Distinguish of Red Ginseng from Different Origins by HPLC–ELSD/PDA Combined with HPSEC–MALLS–RID, Focus on the Sugar-Markers

Qian Cheng [1,2,†], Shuhuan Peng [1,2,†], Fangyi Li [1,2], Pengdi Cui [1,2], Chunxia Zhao [1,2], Xiaohui Yan [2], Tongchuan Suo [1,2], Chunhua Wang [1,2,*], Yongzhi He [1,2] and Zheng Li [1,2,*]

1 College of Pharmaceutical Engineering of Traditional Chinese Medicine, Tianjin University of Traditional Chinese Medicine, Tianjin 301617, China; cq15735178293@163.com (Q.C.); peng910617622@163.com (S.P.); lifangyi@tjutcm.edu.cn (F.L.); cuipengdi@icloud.com (P.C.); zhaochunxia199411@163.com (C.Z.); suotc@tjutcm.edu.cn (T.S.); heyongzhi126@126.com (Y.H.)
2 State Key Laboratory of Component-Based Chinese Medicine, Tianjin University of Traditional Chinese Medicine, Tianjin 301617, China; yanxh@tjutcm.edu.cn
* Correspondence: pharmwch@126.com (C.W.); lizheng@tjutcm.edu.cn (Z.L.); Tel.: +86-22-59791811 (C.W.); +86-22-59791815 (Z.L.)
† Both these authors contribute equally to this work.

Abstract: Red ginseng (RG) has been extensively utilized in Asian countries due to its pharmacological effects. For the quality evaluation of RG, small molecules, such as ginsenosides, have been widely considered as candidates of its quality markers (Q-markers), and various analytical techniques have been developed in order to identify these compounds. However, despite the efforts to analyze the hydrophobic constituents, it is worth pointing out that about 60% of the mass of RG is made of carbohydrates, including mono-, oligo- and polysaccharides. Consequently, the quality differentiation and identification of RG from the perspective of sugar-markers should be focused. High performance liquid chromatography and evaporative light scattering detector (HPLC–ELSD) method for the determination of disaccharides in RG was established. Furthermore, high performance size exclusion chromatography–multi-angle laser light scattering–refractive index detector (HPSEC–MALLS–RID) for the determination of molecular weight and high performance liquid chromatography photodiode array (HPLC–PDA) for the determination of compositional monosaccharides in RG polysaccharides were also established. HPLC–ELSD/PDA combined with HPSEC–MALLS–RID could be used to determine the contents of disaccharides, molecular weights, and compositional monosaccharides of RG polysaccharides, which could be used for quality control, and this is a new view on the sugar marker to quality differentiation of various origins of RG.

Keywords: compositional monosaccharides; HPLC–ELSD/PDA combined with HPSEC–MALLS–RID; molecular weight; quality markers (Q-marker); red ginseng polysaccharides

1. Introduction

Red ginseng (RG) is steamed dried root and rhizome of the cultivated product of Panax ginseng Meyer. It is a mild Chinese medicine that tastes sweet and smells light and has the effects of replenishing the vital qi, restoring pulse, and relieving collapse syndrome, supplementing qi, and activating blood. It is used for shallow breathing, shortness of breath, coldness of limbs, profuse sweating, or weakness [1]. Modern research shows that RG has antitumor [2], anti-aging [3,4], antioxidant [5,6], and other physiological activities. The chemical components isolated from RG include mainly ginsenosides, sugars, volatile oils, amino acids, and trace elements [7,8]. Among these, ginsenosides and sugars are the main chemical constituents and are also recognized as the main active ingredients [9–13].

In the meantime, since the chemical ingredients of RG can fluctuate dramatically in response to the environmental variations (e.g., climate, cultivating conditions, etc.), a reasonable and effective strategy of quality control of this herbal product is greatly required in the community. In this respect, the hydrophobic constituents of RG, such as ginsenosides and saponins, have been widely considered as candidates of quality markers (Q-markers), and various analytical techniques have been developed in order to identify these compounds. For instance, Jeong et al. developed an effective high performance liquid chromatography–photodiode array detector (HPLC–PDA) method and demonstrated that the method was useful for the quantification of maltol in various ginseng products [14]. In et al. reported an HPLC-based method for simultaneous quantification of twelve ginsenosides in RG powder and extract and found the method suitable for quality control of ginseng products [15]. Zhou and his coworkers proposed an ultra-fast liquid chromatography coupled with electrospray ionization triple quadrupole tandem mass spectrometry (UFLC–MS/MS) method to evaluate the quality of RG, which could quantify sixty-six saponins and their six aglycones [16]. Kim et al. investigated the time-dependent changes in the crude saponin and the major natural and artifact ginsenosides contents during simmering and recommended (20S)- and (20R)-ginsenoside Rg3 as new reference materials to complement ginsenoside Rb1 and Rg1 [17]. Lee and his colleagues applied ultra-performance liquid chromatography coupled to quadrupole time-of-flight mass spectrometry (UPLC–QTOF/MS)-based metabolomics for the quality evaluation of four types of ginsengs, and their results indicated that the approach was useful for the quality control of processed ginseng products [18]. Wu et al. performed UPLC–QTOF/MS analysis to detect ginsenosides in white ginseng and RG and assigned several chemical markers with the help of multivariate statistical analysis [19].

However, despite the researchers' efforts to analyze the hydrophobic constituents, it is worth pointing out that ca. 60% [20] of the mass of RG is made of carbohydrates, including mono-, oligo- and polysaccharides [21]. Meanwhile, relevant studies have revealed the pharmacological effects of the polysaccharides of RG, such as immune activity [22–25] and anti-aging effects [26]. Most recently, Shin et al. found that RG polysaccharides could inhibit tau aggregation and promote the dissociation of tau aggregates [27]. Hence, it is quite reasonable to include the sugars in the Q-markers of RG.

In addition, polysaccharide is one of the active components in traditional Chinese medicine (TCM). The activity of polysaccharides is closely related to its structure. The molecular weight and compositional monosaccharides of polysaccharides are also the major factors affecting the therapeutic action of polysaccharides. Therefore, the establishment of determination methods for molecular weight and compositional monosaccharides of polysaccharides can provide the reference for the study of polysaccharides.

Hence, it is quite reasonable to include the sugars in the Q-markers of RG. Based on these considerations, we established a methodology combining evaporative light-scattering detector–photodiode array (HPLC–ELSD/PDA) and high performance size exclusion chromatography (HPSEC)–multi-angle laser light scattering (MALLS)–refractive index detector (RID) to identify and characterize the RG disaccharides and polysaccharides. With the help of principle component analysis (PCA) and cluster analysis, we further showed that RG polysaccharides could be used for the quality control of RG products. In short, this is a novel view of sugar marker to distinguish the different origins of RG.

2. Materials and Methods

2.1. Plant Materials and Reagents

Eight batches of RG from different origins were purchased from Anhui Yishengyuan traditional Chinese medicine decoction pieces Technology Co., Ltd. (bozhou, China). The details of the samples are as follows: Jilin Tonghua (batch number: 190301), Liaoning Xinbin (batch number: 190201), Liaoning Huanren (batch number: 190201), Jilin Jingyu (batch number: 190401), Jilin Huichun (batch number: 190101), Jilin Antu (batch number: 190401), Jilin Fusong (batch number: 190501), and Jilin Dunhua (batch number: 190101). All samples were identified by Dr. Chun-Hua Wang, Tianjin University of Traditional Chinese

Medicine, and these samples were kept at the College of Pharmaceutical Engineering of TCM, Poyanghu Road, Jinghai, Tianjin, China.

Sucrose (batch number: S02S6G1, content ≥98%), maltose (batch number: RM0331FC14, content ≥98%), D-galactose (batch number: Z22J9H64187, content ≥98%), D-anhydrous glucose (batch number: S10S9I69833, content ≥98%), L-arabinose (batch number: T05J6C1, content ≥98%), and 1-phenyl-3-methyl-5-pyrazolone (PMP) reagent were purchased from Shanghai Yuanye Biotechnology Co., Ltd. Acetonitrile and methanol (HPLC grade) were purchased from Fisher (Fair Lawn, NJ, USA). Other reagents were analytically pure grade. Na2SO4 (anhydrous sodium sulphate) was purchased from Tianjin North Tianyi Chemical Reagent Factory. Proclin 300 and ammonium acetate were purchased from Beijing Solebo Technology Co., Ltd.

2.2. Instrumentation

HPLC–ELSD data were detected on a Waters ACQUITY HPLC^{TM} System (Waters, Milford, MA, USA) and ELSD detector 2424 (Waters, Milford, MA, USA). The HPLC–RID–MALLS system consisted of the HPLC instrument LC-20AD (Shimadzu, Kyoto, Japan), RID detector RID-20A (Shimadzu, Kyoto, Japan), and MALLS DAWN8 (Wyatt Technology Co., Santa Barbara, CA, USA). HPLC–PDA analyses were performed using the HPLC ACQUITY Arc instrument (Waters, Milford, MA, USA) and the PDA detector 2998 (Waters, Milford, MA, USA). In addition, we used the 5-digit Analytical Balance AB 135 S and the 4-digit Analytical Balance AL 204 electronic analytical balance (Mettler Toledo instruments Co., Ltd., Shanghai, China), the ultrasonic cleaner KQ2200DB (Kunshan Ultrasonic Instrument Co., Ltd., Kunshan, Jiangsu, China), the vacuum freezing dryer FDU-2110 (EYELA, Tokyo, Japan) and the Q-POD ultrapure water machine (Millipore, Illkirch-Graffenstaden, France).

2.3. Simultaneous Determination of Disaccharides in Rg Using the Hplc–Elsd Method

2.3.1. Preparation of Samples

RG samples were pulverized and passed through a 40-mesh sieve. The sample powders (2.0 g) were added to 20 mL of ultrapure water then sonicated for 30 min in an ultrasonic bath.

The mixed samples contain 16.37 mg of sucrose and 26.30 mg of maltose per 1 mL.

2.3.2. Chromatographic Conditions on the Determination of Disaccharides

Chromatographic column: YMC Pack NH$_2$/S-5 µm/12 nm (250 × 4.6 mm I.D.); flow rate: 1.0 mL/min; mobile phase: acetonitrile-water (79:21); column temperature: 30 °C; ELSD: gain 10; drift tube temperature: 60 °C; air pressure: 35 psi; sprayer heating power level: 60%; injection volume: 10 µL.

2.3.3. Methodological Study on the Determination of Disaccharides

In order to verify the feasibility of the method, we have conducted some method validation experiments, including linearity, precision, repeatability, stability, and recovery.

2.3.4. Content Determination of Disaccharides

Eight batches of RG were collected from Jilin Tonghua, Liaoning Xinbin, Liaoning Huanren, Jilin Jingyu, Jilin Huichun, Jilin Antu, Jilin Fusong, and Jilin Dunhua. RG samples were prepared and were determined according to the chromatographic conditions on the determination of disaccharides (10 µL per injection) [28].

2.4. Determination of Molecular Weight and Compositional Monosaccharides of Rg Polysaccharides

2.4.1. Molecular Weight Analysis of Rg Polysaccharides

Extraction of Polysaccharides

RG powders from different areas were weighed (approximately 1.0 g) and heated to reflux for 3 h after adding 10 mL of pure water. The resulting solution was centrifuged (3750 r/min, 10 min) and the supernatant was evaporated to 10 mL, at which point four volumes of 95% ethanol were added. The solution was left overnight at 4 °C and centrifuged (3750 r/min, 10 min) again; the precipitate was retained and the residual ethanol was removed by heating. The dried precipitate was dissolved in 10 mL of hot water, vortexed, and centrifuged. The supernatant was centrifuged (2220 r/min, 22 min) in an ultrafiltration centrifuge tube (molecular weight cut-off of 3 kDa). Finally, the solution was freeze-dried to obtain the RG polysaccharide's samples.

Preparation of the RG Polysaccharide's Samples

The RG polysaccharides dissolved in the mobile phase were configured into a 1 mg/mL solution through 0.22 µm microporous filter membrane filtration in order to obtain the subsequent filtrate.

Preparation of the Glucan Solution

The MALLS is normalized with a 3 mg/mL control solution of glucan (40 kDa).

Chromatographic Conditions on the Determination of Molecular Weight

Chromatographic column: TSKgel GMPW$_{XL}$ (7.8 mm I.D. × 30 cm, 13 µm); flow rate: 0.6 mL/min; mobile phase: 0.7% Na_2SO_4 (contains 0.02% Proclin 300 antibacterial agent); column temperature: 35 °C; detector: MALLS combined with RID; injection volume: 100 µL.

2.4.2. Analysis of Compositional Monosaccharides of RG Polysaccharides
Preparation of Hydrolyzed Polysaccharides

The polysaccharide was completely dissolved in 2 mol/L TFA (trifluoroacetic acid) in the sealed tube (m polysaccharide: v TFA = 2:1). The tube was kept in boiling water for 6 h to hydrolyze the polysaccharides to monosaccharides. The acid was removed through co-distillation with methanol. The dried products were dissolved in distilled water (m polysaccharides: v H_2O = 2:1) to obtain the hydrolyzed products [29]. The hydrolyzed products were modified with PMP. The monosaccharide aqueous solution (200 µL) was mixed with 200 µL of 0.3 mol/L NaOH, and then 200 µL of 0.5 mol/L PMP methanol solution was added. The reaction was allowed to run for 1 h at 70 °C and then cooled to room temperature and neutralized with a 200 µL 0.3 mol/L HCl solution. The sample solution was extracted with 1 mL of chloroform; the process was repeated three times. Finally, the sample solution was centrifuged (8000 r/min, 10 min) and the supernatant was injected for analysis.

Preparation of the Mixed Monosaccharides Solution

Amounts of 5.42 mg D-mannose, 10.36 mg D-galactose, 25.72 mg D-anhydrous glucose, 5.60 mg L-arabinose, and 4.97 mg D-galacturonic acid were placed in a 10 mL brown volumetric flask. The solution was then diluted with pure water and the method referred to in Preparation of Hydrolyzed Polysaccharides was used for derivation in order to obtain the mixed reference solution.

Chromatographic Conditions on the Determination of Compositional Monosaccharides

Chromatographic column: Kromasil 100-5-C_{18} (4.6 × 250 mm, 5 µm); flow rate: 1.0 mL/min; mobile phase: 0.1 mol/L Ammonium acetate solution-acetonitrile (79:21); column temperature: 35 °C; detective wave: 250 nm; injection volume: 20 µL.

Methodological Study on the Determination of Compositional Monosaccharides of Polysaccharides

In order to verify the feasibility of the method, we have conducted some method validation experiments, including linearity, precision, repeatability, stability, and recovery.

2.4.3. Content Determination of Molecular Weights and Compositional Monosaccharides of Rg Polysaccharides

Determination of Molecular Weights of Polysaccharides

The RG samples from different origins were weighed for the preparation of hydrolyzed polysaccharides and 100 µL were injected according to the chromatographic conditions on the determination of molecular weights of polysaccharides.

Determination of Compositional Monosaccharides of Polysaccharides

The RG samples from different origins were weighed for the preparation of hydrolyzed polysaccharides and 20 µL were injected according to the chromatographic conditions on the determination of compositional monosaccharides.

3. Results and Discussion

3.1. Conditions Optimization

The extraction solvent (including water, 55%, 75%, 95% ethanol), extraction method (including ultrasonic extraction, reflux extraction), and ultrasonic extraction time were optimized in this experiment. The results showed that the optimal extraction process was extracted with water and ultrasonic for 30 min.

The experiment examines the effect of filter membrane adsorption on the determination of sample solution by filter membrane adsorption test. The RG reference sample solution was collected and centrifuged (4000 r/min, 10 min) to determine the peak area (A1) of the sample, and 0.1 mL was discarded to obtain the peak area (A2) of the continued filtrate sample to calculate the sample recovery (A2/A1 × 100%). The results show that the recoveries of sucrose and maltose were 102.08% and 103.18%, respectively. Additionally, he recoveries were 95–105%, which indicated that the membrane has less adsorption on the samples and less interference on the determination results.

Monosaccharides and disaccharides cannot be directly detected by the UV detector, and the process of derivatization of monosaccharides and disaccharides needs to be converted into substances with UV absorption, and sample derivatization may in turn introduce impurities, causing errors. Therefore, in this experiment, the HPLC–ELSD method was established to directly detect disaccharides, and this method is simple and rapid, which does not need the derivatization treatment of samples. However, the amino column has a large column loss, which easily leads to a decrease in column efficiency and poor durability. In order to achieve a better separation, the parameters were optimized in this experiment, and finally, 79% v/v acetonitrile-water isocratic elution was used with a column temperature of 30 °C.

3.2. Results on the Determination of Sucrose and Maltose

The chromatograms showed that the peak separation of sucrose and maltose was good and easy to be distinguished, thus this method could be used for the determination of disaccharides in RG (Figure 1A,B). Therefore, we performed method validation, and the results are shown in Table S1. We can see from the table that the results of the determination of sucrose and maltose were good, which indicated that this method was suitable for the determination of sucrose and maltose. The results of the contents are shown in Table 1.

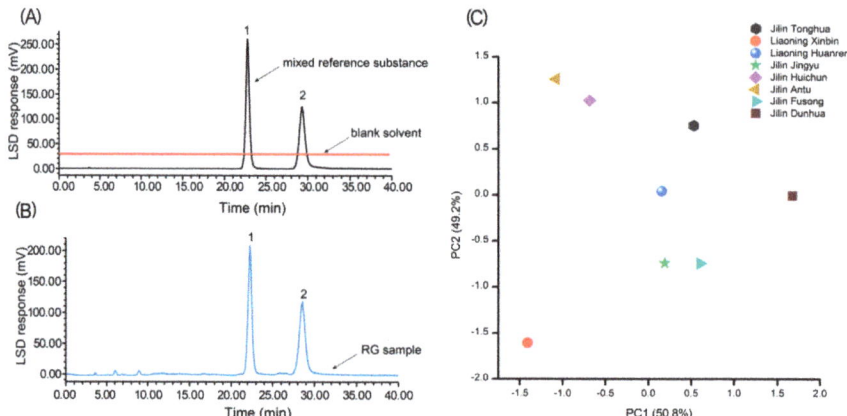

Figure 1. HPLC–ELSD chromatograms of blank solvent, mixed references (**A**) RG sample; (**B**) PCA of RG samples from different origins; (**C**) 1, sucrose; 2, maltose.

Table 1. Results of determination of sucrose and maltose in RG.

Localities	Sucrose (mg/g)	Maltose (mg/g)
Jilin Tonghua	50.37	186.12
Liaoning Xinbin	57.89	92.02
Liaoning Huanren	56.29	162.39
Jilin Jingyu	70.78	145.96
Jilin Huichun	52.46	172.92
Jilin Antu	70.07	179.93
Jilin Fusong	78.22	155.05
Jilin Dunhua	84.10	194.62

3.3. Quantitative Analysis Results

PCA (principal component analysis) can simply express the multivariate information about the samples and makes it intuitive to see the correlation and variability among different samples [30–32].

PCA was performed on the contents of sucrose and maltose present in the eight batches of RG using MATLAB 2018a, and the PCA scatter plots of the eight batches of RG polysaccharide's samples were obtained with principal components 1 and 2, whose cumulative proportion in ANOVA reached 100% (Figure 1C). The distribution of eight batches of RG samples from different areas was relatively scattered, indicating that the contents of sucrose and maltose in RG from different producing regions was quite different.

3.4. Analysis on the Polysaccharides Molecular Weight Determination

HPSEC–MALLS–RID is an efficient and high-quality analysis technique for the molecular weight and distribution of natural polymers [33,34]. In this experiment, the molecular weight and distribution of samples were investigated by HPSEC–MALLS–RID, and dextran standards (40 kDa) were used to verify the accuracy of it. Because the molecular weight of polysaccharides is large, the molecular weight distribution is wide, the separation of gel column is poor, and the samples were washed out of various molecules between 16.65 min and 20 min, the Mws of 3 and 4 could not be precisely determined (Figure 2). Table 2 summarizes the Mw and polydispersity index of polysaccharide fractions (peak 1 and peak 2). The Mws (peak 1 and 2) of the RG polysaccharide fractions of different origins ranged from 7.851×10^2–3.8091×10^3 kDa to 18.7–4.752×10^2 kDa, among which the Mws of the RG polysaccharide fractions (peak 1 and peak 2) from Jilin Jingyu were significantly lower than the others.

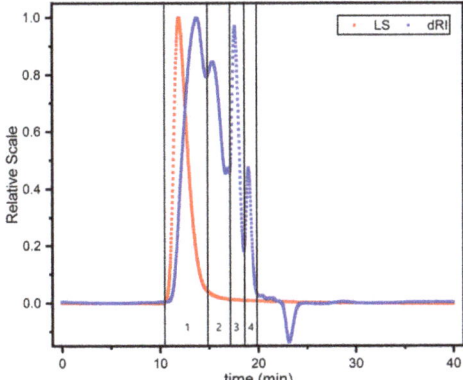

Figure 2. HPSEC–MALLS–RID chromatogram of RG polysaccharides sample.

Table 2. Results of molecular weight determination of RG polysaccharides.

Localities	Mw [a] (kDa) and Error of Peak 1	Mw/Mn [b] and Error of Peak 1	Mw (kDa) and Error of Peak 2	Mw/Mn and Error of Peak 2
Jilin Tonghua	2996.6 (±1.2%)	2.943 (±1.584%)	235.9 (±2.0%)	1.027 (±2.900%)
Liaoning Xinbin	3708.5 (±1.3%)	2.886 (±1.679%)	322.0 (±1.8%)	1.019 (±2.546%)
Liaoning Huanren	3502.4 (±1.5%)	3.158 (±2.097%)	126.3 (±7.6%)	- [c]
Jilin Jingyu	785.1 (±1.9%)	3.577 (±5.558%)	18.7 (±9.2%)	-
Jilin Huichun	3157.7 (±1.1%)	3.187 (±1.305%)	238.0 (±1.2%)	1.020 (±1.681%)
Jilin Antu	3809.1 (±1.3%)	2.532 (±1.645%)	475.2 (±1.7%)	1.016 (±2.417%)
Jilin Fusong	2351.0 (±1.3%)	3.299 (±1.779%)	81.4 (±4.6%)	-
Jilin Dunhua	2463.4 (±1.3%)	3.147 (±1.978%)	83.9 (±6.4%)	-

[a] Mw: molecular weight. [b] Mw/Mn: molecular weight distribution coefficient of polymer. [c] "-"means uncertainty.

3.5. Results of Determinations of the Compositional Monosaccharides of Polysaccharides

The chromatogram showed that the peak separation of monosaccharides was good and easy to distinguish, thus this method could be used for the determination of compositional monosaccharides of polysaccharides (Figure 3A–C). Therefore, we performed method validation, and the results are shown in Table S2. We can see from the table that the results of the determination of compositional monosaccharides were good, which indicated that this method was suitable for the determination of compositional monosaccharides. The results of the contents are shown in Table 3.

Table 3. Results of composition determination of RG polysaccharide.

Localities	Glucose (mg/mL)	Galactose (mg/mL)	Arabinose (mg/mL)
Jilin Tonghua	1.0999	0.0509	0.0410
Liaoning Xinbin	0.7816	0.0519	0.0382
Liaoning Huanren	0.9876	0.0548	0.0430
Jilin Jingyu	1.1636	0.0664	0.0497
Jilin Huichun	0.9833	0.0410	0.0314
Jilin Antu	0.8839	0.0454	0.0302
Jilin Fusong	1.0553	0.0542	0.0396
Jilin Dunhua	1.0366	0.0553	0.0426

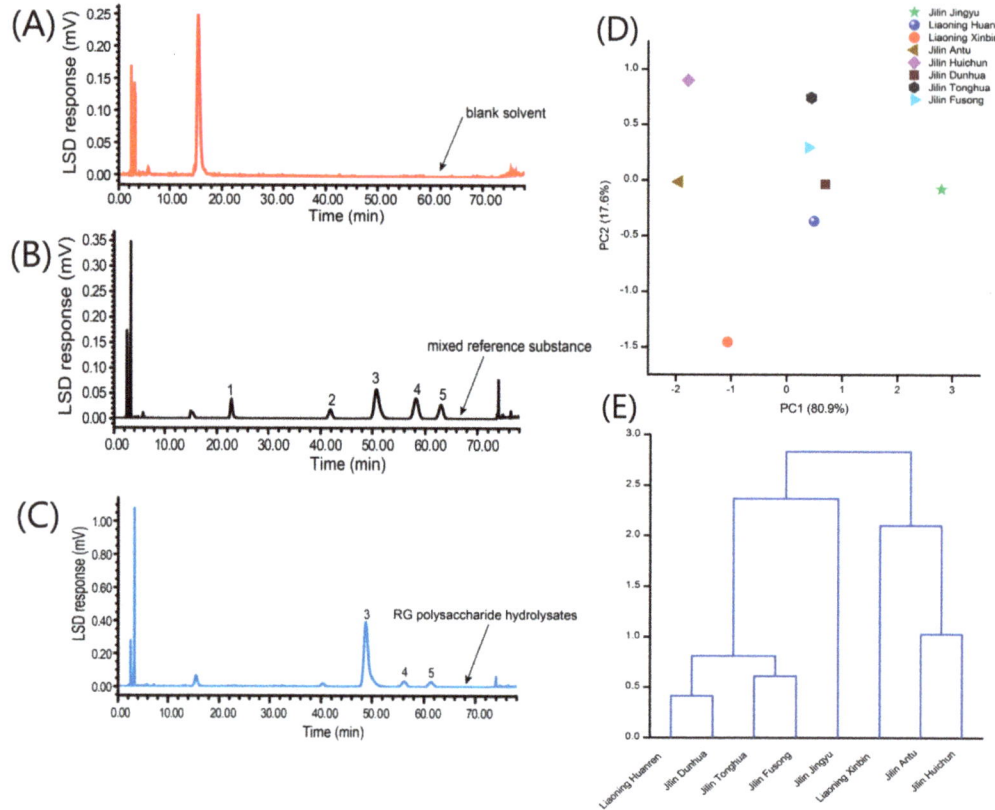

Figure 3. HPLC–PDA chromatograms of PMP derivatized blank solvent (**A**), mixed reference (**B**), RG polysaccharide hydrolysates (**C**); PCA (**D**) and cluster analysis (**E**) results of RG polysaccharide's Scheme 1. mannose; 2, galacturonic acid; 3, glucose; 4, galactose; 5, arabinose.

3.6. Analysis on the Monosaccharide Composition Determination of Polysaccharides
3.6.1. PCA Results

The similarity and difference between different samples can be studied by the PCA analysis. On PCA plots, the more clustered the sample distribution, the more similar the samples are to each other, and the more discrete the sample distribution points are, the greater the variation among samples [35]. PCA was performed on the content of the monosaccharide in the eight batches of RG using MATLAB 2018a. For its dimension-reducing processing, two principal components were obtained whose cumulative proportion in ANOVA reached 100% (Figure 3D). The distribution points of RG samples from different origins are quite discrete.

3.6.2. Results of Systematic Cluster Analysis

The results of compositional monosaccharides determination of polysaccharides were clustered using MATLAB 2018a by the connection method between groups, averaging Euclidean distances. The contents of glucose, galactose, and arabinose were regarded as variables (Figure 3E). The samples from Liaoning Huanren, Jilin Dunhua, Jilin Tonghua, Jilin Fusong, and Jilin Jingyu were included are of a kind. The monosaccharide composition of polysaccharides in RG from these five areas are similar. The samples from Liaoning Xinbin, Jilin Antitu and Jilin Huichun are of another kind. The results of this analysis were identical to the PCA results.

4. Conclusions

In this experiment, the quantitative analysis of disaccharides in RG was established based on the HPLC–ELSD method and its methodology was investigated. The precision of the instrument, the stability, and the repeatability of the sample determination were less than 5%. It is indicated that this method is accurate and reproducible and can be used for the determination of disaccharides in RG. The quantitative results were obtained by PCA analysis, which can intuitively see the similarities and differences of the contents of disaccharides in different sources of RG. The results showed that there are large variations in disaccharides in RG from different areas. This method can be used for the quality control and quality evaluation of raw medicinal materials. In particular, this method is very suitable for the quality control of TCM preparations with high sugar content, such as ginseng products.

Meanwhile, we established the HPSEC–MALLS–RID method to determine the molecular weights of RG polysaccharides from different origins, which can directly measure the molecular weights of polysaccharides without a standard curve. Then, it can provide a reference for other herbal polysaccharides. The PMP pre-column derivatization HPLC method was developed to determine the compositional monosaccharides of RG polysaccharides, and methodological study was carried out. This method is suitable for the analysis of the compositional monosaccharides of RG polysaccharide. RG polysaccharides are mainly composed of glucose, galactose, and arabinose. The experimental results used PCA and cluster analysis to compare the similarities and differences in the compositional monosaccharides of polysaccharides from the RG, and the compositional monosaccharides of the RG samples from Liaoning Xinbin, Jilin Antu, and Jilin Huichun showed marked differences from other locations; thus this study can provide methods and references for the quality control and development and utilization of polysaccharides.

As the literature reported, it is worth pointing out that about 60% of the mass of RG is made of carbohydrates. Our results showed that sugar can be used as a Q-marker to distinguish different origins of RG. However, due to our limited sampling, we have not formed a good cluster analysis. We will continue to collect abundant RG samples according to different places of origins to improve the control system of sugar markers of RG from the perspective of sugars.

Supplementary Materials: The following are available online at https://www.mdpi.com/article/10.3390/separations8110198/s1, Table S1: Results of methodological investigation on the determination of sucrose and maltose; Table S2: Results of methodological investigation on compositional monosaccharides determination of polysaccharides; Table S3: Linear results of content determination of disaccharides. Linear relationship of sucrose; Table S4: Linear relationship of maltose; Figure S1: Linear relationship of sucrose; Figure S2: Linear relationship of maltose.

Author Contributions: Conceptualization, C.W., Z.L., F.L., T.S. and Y.H.; methodology, Q.C., S.P. and C.Z.; software, Q.C. and P.C.; validation, Q.C., S.P. and F.L.; data curation, Q.C. and S.P.; writing—original draft preparation, Q.C. and S.P.; writing—review and editing, C.W., Z.L., F.L., X.Y., T.S. and Y.H.; funding acquisition, Z.L. and X.Y. All authors have read and agreed to the published version of the manuscript.

Funding: This work was supported by the National Key R&D Program of China, Synthetic Biology Research (No. 2019YFA0905300) and the National Natural Science Foundation of China (No. 82074276).

Institutional Review Board Statement: The study did not require ethical approval.

Informed Consent Statement: The study did not require an Informed Consent Statement.

Data Availability Statement: All data is contained within the article.

Conflicts of Interest: The authors have declared no conflict of interest.

References

1. Chinese Pharmacopoeia Commission. *Pharmacopoeia of the People's Republic of China*; China Medical Science Press: Beijing, China, 2020.
2. Zheng, X.; Zhou, Y.; Chen, W.; Chen, L.; Lu, J.; He, F.; Li, X.; Zhao, L. Ginsenoside 20(S)-Rg3 Prevents PKM2-Targeting miR-324-5p from H19 Sponging to Antagonize the Warburg Effect in Ovarian Cancer Cells. *Cell. Physiol. Biochem.* **2018**, *51*, 1340–1353. [CrossRef] [PubMed]
3. Park, M.Y.; Han, S.J.; Moon, D.; Kwon, S.; Lee, J.W.; Kim, K.S. Effects of Red Ginseng on the Elastic Properties of Human Skin. *J. Ginseng Res.* **2020**, *44*, 738–746. [CrossRef] [PubMed]
4. Sun, J.; Jiao, C.; Ma, Y.; Chen, J.; Wu, W.; Liu, S. Anti-Ageing Effect of Red Ginseng Revealed by Urinary Metabonomics Using RRLC-Q-TOF-MS. *Phytochem. Anal.* **2018**, *29*, 387–397. [CrossRef] [PubMed]
5. Kim, M.H.; Lee, E.J.; Cheon, J.M.; Nam, K.J.; Oh, T.H.; Kim, K.S. Antioxidant and Hepatoprotective Effects of Fermented Red Ginseng against High Fat Diet-Induced Hyperlipidemia in Rats. *Lab. Anim. Res.* **2016**, *32*, 217–223. [CrossRef]
6. Liu, J.W.; Yue, C.Y.; Wu, S.; Zhao, X.Q.; Zhang, Y.; Gong, L.H. Study on Antioxidant Ability of Red Ginseng Polysaccharide in Mice. *China Food Addit.* **2019**, *30*, 68–71.
7. Zhou, Q.L.; Xu, W.; Yang, X.W. Chemical constituents of Chinese red ginseng. *Chin. J. Chin. Mater. Med.* **2016**, *41*, 233–249.
8. Zhou, Q.Q.; Ren, W.M.; Wang, Y.H.; Yang, D.; Wang, G.M.; Li, Y.R. Research Progress on Processing Drugs Methods, Chemical Composition and Pharmacological Activity of Red Ginseng. *Shanghai J. Tradit. Chin. Med.* **2016**, *50*, 97–100.
9. Liu, C.L.; Xie, Q.S.; Li, Q.Y.; Gao, T.Y.; Liu, H.X.; Jiang, H. Determination of Total Ginsenosides, Ginsenosides Rg1, Re and Rb1 in Red Ginseng. *J. Pharm. Res.* **2021**, *40*, 87–90.
10. Gao, T.Y.; Jiang, Y.Q.; Li, Q.Y.; Hu, F.D.; Xie, Q.S.; Wang, H. Determination of 12 Ginsenosides in Red Ginseng by High Performance Liquid Chromatography. *J. Food Saf. Food Qual.* **2021**, *12*, 175–181.
11. Da, J.; Wang, Q.R.; Wang, Y.; Yao, S.; Huang, Y.; Wei, W.L.; Liang, J.; Shen, Y.; Franz, G.; Guo, D.A. Quantitative Analysis of Eight Ginsenosides in Red Ginseng Using Ginsenoside Rg1 as Single Reference Standard. *World J. Tradit. Chin. Med.* **2021**, *7*, 1–5. [CrossRef]
12. Lee, D.H.; Cho, H.J.; Kang, H.Y.; Rhee, M.H.; Park, H.J. Total Saponin from Korean Red Ginseng Inhibits Thromboxane A2 Production Associated Microsomal Enzyme Activity in Platelets. *J. Ginseng Res.* **2012**, *36*, 40–46. [CrossRef] [PubMed]
13. Abashev, M.; Stekolshchikova, E.; Stavrianidi, A. Quantitative Aspects of the Hydrolysis of Ginseng Saponins: Application in HPLC-MS Analysis of Herbal Products. *J. Ginseng Res.* **2021**, *45*, 246–253. [CrossRef]
14. Jeong, H.C.; Hong, H.D.; Kim, Y.C.; Rhee, Y.K.; Choi, S.Y.; Kim, K.T.; Kim, S.S.; Lee, Y.C.; Cho, C.W. Quantification of Maltol in Korean Ginseng (Panax Ginseng) Products by High-Performance Liquid Chromatography-Diode Array Detector. *Pharmacogn. Mag.* **2015**, *11*, 657–664. [CrossRef]
15. In, G.; Ahn, N.G.; Bae, B.S.; Han, S.T.; Noh, K.B.; Kim, C.S. New Method for Simultaneous Quantification of 12 Ginsenosides in Red Ginseng Powder and Extract: In-House Method Validation. *J. Ginseng Res.* **2012**, *36*, 205–210. [CrossRef]
16. Zhou, Q.L.; Zhu, D.N.; Yang, X.W.; Xu, W.; Wang, Y.P. Development and Validation of a UFLC-MS/MS Method for Simultaneous Quantification of Sixty-Six Saponins and Their Six Aglycones: Application to Comparative Analysis of Red Ginseng and White Ginseng. *J. Pharm. Biomed. Anal.* **2018**, *159*, 153–165. [CrossRef]
17. Kim, I.W.; Cha, K.M.; Wee, J.J.; Ye, M.B.; Kim, S.K. A New Validated Analytical Method for the Quality Control of Red Ginseng Products. *J. Ginseng Res.* **2013**, *37*, 475–482. [CrossRef] [PubMed]
18. Lee, J.W.; Ji, S.H.; Choi, B.R.; Choi, D.J.; Lee, Y.G.; Kim, H.G.; Kim, G.S.; Kim, K.; Lee, Y.H.; Baek, N.I.; et al. UPLC-QTOF/MS-Based Metabolomics Applied for the Quality Evaluation of Four Processed Panax ginseng products. *Molecules* **2018**, *23*, 2062. [CrossRef]
19. Wu, W.; Sun, L.; Zhang, Z.; Guo, Y.; Liu, S. Profiling and Multivariate Statistical Analysis of Panax Ginseng Based on Ultra-High-Performance Liquid Chromatography Coupled with Quadrupole-Time-of-Flight Mass Spectrometry. *J. Pharm. Biomed. Anal.* **2015**, *107*, 141–150. [CrossRef] [PubMed]
20. Qi, B.; Liu, L.; Zhao, D.Q.; Zhao, Y.; Bai, X.Y.; Zhang, H.M.; Guan, Y.Y.; Zhao, S.N. Comparative Study of Sugar Content in Panax Ginseng, P. quinquefolium and Red Ginseng. *J. China Pharm.* **2013**, *24*, 616–618.
21. Li, L.; Ma, L.; Guo, Y.; Liu, W.; Wang, Y.; Liu, S. Analysis of Oligosaccharides from Panax Ginseng by Using Solid-Phase Permethylation Method Combined with Ultra-High-Performance Liquid Chromatography-Q-Orbitrap/Mass Spectrometry. *J. Ginseng Res.* **2020**, *44*, 775–783. [CrossRef]
22. Lee, S.J.; In, G.; Han, S.T.; Lee, M.H.; Lee, J.W.; Shin, K.S. Structural Characteristics of a Red Ginseng Acidic Polysaccharide Rhamnogalacturonan I with Immunostimulating Activity from Red Ginseng. *J. Ginseng Res.* **2020**, *44*, 570–579. [CrossRef] [PubMed]
23. Lee, Y.Y.; Kim, S.W.; Youn, S.H.; Hyun, S.H.; Kyung, J.S.; In, G.; Park, C.K.; Jung, H.R.; Moon, S.J.; Kang, M.J.; et al. Biological Effects of Korean Red Ginseng Polysaccharides in Aged Rat Using Global Proteomic Approach. *Molecules* **2020**, *25*, 3019. [CrossRef] [PubMed]
24. Youn, S.H.; Lee, S.M.; Han, C.K.; In, G.; Park, C.K.; Hyun, S.H. Immune Activity of Polysaccharide Fractions Isolated from Korean Red Ginseng. *Molecules* **2020**, *25*, 3569. [CrossRef] [PubMed]
25. Park, D.H.; Han, B.; Shin, M.S.; Hwang, G.S. Enhanced Intestinal Immune Response in Mice after Oral Administration of Korea Red Ginseng-Derived Polysaccharide. *Polymers* **2020**, *12*, 2186. [CrossRef] [PubMed]

26. Shin, S.J.; Nam, Y.; Park, Y.H.; Kim, M.J.; Lee, E.; Jeon, S.G.; Bae, B.S.; Seo, J.; Shim, S.L.; Kim, J.S.; et al. Therapeutic Effects of Non-Saponin Fraction with Rich Polysaccharide from Korean Red Ginseng on Aging and Alzheimer's Disease. *Free Radic. Biol. Med.* **2021**, *164*, 233–248. [CrossRef]
27. Shin, S.J.; Park, Y.H.; Jeon, S.G.; Kim, S.; Nam, Y.; Oh, S.M.; Lee, Y.Y.; Moon, M. Red Ginseng Inhibits Tau Aggregation and Promotes Tau Dissociation in Vitro. *Oxid. Med. Cell. Longev.* **2020**, *7829842*, 1–12. [CrossRef] [PubMed]
28. Wang, J.; Li, J.X.; Zhang, H.Z.; Xu, P.Y.; Chen, Z.Q.; Yan, Y.Y. Ethanol Fractional Purification and Antioxidant Activities of Polysaccharides from Polygonum Cuspidatum. *Sci. Technol. Food Ind.* **2019**, *40*, 92–95.
29. Ying, X. *Purification and Structural Analysis of Polysaccharides from Red Ginseng*; Northeast Normal University: Changchun, China, 2018.
30. Aa, J. Analysis of Metabolomic Data: Principal Component Analysis. *Chin. J. Clin. Pharm. Ther.* **2010**, *15*, 481–489.
31. Tai, Y.N.; Wu, X.; Fan, L.M.; Wu, Z.N.; Weng, Y.H.; Lin, Q.Q.; Chu, K.D.; Xie, R.H. Simultaneous Determination of Sixteen Components in Alismatis Rhizoma by UPLC-MS/MS. *Chin. J. Pharm. Anal.* **2018**, *38*, 1337–1350.
32. Wu, D.T.; Li, W.Z.; Chen, J.; Zhong, Q.X.; Ju, Y.J.; Zhao, J.; Anton, B.; Li, S.P. An Evaluation System for Characterization of Polysaccharides from the Fruiting Body of Hericium Erinaceus and Identification of ITS Commercial Product. *Carbohydr. Polym.* **2015**, *124*, 201–207. [CrossRef]
33. Xia, Y.G.; Yu, L.S.; Liang, J.; Yang, B.Y.; Kuang, H.X. Chromatography and Mass Spectrometry-Based Approaches for Perception of Polysaccharides in wild and Cultured Fruit Bodies of Auricularia Auricular-Judae. *Int. J. Biol. Macromol.* **2019**, *137*, 1232–1244. [CrossRef] [PubMed]
34. Cheong, K.L.; Wu, D.T.; Deng, Y.; Leong, F.; Zhao, J.; Zhang, W.J.; Li, S.P. Qualitation and Quantification of Specific Polysaccharides From Panax Species Using GC-MS, Saccharide Mapping and HPSEC-RID-MALLS. *Carbohydr. Polym.* **2016**, *153*, 47–54. [CrossRef] [PubMed]
35. Lv, P.; Yu, S.Q.; Zhang, F.; Xiao, L.; Nie, J. Determination and PCA Analysis of 10 Chemical Components in Chysanthemi Flos by UPLC-UV. *China Pharm.* **2018**, *21*, 1374–1378.

Article

The Separation and Purification of Ellagic Acid from *Phyllanthus urinaria* L. by a Combined Mechanochemical-Macroporous Resin Adsorption Method

Zili Guo, Shuting Xiong, Yuanyuan Xie and Xianrui Liang *

Key Laboratory for Green Pharmaceutical Technologies and Related Equipment of the Ministry of Education, College of Pharmaceutical Sciences, Zhejiang University of Technology, Hangzhou 310014, China; guozili@zjut.edu.cn (Z.G.); shutingx12@163.com (S.X.); xyycz@zjut.edu.cn (Y.X.)
* Correspondence: liangxrvicky@zjut.edu.cn

Abstract: Ellagic acid is a phenolic compound that exhibits both antimutagenic and anticarcinogenic activity in a wide range of assays in vitro and in vivo. It occurs naturally in some foods such as raspberries, strawberries, grapes, and black currants. In this study, a valid and reliable method based on mechanochemical-assisted extraction (MCAE) and macroporous adsorption resin was developed to extract and prepare ellagic acid from *Phyllanthus urinaria* L. (PUL). The MCAE parameters, acidolysis, and macroporous adsorption resin conditions were investigated. The key MCAE parameters were optimized as follows: the milling time was 5 min, the ball mill speed was 100 rpm, and the ball mill filling rate was 20.9%. Sulfuric acid with a concentration of 0.552 mol/L was applied for the acidolysis with the optimized acidolysis time of 30 min and acidolysis temperature of 40 °C. Additionally, the XDA-8D macroporous resin was chosen for the purification work. Both the static and dynamic adsorption tests were carried out. Under the optimized conditions, the yield of ellagic acid was 10.2 mg/g, and the content was over 97%. This research provided a rapid and efficient method for the preparation of ellagic acid from the cheaply and easily obtained PUL. Meanwhile, it is relatively low-cost work that can provide a technical basis for the comprehensive utilization of PUL.

Keywords: *Phyllanthus urinaria* L.; ellagic acid; mechanochemical-assisted extraction; macroporous adsorption resin

1. Introduction

Ellagic acid ($C_{14}H_6O_8$, Figure 1) is an acidic hydrolysis product of polymeric ellagitannins. It is a natural polyphenolic compound that is present in various vegetables, fruits, herbs, and nuts, such as raspberries, strawberries, walnuts, grapes, and black currants [1,2]. At present, the most important pharmacological activity of ellagic acid is its antioxidant activity, which mainly depends on its basic structure containing four hydroxyl groups that are responsible for scavenging superoxide and hydroxyl anion free radicals [3–5]. In addition, ellagic acid has been found to have antimutagenic [6], anticarcinogenic [7–11], antifibrosis [9], and anti-inflammatory effects [12,13]. Furthermore, it can also improve Alzheimer's disease-mediated dementia by suppressing oxidative and inflammatory cell damage and improving antioxidant content. Ellagic acid inhibits cognitive abnormality following traumatic brain injury (TBI) in rats through its anti-inflammatory and antioxidant properties [14]. At present, three main methods are utilized for preparing ellagic acid: plant extraction [15,16], chemical synthesis [17], and enzymatic degradation [18]. Among them, the extraction of ellagic acid from natural plant resources is thought to be one of the most important ways due to its relatively high abundant content and environmental friendliness.

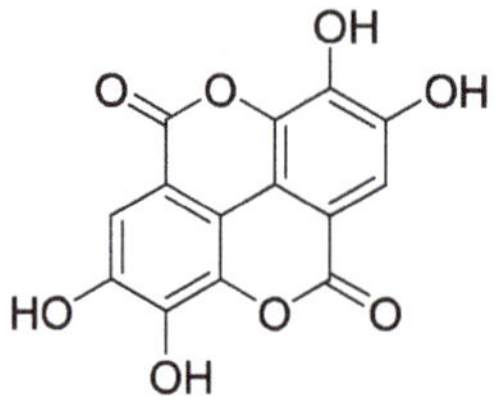

Figure 1. The chemical structure of ellagic acid.

Phyllanthus urinaria L. (PUL), belonging to the *Euphorbiaceae* family, is widely distributed in tropical and subtropical regions of Asian countries [19,20]. The dried whole plant of PUL, also named "pearl grass", "nocturnal grass", and "yin-yang grass", is a traditional Chinese medicine for the treatment of several diseases, including dysentery, jaundice, urinary tract infection, and malnutritional stagnation [21–23]. Pharmacological studies have shown that PUL possesses good biological activities such as anti-viral [24–26], anti-tumor [27], anti-thrombosis [28], hepatoprotective [29], antioxidant [30,31], anti-inflammatory [32] and immunomodulatory activities [33]. Most of the reported ingredients in PUL were tannins, lignans, flavonoids, phenolics, and terpenoids [23,29,30]. Ingredient research also showed that PUL is rich in ellagitannins, which are the main anti-hepatitis B virus active ingredients [22,24] and the main sources of ellagic acid.

In this work, a reliable method based on mechanochemical-assisted extraction (MCAE) and macroporous adsorption resin was developed to extract, separate and purify ellagic acid from PUL. To promote the yield of ellagic acid, the key mechanochemistry ball milling parameters were optimized. Additionally, the different kinds of macroporous resin were further optimized to purify the ellagic acid by the static and dynamic adsorption tests. To date, no separation of ellagic acid from PUL based on mechanochemical-assisted extraction and macroporous adsorption resin has been reported. This research provided both an efficient method to obtain ellagic acid and the comprehensive utilization of cheaply and easily obtained PUL.

2. Materials and Methods

2.1. Materials and Chemicals

The dried PUL was purchased from Tongrentang, Hangzhou City, Zhejiang Province (Hanghzou, China) and stored at room temperature before analysis. The standard ellagic acid (≥98%) was purchased from Shanghai Yuanye Bio-Technology Co., Ltd. (Shanghai, China).

HPLC-grade acetonitrile, methanol and ethanol were supplied by Merck (Darmstadt, Germany). HPLC-grade formic acid was bought from Shanghai Aladdin Bio-Chem Technology Co., Ltd. (Shanghai, China). Ultrapure water (18.2 MΩ) was purified by Barnstead TII super Pure Water System (Boston, MA, USA). Other reagents used in this experiment were all analytical grade and were obtained from Yongda Chemical Reagent Company (Tianjin, China).

2.2. Mechanochemical-Assisted Extraction Procedure

The dried PUL was fully ground and passed through an 80-mesh sifter. The PUL powder (2.0 g) was added into a PM 200 planetary ball mill (grinding media: stainless steel balls of 8 mm diameter; the weight of the balls: 4.2 g; two drums at 50 mL each; the volume of the load/drum ratio: 1:2). Then, the ground mixture (1.0 g) was extracted with 10 mL of 50% ethanol in a 50 mL flask at 30 °C for 30 min under an ultrasonic bath. The supernatant was collected as the ellagitannin extract.

2.3. Acidolysis Experiment

The acid hydrolysis agent (1.5 mL) was added to the ellagitannin extract. The mixture was placed in a 40 °C water bath, heated and stirred for 30 min. The supernatant was concentrated by decompression evaporation and freeze-dried. After removing the solvent, crude ellagic acid was obtained.

2.4. Macroporous Resin Adsorption Experiment

2.4.1. Pretreatment of Adsorbents

Ten macroporous resins named XAD-2, HP-20, AB-8, XDA-8D, LSA-8D, HPD450, HPD826, DA201, LXA-8, and LX-8 were investigated. Their physical properties were listed in Table 1. To remove some water-soluble impurities and suspended material, resins were soaked in deionized water. Then, they were transferred into 95% ethanol for 24 h to swell completely. After washing with deionized water, resins were separately immersed in 5% (w/v) NaOH for 24 h, followed by rinsing with deionized water to a neutral pH. Finally, the macroporous resins were immersed in 5% (v/v) HCl for 24 h and rinsed with deionized water again.

Table 1. Physical properties of the tested resins in this study.

Trade Name	Specific Surface Area (m^2/g)	Particle Size (mm)	Polarity Type
XAD-2	300	0.25–0.84	Non-polarity
HP-20	590	0.25–0.60	Non-polarity
AB-8	480–520	0.30–1.25	Weak polarity
XDA-8D	140	0.20–0.40	Medium polarity
LSA-8D	150	0.30–1.25	Medium polarity
HPD450	500–550	0.30–1.25	Medium polarity
HPD826	500–600	0.30–1.25	Medium polarity
DA201	150–200	0.30–1.25	Polarity
LXA-8	200	0.30–1.25	Polarity
LX-8	1000	0.315–1.26	Polarity

2.4.2. Static Adsorption Tests

Different kinds of resins were compared for their separation capacity through static adsorption tests. The general experimental procedure was as follows: Pretreated resin (1.0 g) was added to a 50 mL flask, then 20.0 mL of 0.09 mg/mL crude ellagic acid solution was added. The flask was shaken in a shaker at 25 °C with 100 rpm for 24 h. The content of the ellagic acid in the adsorption solution was determined and calculated by the UPLC method.

2.4.3. Sorption Kinetics Tests

Sorption kinetics tests were performed to choose the most efficient resin. Similarly, pretreated resin (1.0 g) was added to a 50 mL flask, then 20.0 mL of 0.09 mg/mL crude ellagic acid solution was added. The flask was shaken in a shaker at 25 °C with 100 rpm for 24 h. Every hour, 1 mL of extract was pipetted out to perform UPLC determination.

The adsorption properties including the adsorption capacity, adsorption rate, desorption ratio and recovery rate of each resin were quantified according to the following equations:

$$W \text{ (mg/g dry resin)} = (\rho_o - \rho_e) \times V/m, \tag{1}$$

$$E \text{ (\%)} = (\rho_o - \rho_e)/\rho_o \times 100\%, \tag{2}$$

$$D \text{ (\%)} = \rho_d/(\rho_o - \rho_e) \times 100\%, \tag{3}$$

$$R \text{ (\%)} = (\rho_o - \rho_e) \times V/m, \tag{4}$$

where W was the adsorption capacity at adsorption equilibrium (mg/g dry resin), and ρ_o, ρ_e and ρ_d were the initial, absorption equilibrium and desorption concentrations of analyte in the solutions, respectively (mg/mL). V was the volume of the adsorption solution (mL), and m was the dry weight of the resin (g). E was the adsorption rate (%), D was the desorption rate (%) and R was the recovery rate (%) of the resin.

2.4.4. Dynamic Adsorption Tests

Dynamic adsorption tests were carried out as follows: 60 mL of XDA-8D resins were packed in the column (20 mm × 300 mm), which was loaded with the crude ellagic acid solution, by the wet method. Then, 120 mL of deionized water was loaded to rinse the resins with the flow rate at 1.0 BV/h to make the eluant and extract mix together. After adsorptive equilibrium, different concentrations of ethanol (10%, 20%, 30%, 40%, 50%, 60%, 70%, 80% and 90%) were utilized to desorb the ellagic acid at a constant flow rate of 1.0 BV/h. The elution volume of each concentration was constant by being maintained at 3.0 BV. The contents of ellagic acid in each desorption solution were determined by UPLC.

2.5. Ultra High-Performance Liquid Chromatography (UPLC): Quantitative Analysis and the Characterization of Ellagic Acid

The quantitative analysis of ellagic acid was calculated based on the standard curve of ellagic acid by an ACQUITY UPLC™ system (Waters, Milford, MA, USA) equipped with a binary solvent pump, an autosampler, an integral column heater and a photodiode array detector (PDA eλ Detector). The separation was operated on a Waters ACQUITY UPLC HSS T3 column (2.1 mm × 100 mm, 1.8 µm, Waters, Milford, MA, USA). The mobile phase consisted of 0.1% (v/v) formic acid solution (A) and acetonitrile (B) with the following gradient: 0~3.5 min, 3–4% B; 3.5~5.0 min, 4–8% B; 5.0~9.0 min, 8–10% B; 9.0~14.0 min, 10–11% B; 14.0~38.0 min, 11–15% B; 38.0~46.0 min, 15–20% B; 46.0~50.0 min, 20–30% B; 50.0~55.0 min, 30–40% B; 55.0~57.0 min, 40–90% B; 57.0~60.0 min, 90% B. The flow rate of the mobile phase was set at 0.2 mL/min. The injection volume was 1.0 µL, and the column temperature was maintained at 30 °C. The wavelength at 250 nm was set as the monitoring wavelength.

The proton ^{1}H and carbon ^{13}C NMR spectra of the ellagic acid were obtained at 308 K using a Bruker AVANCE III 600 MHz NMR spectrometer (Bruker, Billerica, MA, USA). The chemical shifts (δ) in the NMR spectra were recorded in ppm with the solvent peak as the reference. The MS analysis was carried out on a micrOTOF-Q II mass spectrometer (Bruker Daltonics, Bremen, Germany) equipped with an ESI source. The ESI source parameters were as follows: the dry gas (N_2) flow rate was 6.0 L/min, the nebulizer gas (N_2) pressure was 0.8 bar, the dry gas temperature was 200 °C, and the capillary voltage was 2800 V in the negative mode.

3. Results and Discussion

3.1. Optimization of the MCAE Procedure

The mechanical force could reduce the particle size of herbal plant powder and destroy the cell wall structure, which increases the release of chemical constituents. In this research, the mechanochemistry ball milling method was applied to the extraction of ellagitannin from PUL. Compared with the traditional extraction method, the ball milling process could effectively increase the peak area of ellagitannin compounds.

To obtain a higher yield of ellagitannin, the key ball milling parameters including milling time, milling speed, and mill filling rate were optimized. To evaluate the influence of milling time on the yield of ellagitannin, the milling time was investigated at 5 min, 10 min and 15 min, respectively. With the extension of milling time, the peak areas of three compounds (corilagin, geraniin, and ellagic acid) generally showed a downward trend, shown in Figure 2A. The possible reason was that the longer milling time resulted in the accumulation of heat in the ball milling tank, which destroyed the compound structural units and led to the oxidation and partial decomposition of the ellagitannin compound. As a result, the milling time of 5 min was preferred.

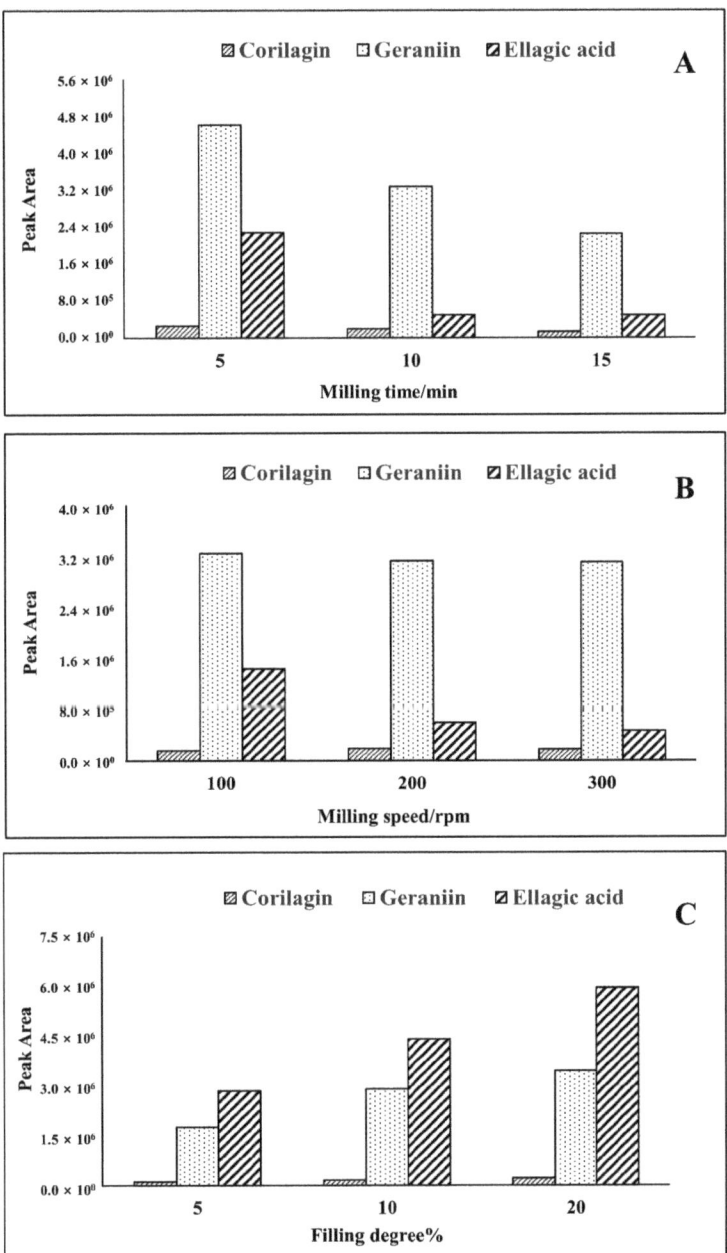

Figure 2. Effect of the MCAE procedure on the yield of ellagitannins and ellagic acid. (**A**) Effect of milling time; (**B**) Effect of milling speed; (**C**) Effect of filling degree.

Ball milling speed is a crucial factor for MCAE. The effect of ball milling speed on the extraction yield of ellagitannin from PUL was shown in Figure 2B. Three speeds including 100 rpm, 200 rpm and 300 rpm were investigated. It was clearly seen that at the ball milling speed of 100 rpm, the yield of the three compounds (corilagin, geraniin, and ellagic acid) reached the maximum value. With increases in ball milling speed, the peak areas of corilagin and geraniin did not change significantly. Generally speaking, the MCAE of the ball mill can continuously destroy cell walls and promote the reaction of bioactive substances with solid phase reagents, thereby greatly improving the extraction efficiency. However, the ball milling speed did not further improve the entire extraction process at the speed of 200 rpm and 300 rpm. Hence, 100 rpm was chosen for the optimal ball milling parameter.

The ball mill filling rates were another factor that influenced the yield of ellagitannin. The filling rates at 5.2%, 10.5% and 20.9% were investigated and the results were shown in Figure 2C. Under the optimal milling time and ball milling speed, with the increase of the filling rate, the peak areas of geranium and ellagic acid showed an overall upward trend, while the peak area of ellagic acid increased significantly. It indicated that the mechanical force in the ball milling tank increased while the increased filling rate correspondingly increased the wall-breaking effect of the plant cells, and more ellagic acid was released. Moreover, the filling rate of 20.9% was used for the ball milling work.

3.2. Optimization of Acid Hydrolysis Conditions

Considering that ellagic acid is an acid hydrolysate of polymerized ellagitannin, the parameters of acid hydrolysis conditions, including acid hydrolysis reagent, acid hydrolysis temperature, and acid hydrolysis time were optimized in this work. Different acid hydrolysis reagents such as sulfuric acid, hydrochloric acid and formic acid were used. The yields of ellagic acid showed that sulfuric acid had a better acidolysis effect. Then, the concentration of sulfuric acid was further optimized. As shown in Figure 3A, the yield of ellagic acid increased with the increasing concentration of sulfuric acid, and the yield of ellagic acid was the highest when the concentration of sulfuric acid was 0.552 mol/L. When the concentration of sulfuric acid is higher than 0.552 mol/L, the yield of ellagic acid decreases. Hence, the concentration of sulfuric acid was selected as 0.552 mol/L.

The acid hydrolysis temperature exerted a greater impact on the yield of ellagic acid. Different acid hydrolysis temperatures at 20 °C, 30 °C, 40 °C, 50 °C, 60 °C, 70 °C, 80 °C and 90 °C were investigated as shown in Figure 3B. The highest peak area of ellagic acid was obtained when the acidolysis temperature was set at 40 °C. As the temperature increased, the yield of ellagic acid increased. This was because high temperatures may increase the degradation rate of ellagitannin. However, when the temperature rose to a certain level (40 °C), the yield of ellagic acid decreased. It was speculated that excessive temperature might cause the oxidative damage of ellagic acid.

Acid hydrolysis time showed little effect on the yield of ellagic acid, shown in Figure 3C. Ellagitannin could be degraded into ellagic acid within 30 min. When the acid hydrolysis time was longer than 30 min, it remained stable. 30 min was thought to be the better acid hydrolysis time.

A low concentration of sulfuric acid leads to inadequate acidolysis and a low yield of ellagic acid. However, the yield of ellagic acid decreased and the energy consumption increased with the high concentration of sulfuric acid. Considering the sensitivity of ellagic acid to air, the excessive temperature may accelerate the oxidation of ellagic acid. Finally, the acid hydrolysis conditions are determined as follows: the concentration of sulfuric acid was 0.552 mol/L, the acidolysis time was 30 min, and the acidolysis temperature was 40 °C.

The yield and content of crude ellagic acid could be improved significantly under the optimized acidolysis conditions. As the content of ellagic acid increased, the contents of corilagin and geraniin decreased. It may be speculated that ellagitannins were more likely to be broken down into ellagic acid in an acid environment [34,35]. In previous reports, Wei [36] and Li [37] studied the extraction process of ellagic acid from muscadine and red raspberry, respectively. The contents of ellagic acid were 616.21 µg/g and 322 µg/g, which were much lower than the 10.2 mg/g in this work.

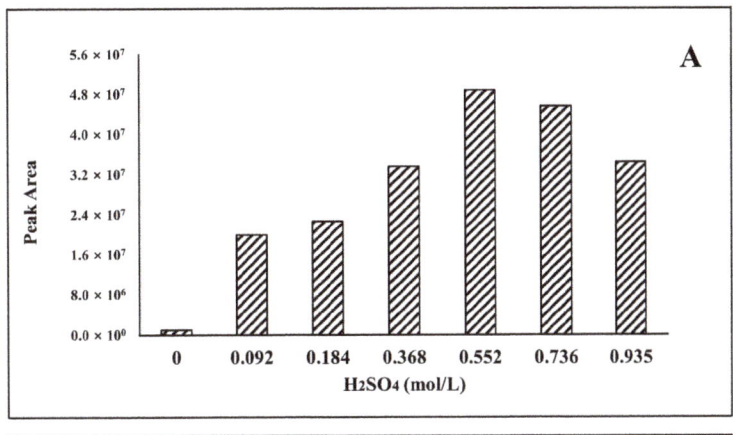

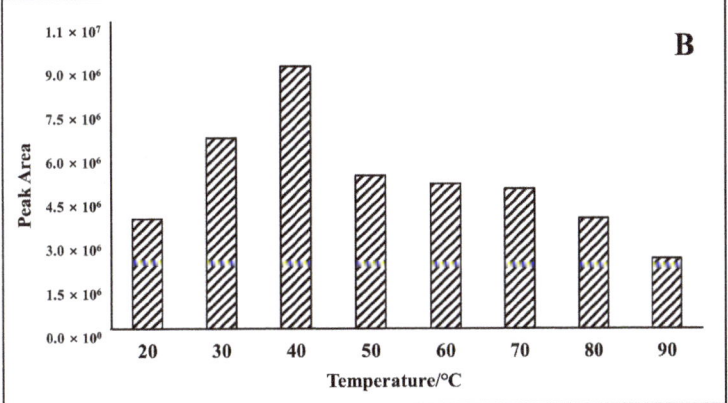

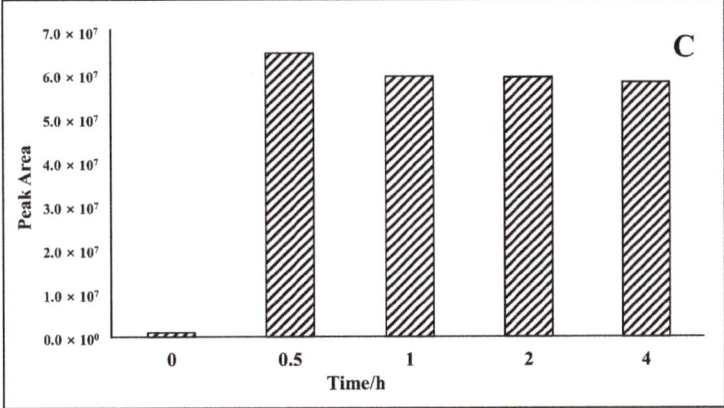

Figure 3. Effect of acid hydrolysis conditions. (**A**) Effect of H_2SO_4 concentration; (**B**) Effect of acid hydrolysis temperature; (**C**) Effect of acid hydrolysis time.

3.3. Screening of Optimum Resin

The adsorption and desorption of ellagic acid by different types of macroporous resins were shown in Table 2 and Figure 4. The XDA-8D type macroporous resin showed the highest adsorption rate (78.03%) of ellagic acid. Although its desorption rate was inferior to HPD450 and LXA-8 resins when the ethanol was selected as the adsorbent, XDA-8D had a higher recovery rate than other resins. As a result, XDA-8D was selected for the separation and purification of crude ellagic acid.

Table 2. Results of the static adsorption and desorption of macroporous resins.

Trade Name	Adsorbent Concentration (mg/mL)	Desorption Solution Concentration (mg/mL)	Adsorption Rate (%)	Desorption Rate (%)	Recovery Rate (%)
XAD-2	0.078	0.007	13.60	60.64	8.25
HP-20	0.076	0.010	16.03	66.68	10.69
AB-8	0.067	0.010	25.80	42.59	10.99
XDA-8D	0.021	0.016	78.03	74.47	58.11
LSA-8D	0.066	0.012	26.85	47.78	12.83
HPD450	0.081	0.009	0.00	92.77	9.42
HPD826	0.045	0.015	39.92	67.98	6.73
DA201	0.062	0.017	31.00	62.44	19.36
LXA-8	0.078	0.009	12.81	76.70	9.82
LX-8	0.060	0.020	33.14	67.62	22.41

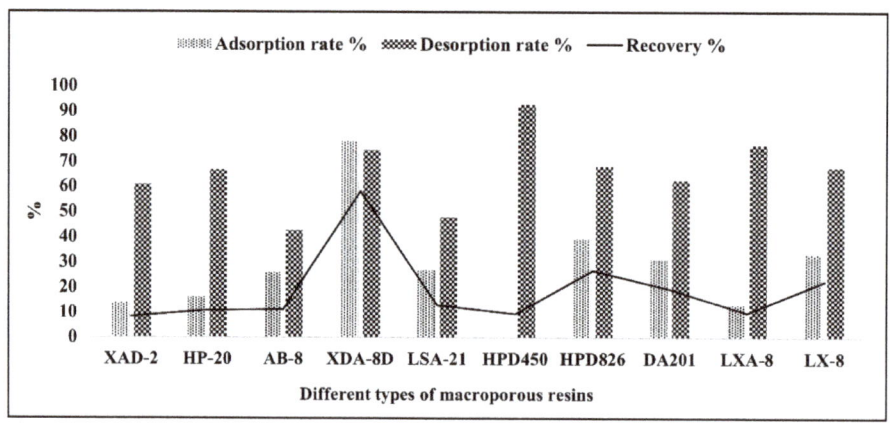

Figure 4. Effect of resin type on the static adsorption and desorption of ellagic acid.

3.4. Static Adsorption Kinetics and Adsorption Isotherms

To enrich ellagic acid from PUL extracts using microporous resins effectively, the optimum type of microporous resin was screened first. Because the external factors have a significant effect, the resin adsorption behavior is significantly influenced by many external factors. To evaluate these effects on the resin adsorption capacity, the adsorption kinetics tests and adsorption isotherms of XDA-8D microporous resin for crude ellagic acid were carried out (Figures 5 and 6).

According to Figure 5, the adsorption rate of XDA-8D for crude ellagic acid increased to the biggest within 1 h. Then, it still maintained a fast adsorption rate within 1–5 h. However, it proceeded slowly between 5–10 h and reached equilibrium after 10 h. The results showed that the adsorption capacity was about 1.2 mg/g of dry resin.

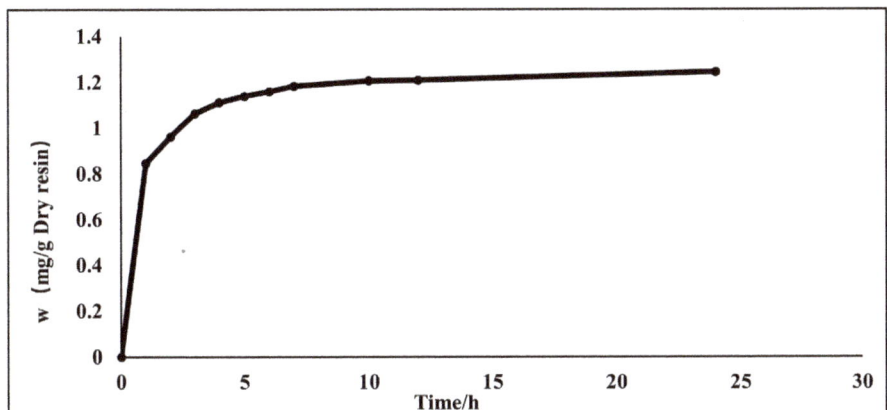

Figure 5. Static adsorption kinetics.

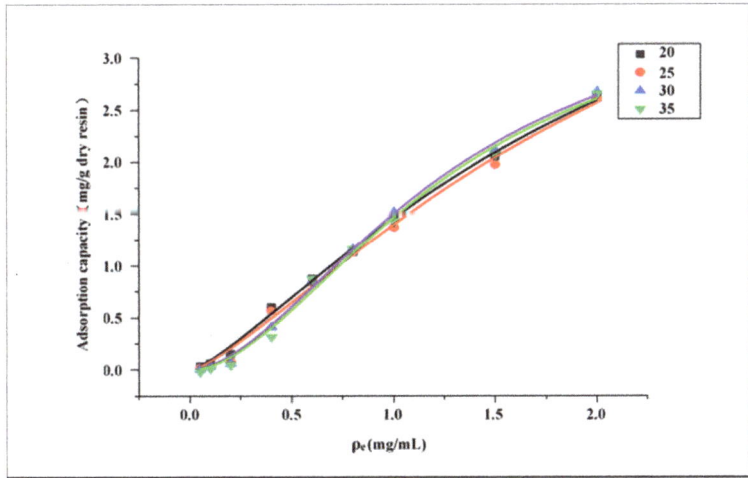

Figure 6. Static adsorption kinetics.

To study the effect of temperature on static adsorption, the adsorption isotherms were investigated at 20 °C, 25 °C, 30 °C and 35 °C. According to the typical adsorption isotherm classified by Brunauer [38] and the preferential equilibrium curves, the adsorption isotherm of XDA-8D belonged to the convex preferential adsorption isotherm (shown in Figure 6). This kind of adsorption isotherm is convex upward along the coordinate direction of adsorption capacity and is called preferential adsorption isotherm. According to Brunauer-Deming-Deming-Teller (BDDT) classification, this means that monolayer molecular adsorption occurs when the pore size of the adsorbent capillary is slightly larger than that of the adsorbent molecule. The experiment data of adsorption isotherms shown in Table 3 were well fitted to a Langmuir model [39]. It showed that the temperature had a certain influence on the adsorption process; 30 °C was the best, with an adsorption capacity of 2.462 mg/mL.

Table 3. Adsorption isotherm equation.

T/°C	Langmuir Isotherm Equation	R^2
20	$Y = 1.994\ X^{1.320}/(1 + 0.369\ X^{1.320})$	0.9970
25	$Y = 1.853\ X^{1.346}/(1 + 0.329\ X^{1.346})$	0.9932
30	$Y = 2.462\ X^{1.773}/(1 + 0.640\ X^{1.773})$	0.9970
35	$Y = 2.392\ X^{1.825}/(1 + 0.636\ X^{1.825})$	0.9927

3.5. Dynamic Adsorption and Elution

The crude ellagic acid product was adsorbed by XDA-8D macroporous resin and eluted with ethanol aqueous solution in gradients, collecting the 30% ethanol to 80% ethanol elution fraction. After removing the solvent, it was characterized by ^{1}H NMR, ^{13}C NMR and MS, shown in Figures S1–S3 (Supplementary Materials), respectively. Meanwhile, the content measured by UPLC was 97% (shown in Figure S4).

3.6. UPLC Quantitative Analysis

3.6.1. Linearity and Limits of Detection and Quantification

The calibration curves were plotted with a series of concentrations of standard solutions. Each analyte curve was made at six levels. Acceptable linear correlation and high sensitivity at these conditions were confirmed by the correlation coefficients (R^2, 0.9989–0.9999). The limits of detection (LODs) and limits of quantification (LOQs) for standards were estimated at signal-to-noise ratios (S/N) of three and ten, respectively, by injecting a series of dilute solutions with known concentrations. The detailed information regarding calibration curves, linear ranges, LODs and LOQs are displayed in Table S1.

3.6.2. Precision, Repeatability, Stability and Recovery

The precisions calculated as relative standard deviation (RSD) were within the range of 0.42–1.28%. The RSD values of three compounds were within the range from 4.67 to 6.15%, which revealed a high repeatability of the method. Stability of the sample solution was tested at room temperature in 12 h. The RSD values of three compounds were all within 5.37%, which demonstrated a good stability within the tested period.

The data for precision, repeatability, stability and recovery were also listed in Table S1. As shown in Table S1, the mean recovery rates of three compounds varied from 93.70 to 107.00% (RSD $\leq$ 7.10%).

4. Conclusions

In this study, a combined mechanochemical-macroporous resin adsorption method was established to separate and purify ellagic acid from PUL. The mechanochemistry ball milling coupled with the ultrasonic-assisted solvent extraction method was utilized to increase the extraction yield of ellagitannin. Under the optimized ball-milling conditions, the yield of ellagitannin was increased as compared to the traditional extraction method. The ellagitannin in the PUL extract could be converted into ellagic acid under sulfuric acid hydrolysis. The optimal reaction conditions were as follows: the sulfuric acid amount was 0.552 mol/L, the acidolysis time was 30 min, and the acidolysis temperature was 40 °C. Finally, the crude ellagic acid was separated and purified by XDA-8D macroporous resin to obtain ellagic acid. Under the optimal technological conditions, the yield of ellagic acid was 10.2 mg/g, and the content was over 97%. It is rapid and efficient for the preparation of ellagic acid. Meanwhile, it can also provide a technical basis for the comprehensive utilization of PUL.

Supplementary Materials: The following are available online at https://www.mdpi.com/article/10.3390/separations8100186/s1, Figure S1: The ^{1}H NMR spectrum of ellagic acid. Figure S2: The ^{13}C NMR spectrum of ellagic acid. Figure S3: The MS spectrum of ellagic acid. Figure S4: UPLC chromatogram of ellagic acid. Table S1: Calibration curves, linear ranges, LODs, LOQs, precision, repeatability, stability and recovery of 5 standards compounds.

Author Contributions: Conceptualization, X.L.; methodology, S.X.; software, Z.G. and S.X.; validation, Z.G.; formal analysis, Z.G. and S.X.; investigation, Y.X.; resources, S.X.; data curation, S.X.; writing—original draft preparation, Z.G.; writing—review and editing, Z.G., Y.X. and X.L.; project administration, Z.G. and X.L. All authors have read and agreed to the published version of the manuscript.

Funding: This research received no external funding.

Institutional Review Board Statement: Not applicable.

Informed Consent Statement: Not applicable.

Acknowledgments: This work was supported by the cooperative project with Zhejiang Hisoar Pharmaceutical Co., Ltd. (KYY-HX-20180525) and the scientific research project of the Zhejiang Provincial Department of Education (Y202043200).

Conflicts of Interest: The authors declare no conflict of interest.

References

1. Pilar, Z.; Federico, F.; Francisco, A.T.-B. Effect of processing and storage on the antioxidant ellagic acid derivatives and fla-vonoids of red raspberry (*Rubus idaeus*) Jams. *J. Agric. Food Chem.* **2001**, *49*, 3651–3655.
2. De Ancos, B.; González, E.M.; Cano, M.P. Ellagic Acid, Vitamin C, and Total Phenolic Contents and Radical Scavenging Capacity Affected by Freezing and Frozen Storage in Raspberry Fruit. *J. Agric. Food Chem.* **2000**, *48*, 4565–4570. [CrossRef] [PubMed]
3. Hassoun, E.A.; Vodhanel, J.; Abushaban, A. The modulatory effects of ellagic acid and vitamin E succinate on TCDD-induced oxidative stress in different brain regions of rats after subchronic exposure. *J. Biochem. Mol. Toxicol.* **2004**, *18*, 196–203. [CrossRef] [PubMed]
4. Seeram, N.P.; Adams, L.S.; Henning, S.M.; Niu, Y.; Zhang, Y.; Nair, M.G.; Heber, D. In vitro antiproliferative, apoptotic and antioxidant activities of punicalagin, ellagic acid and a total pomegranate tannin extract are enhanced in combination with other polyphenols as found in pomegranate juice. *J. Nutr. Biochem.* **2005**, *16*, 360–367. [CrossRef] [PubMed]
5. Zheng, Y.-Z.; Fu, Z.-M.; Deng, G.; Guo, R.; Chen, D.-F. Free radical scavenging potency of ellagic acid and its derivatives in multiple H+/e—Processes. *Phytochemistry* **2020**, *180*, 112517. [CrossRef]
6. Guadalupe, L.-P.; Paul, A.K.; Elvira, G.M.; Norman, Y.K. Inhibitory effects of ellagic acid on the direct-acting mutagenicity of aflatoxin B1 in the Salmonella microsuspension assay. *Mutat. Res.* **1998**, *398*, 183–187.
7. Larrosa, M.; Tomás-Barberán, F.A.; Espín, J.C. The dietary hydrolysable tannin punicalagin releases ellagic acid that induces apoptosis in human colon adenocarcinoma Caco-2 cells by using the mitochondrial pathway. *J. Nutr. Biochem.* **2006**, *17*, 611–625. [CrossRef]
8. Al-Shar'I, N.A.; Al-Balas, Q.A.; Hassan, M.A.; El-Elimat, T.M.; Aljabal, G.A.; Almaaytah, A.M. Ellagic acid: A potent glyoxa-lase-I inhibitor with a unique scaffold. *Acta Pharm.* **2021**, *71*, 115–130. [CrossRef]
9. Han, D.H.; Lee, M.J.; Kim, J.H. Antioxidant and apoptosis-inducing activities of ellagic acid. *Anticancer Res.* **2006**, *26*, 3601–3606.
10. Bell, C.; Hawthorne, S. Ellagic acid, pomegranate and prostate cancer—A mini review. *J. Pharm. Pharmacol.* **2008**, *60*, 139–144. [CrossRef]
11. Losso, J.N.; Bansode, R.R.; Trappey, A.; Bawadi, H.A.; Truax, R. In vitro anti-proliferative activities of ellagic acid. *J. Nutr. Biochem.* **2004**, *15*, 672–678. [CrossRef] [PubMed]
12. Chao, P.-C.; Hsu, C.-C.; Yin, M.-C. Anti-inflammatory and anti-coagulatory activities of caffeic acid and ellagic acid in cardiac tissue of diabetic mice. *Nutr. Metab.* **2009**, *6*, 33. [CrossRef]
13. Rogerio, A.P.; Fontanari, C.; Borducchi, E.; Keller, A.C.; Russo, M.; Soares, E.G.; Albuquerque, D.A.; Faccioli, L.H. An-ti-inflammatory effects of *Lafoensia pacari* and ellagic acid in a murine model of asthma. *Eur. J. Pharmacol.* **2008**, *580*, 262–270. [CrossRef]
14. Gupta, A.; Singh, A.K.; Kumar, R.; Jamieson, S.; Pandey, A.K.; Bishayee, A. Neuroprotective Potential of Ellagic Acid: A Critical Review. *Adv. Nutr.* **2021**, *12*, 1211–1238. [CrossRef]
15. Kang, C.; Duan, Z.H.; Luo, Y.H.; Li, Y.; Wu, S.J.; Mo, F.W.; Deng, N.F.; Xie, W.; Su, H.L.; Shuai, L. Method for Extracting Ellagic Acid from Mango Stone. CN Patent 107163059A, 15 September 2017.
16. Kim, J.H.; Kim, Y.S.; Kim, T.I.; Li, W.; Mun, J.-G.; Jeon, H.D.; Kee, J.-Y.; Choi, J.-G.; Chung, H.-S. Unripe Black Raspberry (*Rubus coreanus* Miquel) Extract and Its Constitute, Ellagic Acid Induces T Cell Activation and Antitumor Immunity by Blocking PD-1/PD-L1 Interaction. *Foods* **2020**, *9*, 1590. [CrossRef] [PubMed]

17. Luo, Z.J.; Feng, B.; Xiao, Z.H.; Luo, B.; Wang, S.J.; Li, J.B.; Zhang, Y.M. One-Step Preparation of Ellagic Acid from Gallate De-Rivatives. CN Patent 110066284A, 30 July 2019.
18. Zhang, Y.; Jiang, B.; Hu, X.S.; Liao, X.J. Extraction of Ellagic Acid from Pomegranate Peels by Biological Enzymic Hydrolysis. CN Patent 101701234A, 5 May 2010.
19. Geethangili, M.; Ding, S.-T. A Review of the Phytochemistry and Pharmacology of *Phyllanthus urinaria* L. *Front. Pharmacol.* **2018**, *9*, 1109. [CrossRef] [PubMed]
20. Liang, Q.P.; Wu, C.; Xu, T.Q.; Jiang, X.Y.; Tong, G.D.; Wei, C.S.; Zhou, G.X. Phenolic Constituents with Antioxidant and Antiviral Activities from *Phyllanthus urinaria* Linnea. *Indian J. Pharm. Sci.* **2019**, *81*, 424–430. [CrossRef]
21. Guo, Q.; Zhang, Q.-Q.; Chen, J.-Q.; Zhang, W.; Qiu, H.-C.; Zhang, Z.-J.; Liu, B.-M.; Xu, F.-G. Liver metabolomics study reveals protective function of *Phyllanthus urinaria* against CCl 4-induced liver injury. *Chin. J. Nat. Med.* **2017**, *15*, 525–533. [CrossRef]
22. Wu, Y.; Lu, Y.; Li, S.-Y.; Song, Y.-H.; Hao, Y.; Wang, Q. Extract from *Phyllanthus urinaria* L. inhibits hepatitis B virus replication and expression in hepatitis B virus transfection model in vitro. *Chin. J. Integr. Med.* **2015**, *21*, 938–943. [CrossRef]
23. Cai, J.; Liang, J.Y. Progress of studies on constituents and pharmacological effect of *Phyllanthus urinaria* L. *Strait Pharm.* **2003**, *15*, 1–3.
24. Li, Y.; Jiang, M.; Li, M.; Chen, Y.; Wei, C.; Peng, L.; Liu, X.; Liu, Z.; Tong, G.; Zhou, D.; et al. Compound *Phyllanthus urinaria* L Inhibits HBV-Related HCC through HBx-SHH Pathway Axis Inactivation. *Evid. Based Complement. Altern. Med.* **2019**, *2019*, 1635837. [CrossRef] [PubMed]
25. Yeo, S.-G.; Song, J.H.; Hong, E.-H.; Lee, B.-R.; Kwon, Y.S.; Chang, S.-Y.; Kim, S.H.; Lee, S.W.; Park, J.-H.; Ko, H.-J. Antiviral effects of *Phyllanthus urinaria* containing corilagin against human enterovirus 71 and Coxsackievirus A16 in vitro. *Arch. Pharmacal Res.* **2014**, *38*, 193–202. [CrossRef] [PubMed]
26. Tan, W.C.; Jaganath, I.; Manikam, R.; Sekaran, S.D. Evaluation of Antiviral Activities of Four Local Malaysian *Phyllanthus* Species against Herpes Simplex Viruses and Possible Antiviral Target. *Int. J. Med. Sci.* **2013**, *10*, 1817–1829. [CrossRef] [PubMed]
27. Huang, S.-T.; Yang, R.-C.; Lee, P.-N.; Yang, S.-H.; Liao, S.-K.; Chen, T.-Y.; Pang, J.-H.S. Anti-tumor and anti-angiogenic effects of *Phyllanthus urinaria* in mice bearing Lewis lung carcinoma. *Int. Immunopharmacol.* **2006**, *6*, 870–879. [CrossRef] [PubMed]
28. Shen, Z.Q.; Dong, Z.J.; Wu, L.O.; Chen, Z.H.; Liu, J.K. Effects of fraction from *Phyllanthus urinaria* on thrombosis and coagu-lation system in animals. *Zhong Xi Yi Jie He Xue Bao* **2004**, *2*, 106–110. [CrossRef]
29. Mediani, A.; Abas, F.; Khatib, A.; Tan, C.P.; Ismail, I.S.; Shaari, K.; Ismail, A.; Lajis, N. Phytochemical and biological features of Phyllanthus niruri and *Phyllanthus urinaria* harvested at different growth stages revealed by 1 H NMR-based metabolomics. *Ind. Crop. Prod.* **2015**, *77*, 602–613. [CrossRef]
30. Fang, S.-H.; Rao, Y.K.; Tzeng, Y.-M. Anti-oxidant and inflammatory mediator's growth inhibitory effects of compounds isolated from *Phyllanthus urinaria*. *J. Ethnopharmacol.* **2008**, *116*, 333–340. [CrossRef]
31. Liu, Y.; She, X.-R.; Huang, J.-B.; Liu, M.-C.; Zhan, M.-E. Ultrasonic-extraction of phenolic compounds from *Phyllanthus urinaria*: Optimization model and antioxidant activity. *Food Sci. Technol.* **2018**, *38*, 286–293. [CrossRef]
32. Lai, C.-H.; Fang, S.-H.; Rao, Y.K.; Geethangili, M.; Tang, C.-H.; Lin, Y.-J.; Hung, C.-H.; Wang, W.-C.; Tzeng, Y.-M. Inhibition of Helicobacter pylori-induced inflammation in human gastric epithelial AGS cells by *Phyllanthus urinaria* extracts. *J. Ethnopharmacol.* **2008**, *118*, 522–526. [CrossRef]
33. Tseng, H.-H.; Chen, P.-N.; Kuo, W.-H.; Wang, J.-W.; Chu, S.-C.; Hsieh, Y.-S. Antimetastatic Potentials of *Phyllanthus urinaria* L on A549 and Lewis Lung Carcinoma Cells via Repression of Matrix-Degrading Proteases. *Integr. Cancer Ther.* **2011**, *11*, 267–278. [CrossRef]
34. Ito, H.; Iguchi, A.; Hatano, T. Identification of Urinary and Intestinal Bacterial Metabolites of Ellagitannin Geraniin in Rats. *J. Agric. Food Chem.* **2007**, *56*, 393–400. [CrossRef]
35. Ito, H. Metabolites of the Ellagitannin Geraniin and Their Antioxidant Activities. *Planta Medica* **2011**, *77*, 1110–1115. [CrossRef] [PubMed]
36. Wei, Z.; Zhao, Y.J.; Huang, Y.; Zhang, Y.L.; Lu, J. Optimization of ultrasound-assisted extraction of ellagic acid and total phenols from Muscadine (*Vitis rotundifolia*) by response surface methodology. *Food Sci.* **2015**, *36*, 29–35.
37. Li, X.P.; Xin, X.L.; Liu, Y.H.; Liang, Q. Study on the extraction processing of ellagic acid form the fruit of red raspberry. *Sci. Technol. Food Ind.* **2010**, *1*, 277–279.
38. Lou, S.; Chen, Z.; Liu, Y.; Ye, H.; Di, D. New Way to Analyze the Adsorption Behavior of Flavonoids on Macroporous Adsorption Resins Functionalized with Chloromethyl and Amino Groups. *Langmuir* **2011**, *27*, 9314–9326. [CrossRef] [PubMed]
39. Bolster, C.H. Revisiting a Statistical Shortcoming when Fitting the Langmuir Model to Sorption Data. *J. Environ. Qual.* **2008**, *37*, 1986–1992. [CrossRef] [PubMed]

Article

Studies on the Separation and Purification of the *Caulis sinomenii* Extract Solution Using Microfiltration and Ultrafiltration

Xi Wang [1,2], Huimin Feng [1,2], Halimulati Muhetaer [1], Zuren Peng [3], Ping Qiu [3], Wenlong Li [1,2,*] and Zheng Li [1,2,*]

1. College of Pharmaceutical Engineering of Traditional Chinese Medicine, Tianjin University of Traditional Chinese Medicine, Tianjin 301617, China; 13847122707@163.com (X.W.); FengHM996@163.com (H.F.); hali0322001x@163.com (H.M.)
2. State Key Laboratory of Component-Based Chinese Medicine, Tianjin University of Traditional Chinese Medicine, Tianjin 301617, China
3. Hunan Zhengqing Pharmaceutical Group Co., Ltd., Huaihua 418005, China; noriaki1009@126.com (Z.P.); qiu_zq@vip.163.com (P.Q.)
* Correspondence: wshlwl@tjutcm.edu.cn (W.L.); lizheng@tjutcm.edu.cn (Z.L.); Tel.: +86-22-5959-6814 (W.L.)

Abstract: The separation and purification process of alkaloids faces great challenges of pollution, high energy consumption and low continuity. In this study, the effects of ceramic microfiltration (MF) membrane (membrane pore size of 0.50 µm, 0.20 µm, 0.05 µm) and organic ultrafiltration (UF) membrane (membrane molecular weight cut-off of 10 KDa and 1 KDa) on the separation and purification of *Caulis sinomenii* extract solution in pilot scale were studied. The cleaning effects of different cleaning methods (pure water, 1% HCl-NaOH, 1% sodium hypochlorite) were investigated. The experimental results indicated that 0.05 µm ceramic membrane and 1 kDa UF membrane have higher sinomenine hydrochloride (SH) permeabilities and total solids (TS) removal rates. The ceramic membrane was circulating cleaned by 1% sodium hypochlorite solution for 1 h; the membrane flux can be restored to more than 90% of the original, the membrane flux of 1 kDa UF membrane can be restored to 99.2% of the original by pure water washing. From the above study, the optimal technic parameters was determined in which 0.05 µm ceramic MF membrane and 1 kDa UF membrane were used to separate and purify the *Caulis sinomenii* extract solution to remove the invalid ingredients, and the two kinds of membranes were cleaned with 1% sodium hypochlorite solution and pure water, respectively, to keep satisfactory membrane fluxes. The study provided an environment-friendly alternative for the separation and purification of alkaloids in natural products, which has a good prospect for the industrial application.

Keywords: *Caulis sinomenii*; microfiltration; ultrafiltration; sinomenine hydrochloride; separation and purification

1. Introduction

There are many problems in The Traditional Chinese Medicine (TCM) industry which need urgent solutions, such as complicated process, time consuming, high energy consumption, consumption of toxic reagents and so on. Because there is a large number of soluble macromolecular impurities in the TCM materials, such as starchs, pectins [1], proteins [2], tannins [3], etc., These macromolecules are easily form suspended hydrosol in solvent, which is typical thermodynamic unstable system. In the subsequent drying, impurity removal and crystallization processes, caking and adhesion can easily occur and ultimately affect the stability and security of the TCM products [4]; this requires removal with a large number of organic reagents in the industrial production. In the industrial production of sinomenine hydrochloride (SH), a series of unit operations, such as percolation, pH adjustment, suspension centrifugation, extraction twice with chloroform, alcohol precipitation and crystallization are needed, to obtain a high-quality active pharmaceutical ingredient (API).

At present, membrane separation technology is widely used in food [5–7], wastewater treatment [8,9], seawater desalination, agricultural waste treatment [10,11], and other areas. Membrane separation technology can retain effective components and remove ineffective components by selecting membrane pore size and molecular weight cut-off (MWCO). It has the advantages of high efficiency, no phase changes, low energy consumption and convenient operation [4]. According to the different separation demands, membrane separation technology can be divided into microfiltration (MF), ultrafiltration (UF), nanofiltration (NF) and reverse osmosis (RO). The MF membrane has a pore size range of 0.01–10 μm is often used as the first step in the membrane separation process to remove microparticles such as bacteria, colloids and insoluble substances in the fluid [12]. UF membrane can generally retain more than 90% of the substances with relative molecular weight of 1000–300,000 Da, and can separate macromolecular organics (such as proteins, bacteria), colloids, suspended solids, etc. [4]. NF is between UF and RO. The MWCO of NF is 80–1000 Da, and the pore size is nanometer. It can intercept more than 90% of organic solutes with a MWCO greater than 300 Da and the organic matter with relatively small molecular weight that passing through the UF membrane and can be used for dialysis of inorganic salts trapped by RO membrane [13,14]. In the single TCM ingredients, such as sophora flavescens [15], ephedra [16], and compound Qingluotongbitang [17], the membrane separation method has a significant effect on the removal of impurities and the enrichment of effective components in the above water extract, and also in the separation and purification of volatile TCM components, such as volatile oil in patchouli [18], asarum [19], forsythia [20]. In addition, the TCM pharmaceutical industry has begun to use UF-MD hybrid system for resource utilization of TCM wastewater [21]. The National Development and Reform Commission (NDRC) of China has proposed in The guidance catalogue of industrial structure adjustment (2019 edition) to encourage the development and application of membrane separation technology in the process of drug production, the development and application of energy saving and emissions reduction technology in the production of API, new technology for quality control of TCM and process technology of modern dosage forms of TCM to solve the above problems. There is a need to produce high efficiency extraction equipment and continuous production technology and equipment.

Chinese Pharmacopoeia regulated that *Caulis sinomenii* (Qingfengteng) derived from the dry cane stem of Menispermaceae plants of *Sinomenium acutum* (Thunb.) Rehd. et Wils. and *Sinomenium acutum* (Thunb.) Rehd. et Wils. Var. cinereum Rehd. et Wils, which are distributed in the Yangtze River Basin and the south provinces of China. As a TCM, *Caulis sinomenii* has a long history in the treatment of rheumatism [22]. *Caulis sinomenii* was firstly recorded in the *Illustrations of Materia Medica* of the Song Dynasty, and later recorded in the *Compendium of Materia Medica* of the Ming Dynasty in ancient China. It is described in the *Compendium of Materia Medica* as being added to wine for the treatment of rheumatism, gout, tuberculous arthritis, pruritus, trauma and ulcers. *Caulis sinomenii* is rich in alkaloids. 91 kinds of alkaloids have been screened out in *Caulis sinomenii*, in which sinomenine (SIN) is a monomer alkaloid. It accounts for about 2% in *Caulis sinomenii*, and its hydrochloride form, i.e., SH, is commonly used in clinic [23]; which has good analgesic and immunosuppressive effects [24], and its different dosage forms, such as enteric coated capsules, sustained release tablets, and injections, were widely used in clinical for the treatment of rheumatoid arthritis, chronic nephritis, gout, ankylosing spondylitis and other diseases [25]. The traditional extraction process of SH, liquid-liquid extraction, in which a large amount of benzene, ethyl acetate or chloroform used, is a commonly used purification method, which could seriously damage the health of operators and cause serious environmental pollution [26]. Therefore, it is urgent to develop a green and efficient separation techniques to address this issue [27].

In this paper, the separation and purification of the extract solution of *Caulis sinomenii* by ceramic MF membrane, an organic UF membrane and under given operating conditions were studied, the aim is to achieve enrichment of the active ingredient, SH, and the removal of impurities, such as biological macromolecules and other alkaloids with similar structures. The technology developed in this paper can provide reference for the development of separation and purification methods of alkaloids in natural products.

2. Materials and Methods

2.1. Materials and Chemicals

Caulis sinomenii (No. 20200601) is purchased from Anhui Yuankang Herbal Pieces Co., Ltd., (Bozhou, Anhui, China) and its origin area is Shangluo City, Shaanxi Province. SH reference substance (No. S27281, ≥98%) is provided by Shanghai Yuanye Biotechnology Co., Ltd., (Shanghai, China). Hydrochloric acid is purchased by Tianjin Damao chemical reagent factory, (Tianjin, China). Calcium hydroxide is obtained by Shanghai Meryer Chemical Technology Co., Ltd., (Shanghai, China). Sodium dihydrogen phosphate is obtained by Tianjin Kermel Chemical Reagent Co., Ltd., (Tianjin, China). Acetonitrile is purchased by Fisher chemical. Sodium hypochlorite is purchased by Tianjin Bohua Chemical Reagent Co., Ltd., (Tianjin, China). Hydrochloric acid, calcium hydroxide, sodium dihydrogen phosphate and sodium hypochlorite are analytical pure and acetonitrile is chromatographic pure.

2.2. Preparation of Caulis sinomenii Extract Solution

1000 g of *Caulis sinomenii* was put into the percolation tube and soaked in 20 L of 0.1 mol·L^{-1} hydrochloric acid solution for 24 h. Medicinal materials were percolated at a flow rate of 15–20 mL·min^{-1}, the pH of the extract solution was adjusted to 11–12 with calcium hydroxide after percolated, then adjusted to pH 8–9 with 6 mol·L^{-1} hydrochloric acid solution after filtrated.

2.3. Filtrating and Cleaning Procedures

Ceramic MF membrane equipment is purchased by Jiangsu JiuWu Hi-Tech Co., Ltd. (Nanjing, China) and UF membrane equipment is purchased by Jinan Bona Biotechnology Co., Ltd. (Jinan, China) The equipments information is listed in Table 1.

Table 1. Equipment parameters of ceramic membrane and organic membrane.

Title	MF Membrane	UF Membrane
device model	JWCMF-0.1	DMJ60-3
membrane type	30 × L500 mm	51 × L306 mm
equipment type	900 × 500 × 1105 mm	1220 × 510 × 1330 mm
temperature range	10–55 °C	10–55 °C
pH range	0–14	2–12
pressure	≤0.2 MPa	≤4 MPa
dead volume	2.4 L	0.3 L

The ceramic membrane material is Al_2O_3, and the UF membrane material is polyethersulfone (PES). The pore sizes and MWCOs of the membranes were 0.5 μm, 0.2 μm, 0.05 μm and 10 kDa 1 kDa respectively. Ceramic membrane equipment operating transmembrane press (TMP): 0.05–0.2 MPa, temperature: 25 ± 5 °C, flow rate: 380–750 L·h^{-1}. UF membrane equipment parameters: operating pressure ≤ 0.1 MPa. The operation flow of the experiment is shown in Figure 1. The extract solution of *Caulis sinomenii* described in Section 2.2 was added into the feed tank, driven by centrifugal pump and circulated in the membrane separation equipment. TMP was controlled through adjusting valves P1 and P2, and the flow rate was recorded with current meter f1. The extracted solution passes through the ceramic membrane, the retention part is introduced into the feed tank for circulating operation, and the permeating solution enters the feed tank of UF module for filtrating, the flux data of ceramic MF membranes were obtained by metal tube rotameter. The flux data of UF membranes were calculated according to the volume of permeate discharged per minute.

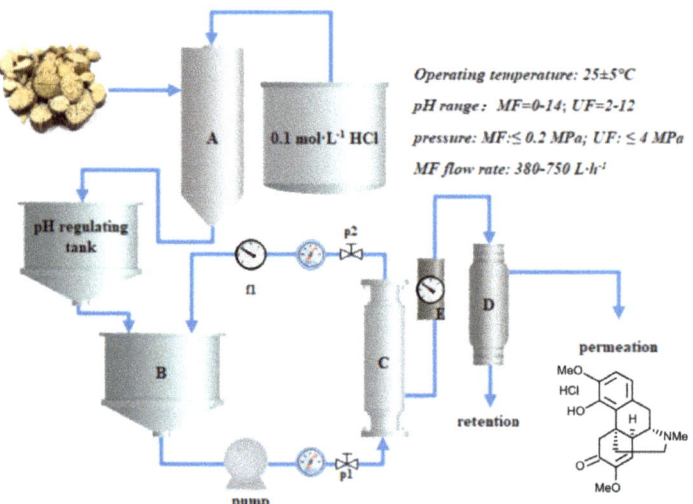

Figure 1. Separation and purification flow chart of *Caulis sinomenii* extract solution. (**A**) percolation tank; (**B**) feed tank; (**C**) ceramic MF membrane module; (**D**) UF membrane module; (**E**) metal tube rotameter.

The ceramic membranes were cleaned with clean water, 1% HCl-NaOH (It refers to cleaning with 1% hydrochloric acid solution for 1 h, followed by 1% sodium hydroxide solution for 1 h) and 1% sodium hypochlorite solution for 1 h at the end of the experiment respectively, and the cleaning volume was 20 L. After cleaning, the ceramic membrane was washed with clean water twice, 30 min each time. After the UF testing, clean membrane with clean water until the membrane flux is restored.

2.4. Analytical Methods

The HPLC method of SH content refers to the first part of *Chinese Pharmacopoeia* 2020 edition [28]. An HPLC (ACQuity Arc, Waters, MI, USA) system equipped with a UV detector was used. Analyses were conducted on a Waters Symmetry-C18 column (4.6 mm × 250 mm, 5 µm) with the column temperature controlled at 25 ± 5 °C. The flow rate of solvent was maintained at 1.0 mL·min^{-1}, while the injection volume of sample was 5 µL. Isocratic elution of the mobile phase containing acetonitrile and 0.78% sodium dihydrogen phosphate (12:88, v/v) was used. The detection wavelength was set at 262 nm. Finally, the regression equation (y = 3.837 × 10^{3}x − 4.666 × 10^3, R^2 = 1) was obtained, and the precision RSD of the method was 0.3%; the repeatability RSD was 2.2%; the stability RSD was 1.8%; the average recovery was 99% (RSD = 0.2%). The test samples are the extract solution and membrane separating permeates in Sections 2.2 and 2.3.

Total solids (TS) contents: 100 mL permeating solution was placed in an evaporating dish, evaporated in water bath, dried at 105 °C for 3 h, cooled in a dryer and weighed [29]. Calculation formula of the TS removal rates and SH permeabilities as follow:

$$\text{TS removal rate} = (Cs_0 - Cs)/Cs_0 \times 100\%, \quad (1)$$

where Cs is the TS in the permeating solution; Cs_0 is the TS in the extract solution of *Caulis sinomenii*.

$$\text{SH permeability} = C/C_0 \times 100\%, \quad (2)$$

where C refers to the SH content of permeating solution; C_0 refers to the SH content of the extract solution.

Evaluation of degree membrane pollution: if the membrane pollution degree reaches more than 30%, it means that the membrane is seriously polluted and the membrane performance is seriously degraded, so it needs to be cleaned. The cleaning effect is expressed by the recovery degree of membrane flux. The formulas of membrane fouling degree and membrane flux recovery rate are (3) and (4) respectively:

$$Jd = (1 - Jp)/Jw \times 100\%, \tag{3}$$

$$Jr = (Ji - Jp)/(Jw - Jp) \times 100\%, \tag{4}$$

where Jd refers to the membrane pollution degree, Jw is the water flux before fouling and Jp is the water flux after fouling, and Ji is the pure water flux after cleaning. When the membrane flux restored over 80%, it can be considered that the membrane has been cleaned.

2.5. Statistical Analysis

Each assay was performed in triplicate, and results were expressed as mean ± SD. Pearson's coefficient (r) was used for the correlational analyses. The SPSS 25.0 (SPSS Inc., Chicago, IL, USA) was used to perform the one-way analysis of variance (ANOVA) with Scheffe as post hoc test. Statistical significance level, except for the ones specially marked, was set at $p < 0.05$.

3. Results

3.1. Effect of Microfiltration

The content of SH in the extract solution was 0.80 mg·mL^{-1} and TS content was 2.28 g·100 mL^{-1} determined with the analysis methods described in Section 2.4. In the extraction process of *Caulis sinomenii*, due to the large amount of medicinal materials used in one test, even for the same batch of medicinal materials, the quality of medicinal materials in each package also has difference, resulting in higher RSD value. The results of SH permeabilities and TS removal rates of *Caulis sinomenii* extract solution separated and purified by three different pore size ceramic membranes are listed in Table 2. It can be seen from the results that the SH permeability and TS removal rate of 0.05 μm ceramic membrane are the highest. The ANOVA results are listed in Table 3, the SH permeabilities were not significantly different among the three different pore size ceramic membranes, but there was a significant difference in the TS removal rate, the clarity of *Caulis sinomenii* extract solutions were significantly improved. There was no significant difference in the SH permeability between the 0.5 μm and 0.2 μm ceramic membranes, 0.05 μm ceramic membrane can enrich SH to some extent, and the TS removal rate is the highest. The 0.05 μm ceramic membrane was selected to study the effect of different TMP and flow rate on the filtration of *Caulis sinomenii* extract solution.

Table 2. SH permeabilities and TS removal rates ($n = 3$).

Pore Sizes	SH Permeabilities	TS Removal Rates
0.5 μm	79.63 ± 5.2%	50.06 ± 1.1%
0.2 μm	73.56 ± 3.1%	55.37 ± 1.3%
0.05 μm	82.61 ± 3.7%	58.99 ± 1.0%

Table 3. ANOVA results of SH permeabilities and TS removal rates.

Item	Source	SS	D	MS	F	p
SH permeabilities	pore sizes	0.013	2	0.006	3.785	>0.05
	error	0.010	6	0.002		
	total	0.023	8			
TS removal rates	pore sizes	0.012	2	0.006	44.985	<0.05
	error	0.001	6	0.000		
	total	0.013	8			

The results of SH permeabilities and TS removal rates of 0.05 μm ceramic membrane under different TMP and flow rates are listed in Table 4. Under the TMP of 0.05 MPa and 0.1 MPa, the SH permeabilities are higher, the TS removal rates are higher under the TMP of 0.15 MPa and 0.2 MPa. The flow rates are 380 L·h^{-1}, 550 L·h^{-1} and 640 L·h^{-1}, the SH permeabilities are higher, but the TS removal rates has no significant difference. The flow rates has little effects on the SH permeabilities and has limited effects on the TS removal rates. High flow velocity can produce wall shear force and reduce the formation of deposition layer on the membrane surface [30], however, excessive flow rate will not only reduce membrane flux [31], but also cause a lot of difficult to remove foam, which will affect subsequent cleaning steps. The results of ANOVA are listed in Table 5, there was no significant difference between the SH permeabilities and the TS removal rates in 0.05 μm ceramic membrane by controlling different TMP and flow rates. Higher TMP causes the concentration polarization of the surface of the membrane [32], which results of the increase in the pollution grade of the ceramic membrane. The large flow rates also causes a large number of difficult to eliminate bubbles in the circulation process, it results in volume loss of the extract solution, which makes it difficult for subsequent cleaning steps. Under different TMP and flow rates, when there is no significant difference in the indexes, the most appropriate operating parameters, 0.1 MPa, 550 L·h^{-1} should be selected.

Table 4. SH permeabilities and TS removal rates under TMP and flow rates (n = 3).

Item	TMP (MPa)			
	0.05	0.10	0.15	0.20
SH permeabilities	84.54 ± 2.4%	80.85 ± 6.4%	76.69 ± 7.3%	75.91 ± 6.5%
TS removal rates	58.47 ± 1.4%	59.28 ± 1.1%	61.17 ± 2.5%	61.33 ± 1.3%
	flow rates (L·h^{-1})			
	380	550	640	750
SH permeabilities	78.96 ± 7.7%	78.96 ± 3.9%	79.61 ± 7.2%	76.61 ± 5.7%
TS removal rates	59.85 ± 1.8%	59.74 ± 0.8%	61.69 ± 1.9%	60.03 ± 2.4%

Table 5. ANOVA results of SH permeabilities and TS removal rates.

Item	Source	SS	D	MS	F	p
SH permeabilities	TMP	0.012	3	0.004	1.080	>0.05
	error	0.031	8	0.004		
	total	0.043	11			
TS removal rates	TMP	0.002	3	0.001	1.896	>0.05
	error	0.002	8	0.000		
	total	0.004	11			
SH permeabilities	flow rates	0.002	3	0.001	0.130	>0.05
	error	0.032	8	0.004		
	total	0.034	11			
TS removal rates	flow rates	0.001	3	0.000	0.779	>0.05
	error	0.003	8	0.000		
	total	0.001	11			

3.2. Effect of Ultrafiltration

After 0.05 μm ceramic membrane separation and purification, *Caulis sinomenii* extract solution would be treated by UF membrane. After 10 kDa UF separation and purification, the SH permeability was 56.95 ± 2.1% and the TS removal rate was 63.46 ± 0.3%. After 1 kDa UF treatment, the SH permeability was 60.77 ± 2.7%, and the TS removal rate was

64.24 ± 0.2%. Only from the results, the treatment effect of 1 kDa UF is better, the SH permeabilities and the TS removal rates are higher than that of 10 kDa UF. According to the ANOVA results listed in Table 6, the removal efficiency of 1 kDa UF membrane is better than that of 10 kDa in terms of TS remove; there was no significant difference in the SH permeabilities. Therefore, 1 kDa UF membrane can be used as the second treatment step of *Caulis sinomenii* extract solution.

Table 6. ANOVA results of SH permeabilities and TS removal rates.

Item	Source	SS	D	MS	F	p
SH permeabilities	MWCO	0.002	1	0.002	3.809	>0.05
	error	0.002	4	0.001		
	total	0.004	5			
TS removal rates	MWCO	0.000	1	0.000	12.086	<0.05
	error	0.000	4	0.000		
	total	0.000	5			

3.3. Effect of Membrane Cleaning

TCM extract solution contain a lot of carbohydrate, protein and other nutrients that can be used for the growth and reproduction of bacteria. If the membrane modules are not cleaned in time, or the cleaning effect is poor, there will be scaling phenomenon on the membrane surface and in the pipeline: bacterial adhesion, biofilm formation [33]. As shown in Figure A1 of Appendix A, 0.5 μm, 0.2 μm, 0.05 μm with TMP = 0.1 MPa and the flow rate was 550 L·h^{-1}, the flux attenuation curves of *Caulis sinomenii* extract solution were obtained. During MF processes, the flux of 0.5 μm ceramic membrane decreased significantly in 6 min, and stabilized at 32 L·h^{-1} after 15 min. The flux of 0.2 μm ceramic membrane decreased rapidly in 5 min and slowly in 14 min, was stable at 51 L·h^{-1} after 15 min. The flux of 0.05 μm ceramic membrane decreased to 44 L·h^{-1} in 5 min, reaching half of the membrane performance, and the flux stabilized to 20 L·h^{-1} after 14 min. After the percolation process, the pH of extract solution was adjusted by calcium hydroxide, a large amount of calcium chloride was produced, and precipitated in the extract to form suspension. Therefore, calcium chloride is the most direct cause of ceramic membrane pollution, followed by the soluble non effective components. The pollution degree of these three kinds of ceramic membranes is more than 60%, as listed in Table 7.

Table 7. Flux and pollution degrees of ceramic membranes (n = 3), (0.1 MPa, 550 L·h^{-1}).

Pore Sizes	Jw (L·h^{-1})	Jp (L·h^{-1})	Jd (%)
0.5 μm	110 ± 1.0	32 ± 3.8	71 ± 2.8%
0.2 μm	150 ± 1.0	51 ± 2.0	66 ± 3.5%
0.05 μm	100 ± 0.6	15 ± 1.3	85 ± 2.3%

In this experiment, pure water, 1% HCl-NaOH and 1% sodium hypochlorite were used to clean the membranes to compare the recovery rates of membrane flux by the above methods. The fouling degrees of these membranes were more than 30%, and their performances were seriously reduced. The fouling degrees of 0.5 μm, 0.2 μm and 0.05 μm ceramic membranes were 71%, 66% and 85%, respectively. The flux recovery effects of ceramic membranes are shown in Figure 2. The Flux of 0.5 μm ceramic membrane was restored to 34.5%, 58.2% and 92.7% after cleaning with pure water, 1% HCl-NaOH and 1% sodium hypochlorite solution, respectively; the flux of 0.02 μm ceramic membrane was restored to 50.6%, 53.3% and 97.3% after cleaning with pure water, 1% HCl–NaOH and 1% sodium hypochlorite solution, respectively. The flux of the 0.05 μm ceramic membrane was restored to 44.0%, 56.0% and 92.0% after initial cleaning with pure water, 1% HCl-NaOH and 1% sodium hypochlorite solution, respectively, and 1% sodium hypochlorite is the most effective of the three cleaning methods.

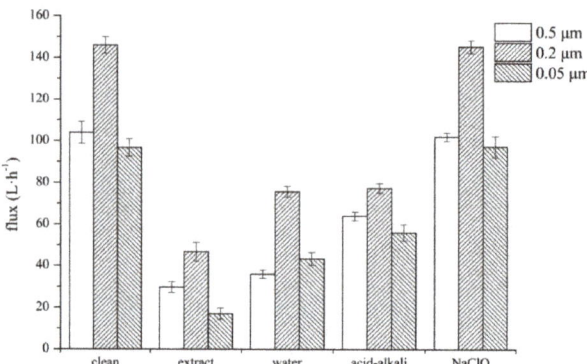

Figure 2. Flux changes of ceramic membranes under different treatment conditions, (0.1 MPa, 550 L·h^{-1}).

PES, polyvinylidene difluoride (PVDF), polyethylene (PE) and other organic materials are the main materials to produce UF membranes, which can not withstand the scouring of strong acid, alkali and oxidation reagents for a long time [34]. On the other hand, sodium hypochlorite is an important reason for the aging of organic membranes. Sodium hypochlorite is a common membrane cleaning agent and swelling agent in drink water and wastewater treatment [35]. Long term exposure to sodium hypochlorite may change the polymer membranes properties, resulting in the decrease of tensile strength and retention capacity, the permeability increase and finally shortening the membrane life [36]. As the extract solution was pretreated by ceramic membrane, the UF membrane was not polluted by large particles, and only macromolecular substances attached to the membrane surface, the flux of 10 kDa and 1 kDa UF membranes could be restored to 96.4% and 99.2% of the original, respectively, after washing with pure water, the cleaning results are listed in Table 8. In this section, the best cleaning method of ceramic membrane is 1% sodium hypochlorite solution circulating cleaning for 1 h, the organic membrane is pure water continuous circulating cleaning.

Table 8. Recovery effects of membrane flux (L·h^{-1}).

MWCO	Initial	Fouling	Water Cleaning	Recovery Rate (%)
10 kDa	9.30 ± 0.3	1.74 ± 0.05	8.97 ± 0.05	>95.0
1 kDa	13.20 ± 0.7	1.98 ± 0.1	13.10 ± 0.8	>95.0

4. Discussion

This paper is based on the exploration of pilot scale production experiment. In three parallel experiments, the SH content of the extract solution casused by each batch of *Caulis sinomenii* were different; the control of the volume of percolate, pH adjustment and the judgment of the separation end point may also cause fluctuations in the results. Therefore, it is very important to narrow the differences between groups. A researcher have investigated 73 TCM pharmaceutical enterprises [37]; the demand rate of membrane separation technology was more than 80%, but only half of the companies had undertaken process exploration and practical application, which shows that a large amount of researches on the membrane separation of TCM products were carried out on the laboratory scale, but have not taken proceeded to the pilot and industrial production scale-up and verification [38]. The later further maintenance of membrane separation equipment requires comes at a large cost, and a large amount of hypochlorous acid solution is also required for membrane cleaning. In the laboratory environment, researchers can try to ignore the loss of time, electricity and water resources to achieve more ideal results. However, in industrialized production, TCM pharmaceutical companies need to take full consideration of the production cost, organic solvent residues and compliance with the regulations on drug production supervision to avoid the problems caused by process changes and environmental issues. At present, the application of membrane separation technology in TCM pharmaceutical industry is still in the exploratory stage and cannot be compared. This research is committed to developing membrane separation technology of SH to replace the traditional chloroform extraction process, which can provide reference for the solution of similar problems in the industrialized production of TCM.

5. Conclusions

In this study, a combination of MF and UF method was established for the separation and purification of *Caulis sinomenii* extract solution. On the premise purpose of protecting the subsequent organic membrane from being polluted by large particles, the most suitable pore size of ceramic membrane was selected. The optimal pore size of ceramic membrane is 0.05 μm, whose SH permeability and TS removal rate are the highest. Under optimized process conditions, the TMP is 0.1 MPa and the flow rate is 550 L·h^{-1}, the TS removal rate was 59.74 ± 0.8%, and the SH permeability reached 80.85 ± 6.4%. The SH permeability and TS removal rate of 1 kDa UF membrane are higher than 10 kDa membrane. The SH permeability of 1 kDa UF membrane was 60.77 ± 2.7%, and the removal rate of TS was 64.24 ± 0.2%. In the process of membrane cleaning, a washing method of ceramic membrane with 1% sodium hypochlorite solution for 1 h was accepted. The ceramic membrane flux can be restored to more than 90%, and the UF membrane can be restored more than 99% after pure water washing. In the context of increasingly stringent regulations, membrane separation technology can be used as a technical reservation for the green production of TCM.

Author Contributions: Conceptualization, W.L., Z.L., Z.P. and P.Q.; methodology, Z.P., P.Q., X.W., H.M. and H.F.; software, X.W. and H.M.; validation, X.W., H.F. and W.L.; formal analysis, X.W. and W.L.; investigation, X.W., H.F. and H.M.; resources, Z.P., P.Q., W.L. and Z.L.; data curation, X.W., H.F. and H.M.; writing—original draft preparation, X.W.; writing—review and editing, X.W., Z.P., P.Q. and W.L.; visualization, X.W., H.F. and H.M.; supervision, W.L. and Z.L.; project administration, Z.L. and W.L.; funding acquisition, Z.L. and W.L. All authors have read and agreed to the published version of the manuscript.

Funding: This research was funded by Tianjin Science and technology project (No. 20ZYJDJC00090) and the National Nature Science Foundation of China (No. 82074276).

Institutional Review Board Statement: Not applicable.

Informed Consent Statement: Not applicable.

Data Availability Statement: All data is contained within the article.

Conflicts of Interest: The authors declare no conflict of interest.

Abbreviations

MF	microfiltration
UF	ultrafiltration
NF	nanofiltration
RO	reverse osmosis
SH	sinomenine hydrochloride
TS	total solids
TCM	tradition Chinese medicine
SIN	sinomenine
API	active pharmaceutical ingredient

MWCO	molecular weight cut-off
NDRC	national development and Reform Commission
PES	polyethersulfone
TMP	transmembrane press
ANOVA	analysis of variance
PVDF	polyvinylidene difluoride
PE	polyethylene

Appendix A

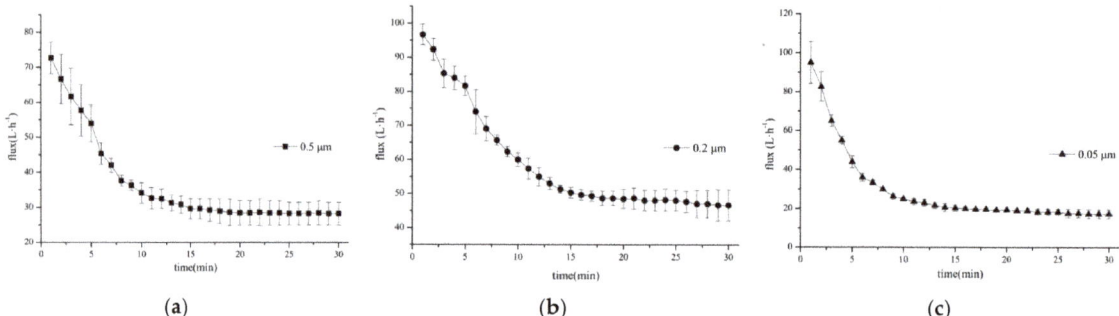

Figure A1. (**a**) 0.5 µm flux decay curve; (**b**) 0.2 µm flux decay curve and (**c**) 0.05 µm flux decay curve.

References

1. Shao, F.L.; Xu, J.T.; Zhang, J.Y.; Wei, L.Y.; Zhao, C.C.; Lu, C.X.; Fu, Y.Z. Study on the Influencing Factors of Natural Pectin's Flocculation: Their Sources, Modification and Optimization. *Water Environ. Res.* **2021**. online ahead of print. [CrossRef]
2. Xu, D.H.; Yu, C.L.; Wang, J.J.; Fan, Q.R.; Wang, Z.Z.; Xiao, W.; Duan, J.A.; Zhou, J.; Ma, H.Y. Ultrafiltration strategy combined with nanoLC-MS/MS based proteomics for monitoring potential residual proteins in TCMIs. *J. Chromatogr. B Anal. Technol. Biomed. Life Sci.* **2021**, *1178*, 122818. [CrossRef] [PubMed]
3. Das, I.; Sasmal, S.; Arora, A. Effect of thermal and non-thermal processing on astringency reduction and nutrient retention in cashew apple fruit and its juice. *J. Food Sci. Technol.* **2021**, *58*, 2337–2348. [CrossRef] [PubMed]
4. Guo, L.W.; Zhu, H.X.; Tang, Z.S.; Li, B.; Pan, Y.L. *Chinese Medicine Pharmaceutical Separation Technology Based on Membrane Process: Foundamentals and Applications*, 1st ed.; Science Press: Beijing, China, 2019; pp. 49–137.
5. Pelin, O. Production of high quality clarified pomegranate juice concentrate by membrane processes. *J. Membr. Sci.* **2013**, *442*, 264–271.
6. Echavarría, A.P.; Torras, C.; Pagán, J.; Ibarz, A. Fruit Juice Processing and Membrane Technology Application. *Food. Eng. Rev.* **2011**, *3*, 136–158. [CrossRef]
7. Sjölin, M.; Thuvander, J.; Wallberg, O.; Lipnizki, F. Purification of Sucrose in Sugar Beet Molasses by Utilizing Ceramic Nanofiltration and Ultrafiltration Membranes. *Membranes* **2019**, *10*, 5–22. [CrossRef]
8. Pramanik, B.K.; Pramanik, S.K.; Monira, S. Understanding the fragmentation of microplastics into nano-plastics and removal of nano/microplastics from wastewater using membrane, air flotation and nano-ferrofluid processes. *Chemosphere* **2021**, *282*, 131053. [CrossRef]
9. Vemuri, B.; Xia, L.C.; Chilkoor, G.; Jawaharraj, K.; Sani, R.K.; Amarnath, A.; Kilduff, J.; Gadhamshetty, V. Anaerobic wastewater treatment and reuse enabled by thermophilic bioprocessing integrated with a bioelectrochemical/ultrafiltration module. *Biosour. Technol.* **2021**, *321*, 124406. [CrossRef]
10. Didaskalou, C.; Buyuktiryaki, S.; Kecili, R.; Fonte, C.P.; Gyorgy, S. Valorisation of agricultural waste with an adsorption/nanofiltration hybrid process: From materials to sustainable process design. *Green Chem.* **2017**, *19*, 13–30. [CrossRef]
11. Voros, V.; Drioli, E.; Fonte, C.; Szekely, G. Process Intensification via Continuous and Simultaneous Isolation of Antioxidants: An Upcycling Approach for Olive Leaf Waste. *ACS Sustain. Chem. Eng.* **2019**. [CrossRef]
12. Adam, M.R.; Othman, M.H.D.; Kadir, S.H.S.A.; Sokri, M.N.M.; Tai, Z.S.; Iwamoto, Y.; Tanemura, M.; Honda, S.; Puteh, M.H.; Rahman, M.A.; et al. Influence of the Natural Zeolite Particle Size Toward the Ammonia Adsorption Activity in Ceramic Hollow Fiber Membrane. *Membranes* **2020**, *10*, 63–81. [CrossRef] [PubMed]
13. Wang, X.; Qiu, P.; Peng, X.S.; Li, Z.; Li, W.L. Applications of Integrated Membrane Technology in Pharmaceutical Industry of Traditional Chinese Medicine. *Chin. J. Pharm.* **2020**, *55*, 1836–1841.
14. Hailemariam, R.H.; Woo, Y.C.; Damtie, M.M.; Kim, B.C.; Park, K.D.; Choi, J.S. Reverse osmosis membrane fabrication and modification technologies and future trends: A review. *Adv. Colloid interface Sci.* **2020**, *276*, 102100. [CrossRef] [PubMed]

15. Echavarria, A.P.; Torras, C.; Pagan, J.; Ibarz, A.; Gao, H.N.; Jin, W.Q.; Guo, L.W. Research on refining of radix sophorae flavescentis decoction by microfiltration-ultrafiltration technology. *Tradit. Chin. Drug Res. Clin. Pharm.* **2009**, *20*, 571–573.
16. Shen, R.M.; Hu, B.; Zhang, Y.; Xia, Y.; Chen, Y.; Wan, D.J. Application of membrane separation technology in ephedrine production. *Chem. Bioeng.* **2008**, *7*, 58–60.
17. Cao, Y.T.; Pan, L.M.; Zhu, H.X.; Guo, L.W.; Shi, D.L. The research of different molecular membrane gradient purificate Qingluotongbi decoction. *World Sci-Tech RD* **2009**, *31*, 1000–1002.
18. Wang, H.; Liu, H.B.; Li, B.; Pan, L.M.; Fu, T.M.; Zhang, Y.; Song, Z.X.; Tang, Z.S.; Zhu, H.X. Separation of volatile oil from Pogostemon cablin based on ultrafiltration and vapor permeation membrane methods. *Chin. Tradit. Herb. Drugs* **2021**, *52*, 1582–1590.
19. Zhang, Q.; Zhu, H.X.; Tang, Z.S.; Li, B.; Pan, Y.L.; Yao, W.W.; Liu, H.B.; Fu, T.M.; Guo, L.W. Feasibility of vapor permeation technology for separating of Asari Radix et Rhizoma essential oil-bearing water. *Chin. Tradit. Herb. Drugs* **2019**, *50*, 1795–1803.
20. Zhang, Q.; Zhu, H.X.; Tang, Z.S.; Pan, Y.L.; Li, B.; Fu, T.M.; Yao, W.W.; Liu, H.B.; Pan, L.M. Study on essential oil separation from Forsythia suspensa oil-bearing water body based on vapor permeation membrane separation technology. *Chin. J. Chin. Mater. Med.* **2018**, *43*, 1642–1648.
21. Zhong, W.W.; Zhao, Y.N.; Chen, S.Q.; Zhong, J.L.; Guo, L.W.; Zheng, D.Y.; Xie, C.; Ji, C.; Guo, Y.; Dong, G.X.; et al. Resources recycle of traditional Chinese medicine (TCM) wastewater 1: Effectiveness of the UF-MD hybrid system and MD process optimization. *Desalination* **2021**, *504*, 114953. [CrossRef]
22. Chen, L.G.; Wang, H.J.; Ji, T.F.; Zhang, C.J. Chemoproteomics-based target profiling of sinomenine reveals multiple protein regulators of inflammation. *Chem. Commun.* **2021**, *57*, 5981–5984. [CrossRef]
23. Jiang, Z.M.; Wang, L.J.; Pang, H.Q.; Guo, Y.; Xiao, P.T.; Chu, C.; Guo, L.; Liu, E.H. Rapid profiling of alkaloid analogues in Sinomenii Caulis by an integrated characterization strategy and quantitative analysis. *J. Pharm. Biomed. Anal.* **2019**, *174*, 376–385. [CrossRef]
24. Zhao, Z.Z.; Liang, Z.T.; Zhou, H.; Jiang, Z.H.; Liu, Z.Q.; Wong, Y.F.; Xu, H.X.; Liu, L. Quantification of sinomenine in caulis sinomenii collected from different growing regions and wholesale herbal markets by a modified HPLC method. *Biol. Pharm. Bull.* **2005**, *28*, 105–109. [CrossRef]
25. Wang, X.; Zhang, Z.Y.; Qiu, P.; Peng, X.S.; Li, Z.; Li, Y.X.; Li, W.L. Research progress on Caulis Sinomenii, Sinomenine and related preparations. *Chin. J. Pharm.* **2021**, *56*, 85–93.
26. Milić, M.; Antović, A.; Trandafilović, M.; Zdravković, M. Fatal consequences caused by prolonged chloroform inhalation in a child. *Srp. Arh. za Celok. Lek.* **2019**, *147*, 230–234. [CrossRef]
27. Cheng, Y.Y.; Qu, H.B.; Zhang, B.L. Innovation guidelines and strategies for pharmaceutical engineering of Chinese medicine and their industrial translation. *China. J. Chin. Mater. Med.* **2013**, *38*, 3–5.
28. *Chinese Pharmacopoeia*; The Medicine Science and Technology Press of China: Beijing, China, 2015; Volume 1, p. 195.
29. Pan, J.J.; Shao, J.Y.; Qu, H.B.; Gong, X.C. Ethanol precipitation of Codonopsis Radix concentrate with a membrane dispersion micromixer. *J. Clean. Prod.* **2020**, *251*, 119633. [CrossRef]
30. Qu, P.; Gésan-Guiziou, G.; Bouchoux, A. Dead-end fifiltration of sponge-like colloids: The case of casein micelle. *J. Membr. Sci.* **2012**, *11*, 10–19. [CrossRef]
31. Roland, S.; Florian, S.; Johanna, L.; Ulrich, K. Comparative Assessment of Tubular Ceramic, Spiral Wound, and Hollow Fiber Membrane Microfiltration Module Systems for Milk Protein Fractionation. *Foods* **2021**, *10*, 692–706.
32. Daisuke, S.; Karkhanechi, H.; Matsuura, H.; Matsuyama, H. Effect of operating conditions on biofouling in reverse osmosis membrane processes: Bacterial adhesion, biofilm formation, and permeate flux decrease. *Desalination* **2016**, *378*, 74–79.
33. Vries, H.J.D.; Kleibusch, E.; Hermes, G.D.A.; Brink, P.V.D.; Plugge, C.M. Biofouling control: The impact of biofilm dispersal and membrane flushing. *Water Res.* **2021**, *198*, 117163. [CrossRef] [PubMed]
34. Li, K.; Su, Q.; Li, S.; Wen, G.; Huang, T. Aging of PVDF and PES ultrafiltration membranes by sodium hypochlorite: Effect of solution pH. *J. Environ. Sci.* **2021**, *104*, 444–455. [CrossRef] [PubMed]
35. Woo, Y.C.; Lee, J.J.; Oh, J.S.; Kim, H.S. Effect of chemical cleaning conditions on the flux recovery of MF membrane as pretreatment of seawater desalination. *Desalin. Water Treat.* **2013**, *51*, 6329–6337. [CrossRef]
36. Conidi, C.; Cassano, A.; Drioli, E. A membrane-based study for the recovery of polyphenols from bergamot juice. *J. Membr. Sci.* **2011**, *375*, 182–190. [CrossRef]
37. Zhu, H.X.; Tang, Z.S.; Pan, L.M.; Li, B.; Guo, L.W.; Fu, T.M.; Zhang, Q.C.; Pan, Y.L.; Duan, J.A.; Liu, H.B.; et al. Design, integration, and application of special membrane materials and equipment for new separation process in Chinese materia medica industry. *Chin. Tradit. Herb. Drugs* **2019**, *50*, 1776–1784.
38. Guo, L.W. Thoughts on the systematic research of "separation principle and technology of traditional Chinese medicine"-modern separation science and separation of traditional Chinese medicine. *World Sci. Tech Mod. Tradit. Chin. Med.* **2005**, *4*, 61–88.

Article

An Index for Quantitative Evaluation of the Mixing in Ethanol Precipitation of Traditional Chinese Medicine

Yanni Tai [1,2,†], Jingjing Pan [1,2,†], Haibin Qu [1,2] and Xingchu Gong [1,2,*]

1 Pharmaceutical Informatics Institute, College of Pharmaceutical Sciences, Zhejiang University, Hangzhou 310058, China; taiyn@zju.edu.cn (Y.T.); 21819006@zju.edu.cn (J.P.); quhb@zju.edu.cn (H.Q.)
2 Innovation Center in Zhejiang University, State Key Laboratory of Component-Based Chinese Medicine, Hangzhou 310058, China
* Correspondence: gongxingchu@zju.edu.cn
† Yanni Tai and Jingjing Pan contributed equally to this work.

Abstract: (1) Background: Ethanol precipitation is widely used in the manufacturing traditional Chinese medicines (TCMs). Insufficient mixing of ethanol solution and concentrate usually results in the coating loss of active ingredients. However, there is no index for quantitative evaluation of the mixing in ethanol precipitation. Therefore, this study aimed to define an index for quantitative evaluation of the mixing effect in ethanol precipitation of TCMs. (2) Methods: The concept and requirements of a mixing indicator were proposed. The mass percentage of concentrate fully mixed with ethanol solution (well-mixing ratio, WMR) was used as an index to evaluate the mixing effect. The formula for calculation of WMR was derived. The utility of the WMR was evaluated on stirring devices and a micromesh mixer. (3) Results: Increasing stirring speed, decreasing total solid content of the concentrate, and decreasing the diameter of the ethanol solution droplets all resulted in higher retention rates for lobetyolin and higher WMR. The WMR increased with the increasing flow rate of the concentrate and ethanol solution in the micromesh mixer. The mixing of ethanol solution and concentrate was better when using a micromesh mixer with a smaller internal mixing zone. The results revealed that WMR could be used to quantitatively characterize the mixing of concentrate and ethanol solution, although it has some limitations. (4) Conclusions: The proposed index WMR could guide quality control of the TCM ethanol precipitation process. This study represents a new contribution to improving ethanol precipitation equipment, optimizing process parameters, and enhanced properties of concentrate for TCM enterprises.

Keywords: ethanol precipitation; loss of active ingredients; mixing condition indicator; micromesh mixer

Citation: Tai, Y.; Pan, J.; Qu, H.; Gong, X. An Index for Quantitative Evaluation of the Mixing in Ethanol Precipitation of Traditional Chinese Medicine. *Separations* **2021**, *8*, 181. https://doi.org/10.3390/separations8100181

Academic Editor: Alberto Cavazzini

Received: 7 September 2021
Accepted: 8 October 2021
Published: 12 October 2021

Publisher's Note: MDPI stays neutral with regard to jurisdictional claims in published maps and institutional affiliations.

Copyright: © 2021 by the authors. Licensee MDPI, Basel, Switzerland. This article is an open access article distributed under the terms and conditions of the Creative Commons Attribution (CC BY) license (https://creativecommons.org/licenses/by/4.0/).

1. Introduction

As a simple and effective way to remove impurities, ethanol precipitation has been extensively applied in foods and herbal products for the purification process. Because of its simple operation and solvent safety, approximately 20% of prescription preparations and single-flavor preparations included in the Chinese Pharmacopoeia (first volume, 2020 edition) are subjected to ethanol precipitation to remove impurities [1]. Ethanol solution is introduced into the TCM concentrate, which effectively removes proteins [2], polysaccharides [3], and tannins [4], thus improving the purity of active ingredients in the supernatant. For example, the purity of total chlorogenic acid in Lonicerae Japonicae Flos [5] and hydroxysafflor yellow A [6] in Carthami Flos were improved obviously after ethanol precipitation. In addition, pectins can be purified from food by ethanol precipitation in the food industry [7,8].

The loss of active ingredients often occurs in industrial production [9]. Previous research indicated that a large number of phenolic acids were lost from *Salvia miltiorrhiza* concentrate in the process of industrial ethanol precipitation, and the loss ratios for danshensu, salvianolic acid B, and salvianolic acid D were even more than 50% [10,11]. A considerable number of domestic scholars have found that the active ingredients of siwu decoction [12], shuanghuanglian preparation [13], ganmaoling concentrate [14], huangqi concentrate [15], and biqiu granules [16] were lost to varying degrees in the ethanol precipitation process. To date, few foreign scholars have studied the ethanol precipitation process used with TCMs. However, Koh et al. [17] also demonstrated that rutin and tannic acid were lost in the refinement of sweet tea concentrate by the ethanol precipitation process.

It is generally considered that there are at least three reasons for the loss of active ingredients in ethanol precipitation [18], namely, coating, precipitation, and degradation. Precipitation loss is due to the low solubility of active ingredients in the supernatant [19–21]. In degradation loss, active ingredients generate other ingredients due to chemical reactions occurring during ethanol precipitation. The reason for coating loss is the insufficient mixing of ethanol solution and concentrate. The high viscosity of concentrate, large density difference of concentrate and ethanol solution, and a large amount of solid precipitate are the main factors resulting in the difficulties in mixing [20,21].

It is essential to improve the mixing of concentrate and ethanol solution to reduce losses arising from the coating. To the best of our knowledge, equipment factors [22], process parameters [23], and raw material characteristics [24] influence the mixing effect. Equipment factors include the mixing mode, size and shape of the stirring blade, and so on. The process operation parameter is principally the speed of ethanol solution addition. Generally, slower ethanol solution addition speed is more favorable. The total solid content of concentrate, concentrate pH, ethanol concentration [25,26], and material properties greatly impact the mixing effect. Although extensive research has been carried out on the ethanol precipitation process, there has been little discussion of a quantitative evaluation of the mixing effect. In this situation, it is impossible to judge whether the equipment, process parameters, and raw material properties have been optimized to enable full mixing of the ethanol solution and concentrate. Therefore, it is particularly important to find a method that can evaluate the mixing effect of ethanol solution and concentrate.

This research was designed to put forward an evaluation index, that is, a method for determining the proportion of the concentrate sufficiently mixed with ethanol solution (well-mixing ratio, WMR). This evaluation index was adopted to quantitatively evaluate the mixing of concentrate and ethanol solution in the process of ethanol precipitation. First, the hypothesis presented in this work was based on the principle of mass conservation, and a quantitative formula was derived for the new evaluation index. Second, Codonopsis Radix (dangshen), a widely used herbal medicinal material with the pharmacological roles of antioxidant [27] and antitumor agent [28], was employed as an example. The indicator of mixing condition was determined by the desorption method. Finally, using fundamental data from the research group's previous work [22], single-factor experiments were performed with stirring devices and the micromesh mixer. The influences of ethanol addition mode, the droplet size of the concentrate and ethanol solution, stirring speed, the total solid content of the concentrate, and flow rate when using the micromixer on the mixing effect were investigated. On this basis, the use of the new index to evaluate quantitatively mixing was investigated, results were analyzed, and shortcomings were noted.

2. Derivation of Quantitative Evaluation Index

2.1. Features of Mixing Condition Indicator

Many previous researchers have studied the influence of ethanol content in the supernatant, concentrate density, standing time of the mixture, and ethanol consumption as ethanol precipitation indicators [29–32]. Most of these studies took the retention rate of active ingredients as one of the optimization goals. When the concentrate and the ethanol solution are insufficiently mixed, the rapidly generated precipitate will coat some of the concentrate, which prevent the active ingredients from dissolving into the ethanol solution. Nevertheless, the low retention of active ingredients was not necessarily the result of coating loss. The lower solubilities of the active ingredients in ethanol solution and chemical reactions could also reduce the retention. In addition, it has been experimentally demonstrated that precipitates adsorb small amounts of supernatant. This phenomenon ultimately led to a reduction in the retention of active ingredients. For this reason, there are limitations in the evaluation of coating loss with the retention rate for active ingredients.

A mixing condition indicator can either be a certain component in the concentrate or a component added to indicate the mixing degree of ethanol solution and concentrate. The mixing condition indicator is a component with high solubility in the ethanol solution. Therefore, the precipitation loss will not occur for the mixing condition indicator in the ethanol precipitation process. The mixing condition indicator is also a component chemically stable in the ethanol precipitation process. Therefore, the total amount of a mixing condition indicator should remain constant before and after ethanol precipitation. If the total amount of the mixing condition indicator significantly decreased after ethanol precipitation, the coating loss probably occurs. If a mixing condition indicator can be found in the ethanol precipitation system, the mixing degree of ethanol solution and concentrate can be quantitatively determined.

2.2. Derivation of WMR Calculation Formula

Component A was used as the mixing condition indicator used to derive the formula for the WMR. According to the principle of conservation of mass in the Codonopsis Radix ethanol precipitation system, there are:

$$m_0 + m_1 = m_2 + m_3 \tag{1}$$

where m_0, m_1, m_2, and m_3 represent the concentrate mass, ethanol mass, the total mass of the supernatant, and the total mass of precipitate, respectively. It was assumed that the precipitate consisted of normal precipitate and encapsulated concentrate in the precipitate. Normal precipitate produces dried precipitate and precipitate-adsorbed supernatant. A schematic diagram of the coating phenomenon is shown in Figure 1.

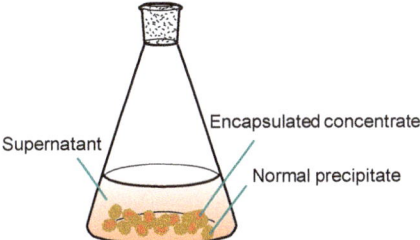

Figure 1. Schematic diagram of the coating phenomenon.

According to mass conservation of the precipitate in Figure 2, there is:

$$m_3 = m_4 + m_5 + m_0 \left(1 - \text{WMR}\right) \tag{2}$$

where m_4 and m_5 represent the mass of dried precipitate and the mass of the precipitate-absorbed supernatant, respectively. WMR is the evaluation index, and it represents the mass percentage of concentrate fully mixed with ethanol solution.

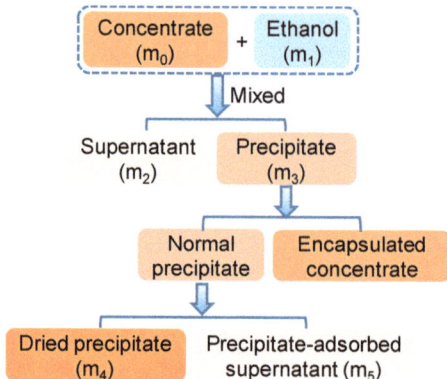

Figure 2. Illustration of mass distribution before and after ethanol precipitation.

According to mass conservation of total solid content in the ethanol precipitation system, there is:

$$m_0 S_0 = m_2 S_2 + m_0 (1 - \text{WMR}) S_0 + m_4 \tag{3}$$

where S_0 and S_2 represent the total solid contents of the concentrate and supernatant, respectively. According to mass conservation for the mixing condition indicator, component A, there is:

$$m_0 C_0 = m_2 C_2 + m_0 (1 - \text{WMR}) C_0 + m_5 C_2 \tag{4}$$

where C_0 and C_2 represent the content of component A in the concentrate and the content of component A in the supernatant, respectively. From Formulas (1)–(4), there is:

$$\text{WMR} = \frac{\frac{m_1}{m_0} + \frac{m_2}{m_0} \cdot S_2}{S_0 + \frac{C_0}{C_2} - 1} \times 100\% \tag{5}$$

The retention rate of component A was defined as η. Based on its physical meaning, η is calculated as:

$$\eta = \frac{m_2 C_2}{m_0 C_0} \tag{6}$$

Formula (6) was substituted into Formula (5), and the method for calculating the WMR with the retention rate was obtained. This method is shown in Formula (7).

$$\text{WMR} = \frac{m_1 + m_2 S_2}{\frac{m_2}{\eta} - m_0(1 - S_0)} \tag{7}$$

If the WMR is 0%, the concentrate is completely encapsulated. At this time, S_2 is zero, and it can be shown that m_1 is zero. The physical meaning is that when ethanol is not consumed, it can be regarded as an indicator of complete coating. The concentrate can be regarded as a completely encapsulated precipitate, and components can be partially dissolved when ethanol solution is added. If the WMR is 100%, the concentrate and ethanol solution are completely mixed. In that case, Formula (8) can be obtained:

$$\eta = \frac{m_2}{m_1 + m_2 S_2 + m_0(1 - S_0)} \tag{8}$$

where m_1 represents the mass of ethanol consumption, $m_0 \times (1-S_0)$ represents the water contained in the concentrate, and $m_1 S_2$ represents total solid contained in the supernatant. If a dried precipitate with no supernatant adsorbed on the surface could be obtained, then the above three items would add up to the theoretical mass of the supernatant. Because m_2 is the mass of the supernatant actually measured, when the concentrate and ethanol solution are completely mixed without coating, the retention rate depends on the amount of precipitate-absorbed supernatant.

3. Materials and Methods

3.1. Chemicals and Reagents

The electronic balance used was a model XS105 (Mettler-Toledo, Greifensee, Switzerland). Syringe needles were purchased from Hangzhou Chengdian Experimental Equipment Co., Ltd. (Hangzhou, Zhejiang, China). Reference substance lobetyolin (batch number: 180307, purity > 98%) used for HPLC analysis was purchased from Shanghai Ronghe Pharmaceutical Technology Co., Ltd. (Shanghai, China). Absolute ethanol and 95% ethanol were purchased from Zhejiang Evergreen Chemical Co., Ltd. (Hangzhou, Zhejiang, China). HPLC-grade acetonitrile was purchased from Merck (Darmstadt, Germany). HPD-100 resin was purchased from Cangzhou Baoen Adsorption Material Technology Co., Ltd. (Cangzhou, Hebei, China). Ultrahigh-purity water was produced using a water purification system (Milli-Q, Millipore, Burlington, MA, USA). Codonopsis Radix (batch number: 191022) was purchased from Anhui Bozhou Yuanfengtang Agricultural and Sideline Products Distribution Co., Ltd. (Bozhou, Anhui, China). Codonopsis Radix was identified as the dried root of *Codonopsis pilosula* (Franch.) Nannf. by Dr. Gong Xingchu and deposited at the Smart Pharmaceutical Laboratory, Pharmaceutical Informatics Institute, College of Pharmaceutical Sciences, Zhejiang University, Hangzhou, China.

3.2. Preparation of Codonopsis Radix Water Extract Concentrate

Codonopsis Radix was refluxed with water at a ratio of 1:8 (m/v) for 0.5 h. The water extracts were collected by filtration with an oil-free vacuum filter pump (DP-01, Shanghai Leigu Instrument Co., Ltd., Shanghai, China). Codonopsis Radix was refluxed with water at a ratio of 1:6 (m/v) for 0.5 h again to obtain another water extract. After filtration, the two filtrates were mixed. The mixtures were concentrated with a rotary evaporator (V-100, BUCHI Labortechnik AG, Flawil, Switzerland). The relative density of the concentrate was measured by weighing 5 mL of concentrate. All experiments were repeated three times. The concentration process was completed when the relative density approximately of the concentrate was 1.2 g/mL (The total solid content was about 60%). The samples with different solid content concentrate were obtained by diluting with water in the study.

3.3. Analytical Methods

The established HPLC method was used to determine lobetyolin content [33]. An HPLC system (1100, Agilent Technologies, Santa Clara, CA, USA) equipped with a variable wavelength detector (G1314C), a quaternary pump (G1311A), a column thermostat (G1316A), an automatic liquid sampler (G1313A), and a degasser (G1322A) was used for all measurements. Chromatographic separation was carried out at 30 °C on a Zorbax SB-C18 column (250 mm × 4.6 mm, 5 μm particle size). The flow rate of the mobile phases containing acetonitrile (B) and water (A) is 1.0 mL/min. The isocratic elution was set performed using 80% A, and the total runtime is 25 min. The sample volume injected was 10 μL, and the detection wavelength was 269 nm. After each run, the chromatographic system was set to 100% B and balanced for 10 min with a 1.0 mL/min flow rate. Representative HPLC chromatograms of the supernatant sample and the reference standard sample are presented in Appendix A, Figure A1. The supernatant (2.0 g) was diluted with 50% methanol-water in a 5 mL volumetric flask. The concentrate (0.5 g) were diluted with 30% methanol-water in a 5 mL volumetric flask.

The total solid content was determined using a gravimetric method. The supernatant or the concentrate was placed into a weighing bottle dried to a constant mass. Samples were dried at 105 °C for 3.0 h using a drying oven (DHG-9146A, Shanghai Jinghong Experimental Equipment Co., Ltd., Shanghai, China) and then incubated in a desiccator for 0.5 h. Their masses were determined, and the total solid content in the samples were calculated.

3.4. Determination of the Mixing Condition Indicator

Previous studies have demonstrated that the retention rate of lobetyolin in the supernatant was higher than 90% when using the optimized ethanol precipitation process [22,34], which indicated that little lobetyolin was lost during the ethanol precipitation process. Therefore, lobetyolin might be regarded as a potential indicator of the mixing situation. However, it was still necessary to confirm that the content of lobetyolin showed little difference before and after ethanol precipitation, i.e., its solubility in the supernatant was large enough to avoid precipitation loss.

A total of 15.0 g of Codonopsis Radix concentrate containing 45% total solid content was placed into a conical bottle, and 37.5 g of 95% (v/v) ethanol solution was added. The mixtures were stirred for 30 min with a magnetic stirrer (JJ-IA, Changzhou Yunhua Electrical Appliance Co., Ltd., Changzhou, Zhejiang, China) and filtered under vacuum. Then the supernatant and a precipitate were obtained. The precipitate was dried to a constant mass at 105 °C for 3.0 h using a drying oven, and the mass of supernatant adsorbed by the precipitate was calculated. The total lobetyolin content in the supernatant and precipitate was analysed referencing the analytical method of Section 3.3, then compared with that in the concentrate. There is no obvious degradation loss in the ethanol precipitation process if the latter is not different from the former.

Direct detection of the solubility of lobetyolin in different concentrations of ethanol requires more reference substances, and the experiments will be expensive. Therefore, this study utilized the desorption method to indirectly determine solubility trends with different ethanol concentrations [35].

The desorption method was as follows: 91.0 g of concentrate was diluted with 454.0 g ultrahigh-purity water and then mixed well by a magnetic stirrer. 70 mL of HPD-100 resin was immersed with 90% (v/v) ethanol solution in a beaker for 12 h. The resin was washed 5 times with water. The diluted Codonopsis Radix concentrate was adsorbed with 70 mL HPD-100 resin. The mixture of concentrate and resin was stirred for 3 h and filtered. After that, the resin was collected by filtration. Then each 10 g of wet resin was contact with 20 mL of desorption solution composed of different ethanol concentrations (0%, 30%, 50%, 70%, and 90%, v/v) for 3 h under stirring. After that, the desorption solutions were collected by filtration. The lobetyolin concentrations of desorption solution were determined. If the lobetyolin amount increased with increasing ethanol concentration, it indirectly illustrated that the solubility of lobetyolin increased with increasing ethanol concentration.

3.5. Single-Factor Experiments with Stirring Devices

Single-factor experiments were performed with the stirring devices to evaluate the utility of the evaluation index. The stirring devices are shown in Figure 3. Two addition modes were used in ethanol precipitation experiments: ethanol solution was added dropwise to the concentrate, and concentrate was added dropwise to the ethanol solution. Different silicone tubes and syringe needles were used to control the droplet sizes of the concentrate and ethanol solution. Their inner diameter (ID), outer diameter (OD), and droplet diameters are listed in Table 1. Droplet shapes were captured by a smartphone camera (iPhone 8, Apple Inc., Cupertino, CA, USA). Droplet diameters were calculated based on the known OD of the syringe needle and silicone hose. Total solid content of the concentrate was 50%, and 95% ethanol solution or the concentrate was pumped into the Erlenmeyer flask with a peristaltic pump (BT300-2J, Changzhou Runhua Electric Co., Ltd., Changzhou, Zhejiang, China). The mass ratio of ethanol solution to concentrate (ECR) was 1.5 (g/g). A magnetic stirrer completed the mixing of ethanol solution and concentrate with a stirring speed of 400 rpm. The total time of ethanol solution addition and stirring after the addition was 30 min. After ethanol precipitation, the mixture was filtered under vacuum, and the supernatant was collected.

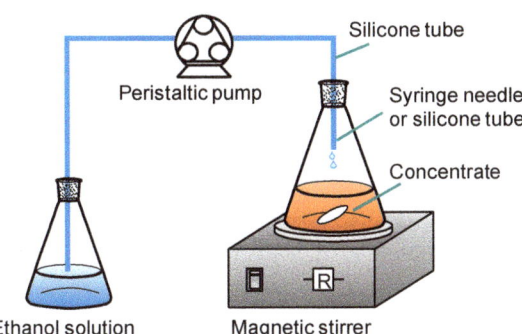

Figure 3. Schematic diagram for adding ethanol solution dropwise into the concentrate under stirring. When adding concentrate dropwise into ethanol solution, the position of ethanol solution and concentrate was exchanged.

Table 1. Conditions for single-factor experiments on stirring devices.

Addition Modes	Droplet Diameters (mm)	Substances or Specifications for Diameter Control
Concentrate added dropwise into ethanol solution	2.09 ± 0.050	Syringe needle, ID: 0.46 mm; OD: 0.70 mm.
	3.20 ± 0.037	Silicone tube, ID: 3.1 mm; OD: 6.2 mm.
	3.52 ± 0.016	Silicone tube, ID: 4.8 mm; OD: 8.0 mm.
Ethanol solution added dropwise into concentrate	2.06 ± 0.031	Syringe needle, OD: 0.45 mm; length: 16 mm.
	3.25 ± 0.041	Syringe needle, ID: 0.46 mm; OD: 0.70 mm.
	3.49 ± 0.034	Silicone tube, ID: 3.1 mm; OD: 6.2 mm.

Droplet diameters are expressed as the mean ± standard deviations, $n = 3$.

3.6. Single-Factor Experiments with the Micromesh Mixer

A schematic diagram of the micromesh mixer device is shown in Figure 4. It mainly consists of two polytetrafluoroethylene (PTFE) plates (40 × 40 × 12 mm^3), one PTFE micromesh (20 × 10 × 1.5 mm^3), and a silicone gasket. The silicone gasket was placed between two PTFE plates. A micromesh with a 0.6 mm diameter was placed in the middle of the micromixer. The mixing chamber was 16 mm in length, 8 mm in width, and 2.5 mm in depth. Ethanol solution was the continuous phase, and the concentrate was the dispersed phase. The continuous phase was pumped into the micromesh mixer by an advection pump (2PB-20005II, Beijing Xingda Technology Development Co., Ltd., Beijing, China). The dispersed phase was pumped into the micromesh mixer by a gear pump (CT3001F, Baoding Reef Fluid Technology Co., Ltd., Baoding, Hebei, China). The ECR was controlled at 1.5 (g/g), and mixtures were collected in an Erlenmeyer flask. 30.0 g of Codonopsis Radix concentrate with a total solid content of 54% was mixed with 45.0 g of 95% (v/v) ethanol solution at flow rates of 40 mL/min, 60 mL/min, and 80 mL/min. Three tests were performed at each flow rate. After the mixtures were collected, they were immediately stirred for 5 min with a stirring speed of 300 rpm to prevent the precipitation from heaping.

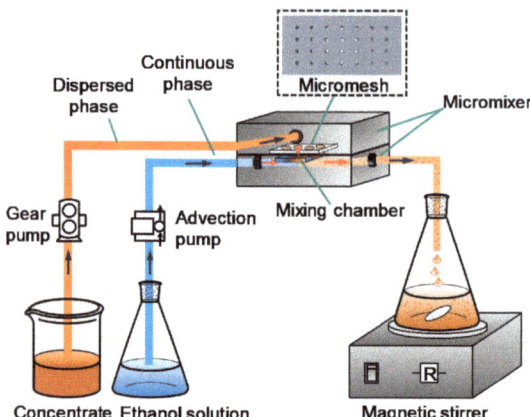

Figure 4. Schematic diagram of the micromesh mixer device.

4. Results

4.1. Results of Determination for Mixing Condition Indicator

The total content of lobetyolin in the supernatant and precipitate after ethanol precipitation was analyzed and then compared with that in the concentrate before ethanol precipitation. It was found that the former was 99.4% of the latter, indicating that the total content of lobetyolin remained unchanged in the process of ethanol precipitation. Therefore, degradation loss could be ignored.

Figure 5 shows lobetyolin concentration data with different ethanol concentrations. The lobetyolin concentration increased with ethanol concentration, which indicated that a higher ethanol concentration caused dissolution of more lobetyolin. The results further demonstrated that the higher the ethanol concentration was, the higher the lobetyolin solubility [36]. Theoretically, precipitation loss of lobetyolin probably did not occur due to an increase of lobetyolin solubility after adding ethanol solution. If a large amount of lobetyolin loss was observed after ethanol precipitation, the loss should have been due to coating loss. Therefore, lobetyolin was regarded as a mixing condition indicator for the Codonopsis Radix ethanol precipitation system in follow-up studies.

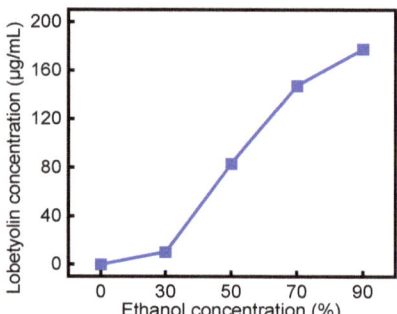

Figure 5. Lobetyolin concentration with varying ethanol concentration.

4.2. Results of Ethanol Precipitation with Stirring Devices

It can be seen from Figure 6 (the raw data can be seen from Supplementary Materials Table S1) that when the ethanol solution was added dropwise to the concentrated solution, the retention rate of lobetyolin in the supernatant was higher than 72.3%, and the calculated WMR value was higher than 87.0%. It suggests that dropwise addition of ethanol solution to the concentrate had a better mixing effect than dropwise addition of concentrate to ethanol solution. This result is consistent with the actual production steps. The explanation for the above result was that when drops of ethanol solution were added to the concentrate, the ethanol content in the system gradually increased, precipitation gradually occurred, and this was conducive to full mixing. On the other hand, when concentrate was added dropwise into the ethanol solution, the initial ethanol content of the system was extremely high. The amount of precipitation generated per unit mass of concentrate was higher at the beginning than at the end of ethanol addition. This situation was more prone to the coating phenomenon. With increasing droplet diameters, the retention rate of lobetyolin and the WMR decreased. Overall, these results showed that reducing the droplet diameter was beneficial for reducing the mixing scale and improving the mixing effect.

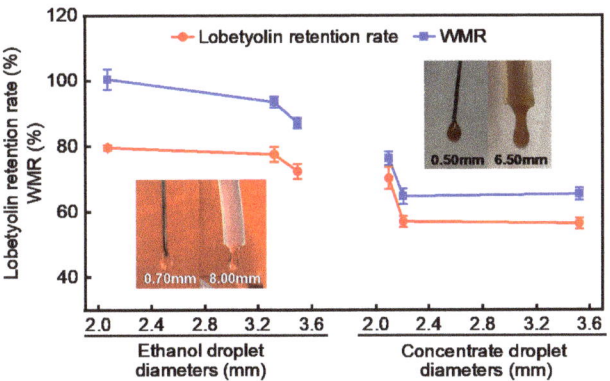

Figure 6. Results of lobetyolin retention rates and WMR values on different addition modes. The numbers below the droplet are the outer diameter sizes of the syringe needle or the silicone tube.

The influences of stirring speed and total solid content of the concentrate on ethanol precipitation process were investigated in the authors' previous research [22]. The ethanol precipitation was carried out by adding ethanol solution dropwise into the concentrate with a syringe needle (ID: 3.1 mm, OD: 6.2 mm). In this section, obtained data previously was remodeled. Data were substituted into Formula (7), and the WMR was obtained with different stirring speeds and concentrate total solid content, as shown in Figure 7 (the raw data can be seen from Supplementary Materials Table S2). The higher the stirring speed was, the higher the WMR and the more sufficient the concentrate mixing with ethanol solution. With decreasing total solid content, the mixing effect was enhanced. When the total solid content of concentrate was 45%, and the stirring speed reached 300 rpm, the WMR value was close to 1.0, which meant that the concentrate was fully mixed with ethanol. At this point, the WMR value was almost unchanged when the stirring speed was increased. These results demonstrated that the WMR well reflected the mixing of concentrate and ethanol solution.

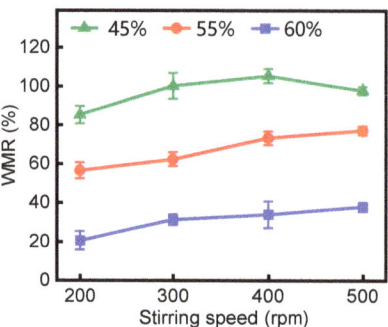

Figure 7. WMR values at different stirring speeds (200, 300, 400 and 500 rpm) and different total solid content of the concentrate (45%, 55%, and 60%) on stirring devices. The results are expressed as mean ± standard deviations, n = 3.

4.3. Results of Ethanol Precipitation with a Micromesh Micromixer

From the experiment results of droplet diameter on the mixing effect, it was known that the mixing effect was improved by decreasing the diameters of ethanol solution droplets, thus reducing coating loss. Therefore, a micromesh was put into the micromixer to carry out ethanol precipitation experiments. WMR values and lobetyolin retention rates under different concentrate flow rate on the micromesh mixer are shown in Figure 8 (the raw data can be seen from Supplementary Materials Table S3). WMR values showed an upward trend with increasing concentrate flow rate, consistent with the result obtained with the stirring devices. The WMR values with the micromesh mixer were approximately 100%, and this indicated that increasing the flow rate improved the mixing effect on contact and reduced the loss of active components simultaneously. Surprisingly, lobetyolin retention rates were approximately 15% lower than the WMR values, indicating that the precipitate possibly adsorbed approximately 15% of the supernatant.

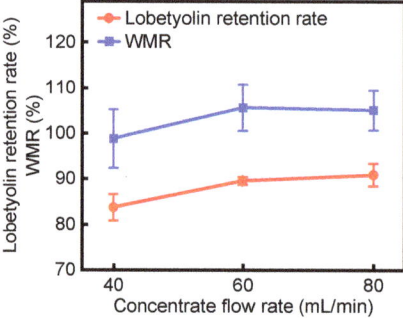

Figure 8. WMR values and lobetyolin retention rates under different concentrate flow rate on the micromesh mixer. The results are expressed as mean ± standard deviations, n = 3.

4.4. The Analysis of WMR Value Calculated with the Previous Dataset

In this study, the previous dataset [22] for the membrane dispersion micromixer was substituted into Formula (7), and the calculated WMR values are shown in Figure 9 (raw data can be seen from Supplementary Materials Tables S4–S7). WMR was improved by reducing the membrane pore size, mixing chamber width and depth, and increasing the concentrate flow rate. With a fixed concentrate flow rate, as Figure 9b shows, no significant differences in WMR were found by increasing ECR. Figure 9c shows a comparison of WMR values for the membrane dispersion micromixer and for stirring devices. With increasing total solid content in the concentrate, the WMR value for the membrane dispersion micromixer was significantly higher than for stirring devices.

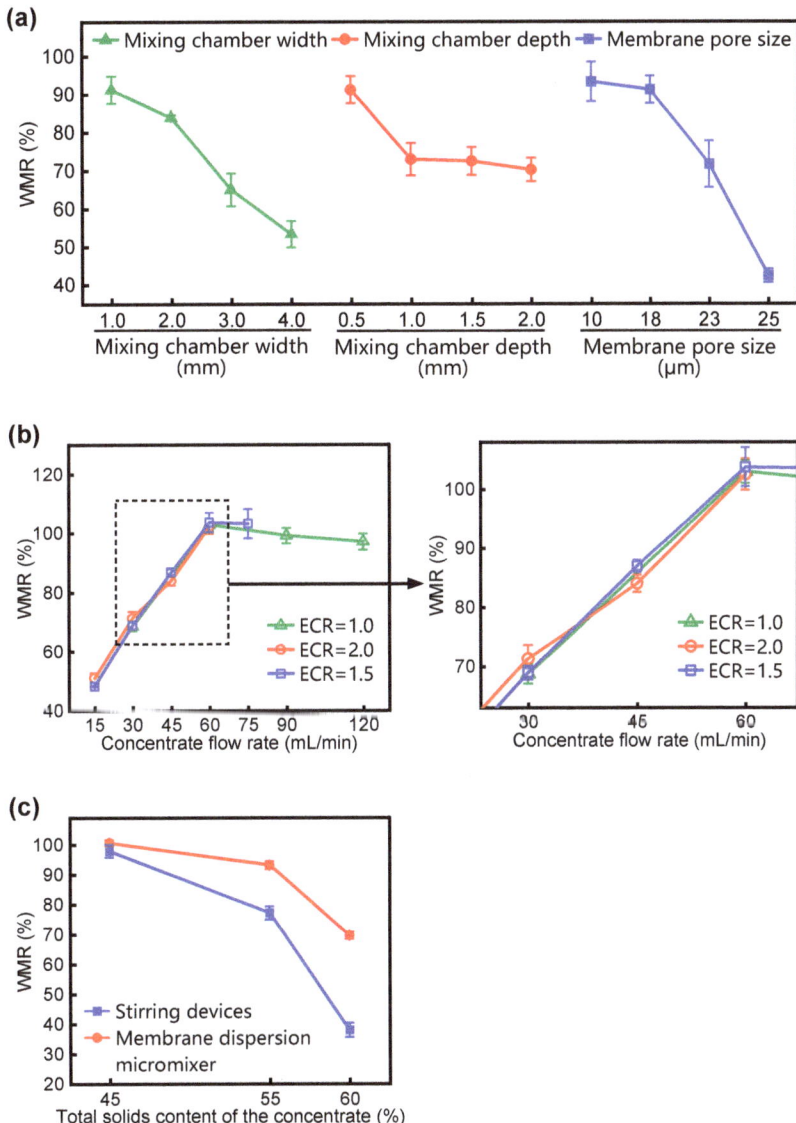

Figure 9. WMR values under different membrane pore size, mixing chamber width, mixing chamber depth (**a**), concentrate flow rate, ECR (**b**), and total solid content of the concentrate (**c**) on a membrane dispersion micromixer. The results are expressed as mean ± standard deviations, $n = 3$.

5. Discussions

5.1. Relationship between WMR Values and Total Solid Removal Rates

The total solid removal rate is also one of the evaluation indexes used with the ethanol precipitation process, and it partially characterizes the ability to remove impurities. The calculation is as shown in Formula (9) [37]:

$$\text{Total solid removal rate} = \left(1 - \frac{m_s \times DM_s}{m_c \times DM_c}\right) \times 100\% \tag{9}$$

where m and DM are the mass and total solid content, respectively. Subscripts S and C represent the supernatant and concentrate, respectively. To better understand the relationship between the WMR value and the total solid removal rate, the linear formula $y = a_1 x + b_1$ was used to fit the linear relationship between them. The dataset to be fitted was from the current studies and previous dataset. According to the fitting results shown in Table 2 and Figure 10, a_1 negative values mean that the higher the WMR was, the lower the total solid removal rate. The fitting results revealed that the total solid were dissolved in the supernatant to the extent possible after the concentrate and ethanol were fully mixed. The fitting results show that the higher the degree of fully mixing the concentrate and ethanol solution, the higher the total solid content in the supernatant, further indicating that more components are dissolved in the supernatant. The R^2 values of the two linear fittings shown were less than 0.75, which indicated that in addition to ECR, other important factors, such as concentrate properties, affected the WMR and total solid removal rate.

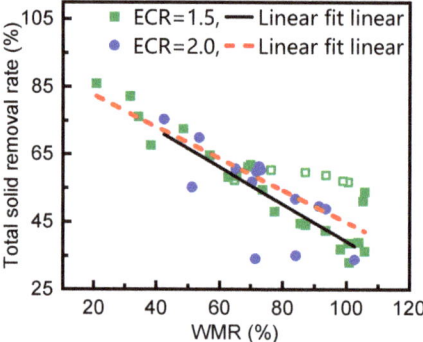

Figure 10. Relationship between the WMR value and total solid removal rate: □ represents the data from this work; • and ■ represent the data from Pan, J.; Shao, J.; Qu, H.; Gong, X, ethanol precipitation of Codonopsis Radix concentrate with a membrane dispersion micromixer; published by J. Clean. Prod, 2020.

Table 2. Linear fitting results for total solid removal rate and WMR.

ECR	a_1	b_1	R^2
2.0	−0.548 ± 0.148	0.940 ± 0.111	0.5334
1.5	−0.471 ± 0.0534	0.920 ± 0.436	0.7424

5.2. Deficiency of the Index

The index WMR was used to quantitatively evaluate the mixing situation of ethanol solution and concentrate, and the proportion of active components lost due to coating. According to the assumptions described in Section 2.2, the WMR should be positively correlated with the retention rate and slightly exceed the retention rate. Experimental data for an ECR of 1.5 were extracted. The linear formula $y = a_2 x + b_2$ was used to describe the linear relationship between lobetyolin retention rate and WMR. According to the results shown in Table 3 and Figure 11, when the WMR was more than 40%, the linear relationship between the retention rate and WMR was good, with $R^2 = 0.9075$. The value of a_2 was less than 1.0, which indicated that the WMR value was higher than the lobetyolin retention rate. As the retention rate did not reflect the content of lobetyolin in the supernatant adsorbed by precipitation, the WMR value was greater than the lobetyolin retention rate, which was consistent with the assumptions used for the WMR calculation derived in this study. When the WMR was less than 40%, the linear relationship between the retention rate and WMR was also good, with $R^2 = 0.9849$. However, the value of a_2 was greater than 1.0, indicating that the WMR was less than the lobetyolin retention rate. This result was inconsistent with the previous assumption. The reasons may be as follows: according to the hypotheses used to define the new evaluation index, precipitation was divided into normal precipitate and the concentrate encapsulated in the precipitate. The supernatant was not included in the precipitation. When the WMR was low, the coating loss was serious. The precipitate may encapsulate part of the supernatant.

Another limitation of this study is that an indicator of mixing conditions must be found for the ethanol precipitation system. Without a suitable indicator, it would be difficult to calculate the WMR.

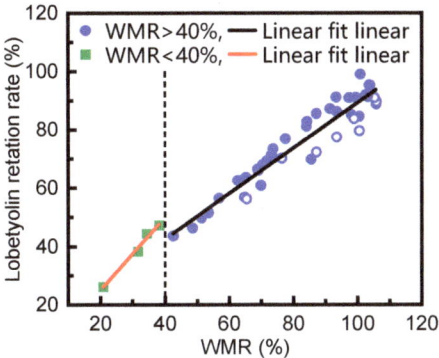

Figure 11. Relationship between WMR and lobetyolin retention rate: ○ represents the data from this work; ● and ■ represent the data from Pan, J.; Shao, J.; Qu, H.; Gong, X, ethanol precipitation of Codonopsis Radix concentrate with a membrane dispersion micromixer; published by J. Clean. Prod, 2020.

Table 3. Linear fitting results for lobetyolin retention rate and WMR.

WMR	a_1	b_1	R^2
>40%	0.777 ± 0.0388	0.116 ± 0.0325	0.9075
<40%	1.25 ± 0.109	0.00114 ± 0.0348	0.9849

6. Conclusions

In this study, a new index was proposed for quantitative evaluation of the mixing of concentrate and ethanol solution in the ethanol precipitation process. The index is WMR, which denotes the mass percentage of concentrate fully mixed with ethanol. The higher the WMR was, the higher the proportion of concentrate and ethanol solution that was fully mixed. The concept and requirements of a mixing condition indicator were put forward, and a formula for the WMR was derived. Lobetyolin was verified as a mixing condition indicator in the ethanol precipitation of Codonopsis Radix concentrate.

It was found that dropwise addition of ethanol solution to the concentrate showed better mixing than the dropwise addition of concentrate to the ethanol solution. Reducing the droplet diameters of concentrate and ethanol improved the mixing effect. When using a micromixer to mix an ethanol solution and a concentrate, reducing the pore size and the size of the mixing chamber improved the WMR. Increasing the two-phase flow rate also led to higher WMR values. Lobetyolin retention rate increased as WMR increased.

The results were consistent with expectations indicating that the WMR could quantitatively characterize the concentrate and ethanol solution mixing. The research also helps to solve the long-standing problem of indistinguishability between coating loss and precipitation loss on ethanol precipitation. It provides a beneficial foundation for improving the quality control of ethanol precipitation process.

Supplementary Materials: The following are available online at https://www.mdpi.com/article/10.3390/separations8100181/s1, Table S1: WMR values, total solid removal rate, lobetyolin retention rates on different addition modes. Table S2: WMR values under different stirring speeds (200, 300, 400, and 500 rpm) and different total solid content of the concentrate (45%, 55%, and 60%) on stirring devices. Table S3: WMR values, total solid removal rate, lobetyolin retention rates under different concentrate flow rates on the micromesh mixer. Table S4: WMR values under different membrane pore sizes on a membrane dispersion micromixer. Table S5: WMR values under different mixing chamber widths on a membrane dispersion micromixer. Table S6: WMR values under different mixing chamber depths on a membrane dispersion micromixer. Table S7: WMR values under different concentrate flow rates, ECR on a membrane dispersion micromixer.

Author Contributions: Conceptualization, X.G.; investigation, J.P. and Y.T.; data curation, J.P. and Y.T.; writing—original draft preparation, Y.T. and J.P.; writing—review and editing, X.G. and Y.T.; supervision, X.G. and H.Q.; funding acquisition, X.G. and H.Q. All authors have read and agreed to the published version of the manuscript.

Funding: This research was funded by the National S&T Major Project of China (2018ZX09201011-002), the Basic Public Welfare Research Program of Zhejiang Province (LGG18H280001), and the National Project for Standardization of Chinese Materia Medica (ZYBZH-C-GD-04).

Data Availability Statement: The data presented in this study are available on Supplementary Materials.

Conflicts of Interest: The authors declare no conflict of interest.

Abbreviations

TCMs: traditional Chinese medicines; HPLC: high performance liquid chromatography; WMR: the mass percentage of concentrate fully mixed with ethanol solution (well-mixing ratio); ECR: the mass ratio of ethanol solution to concentrate; ID: inner diameter; OD: outer diameter.

Appendix A

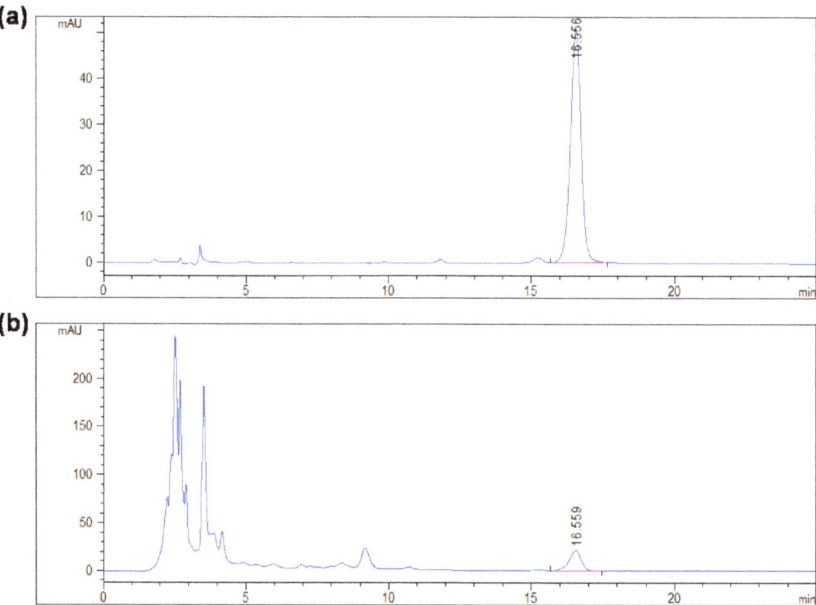

Figure A1. Typical HPLC chromatogram of the reference standard and the supernatant sample. (**a**) HPLC chromatogram of the lobetyolin. (**b**) HPLC chromatogram of the supernatant.

References

1. State Pharmacopoeia Commission. *Pharmacopoeia of the People's Republic of China*; China Medical Science Press: Beijing, China, 2020.
2. Lee, H.; Gupta, R.; Kim, S.; Wang, Y.; Rakwal, R.; Agrawal, G.; Kim, S. Abundant storage protein depletion from tuber proteins using ethanol precipitation method: Suitability to proteomics study. *Proteomics* **2015**, *15*, 1765–1769. [CrossRef] [PubMed]
3. Yan, J.; Wang, C.; Yu, Y.; Wu, L.; Chen, T.; Wang, Z. Physicochemical characteristics and in vitro biological activities of polysaccharides derived from raw garlic (*Allium sativum* L.) bulbs via three-phase partitioning combined with gradient ethanol precipitation method. *Food Chem.* **2021**, *339*, 128081. [CrossRef]
4. Gong, X.; Li, Y.; Qu, H. Removing tannins from medicinal plant extracts using an alkaline ethanol precipitation process: A case study of Danshen injection. *Molecules* **2014**, *19*, 18705–18720. [CrossRef] [PubMed]
5. Lu, X. Studies on Ethanol Precipitation and Countercurrent Extration Refining Processes of Lonicerae Japonicae Flos and Artemisiae Annuae Herba Water Extract. Master's Thesis, Zhejiang University, Hangzhou, China, 2014.
6. Yuan, J. Optimization of Alcohol Precipitation Process for Extract of Carthamus Tinctorius and Salvia Miltiorrhiza and Investigation on the Encapsulated Loss Phenomena. Master's Thesis, Zhejiang University, Hangzhou, China, 2011.
7. Zhang, T.; Guo, X.; Meng, H.; Tang, X.; Ai, C.; Chen, H.; Lin, J.; Yu, S. Effects of bovine serum albumin on the ethanol precipitation of sugar beet pulp pectins. *Food Hydrocoll.* **2020**, *105*, 105813. [CrossRef]
8. Guo, X.; Meng, H.; Zhu, S.; Tang, Q.; Pan, R.; Yu, S. Developing precipitation modes for preventing the calcium-oxalate contamination of sugar beet pectins. *Food Chem.* **2015**, *182*, 64–71. [CrossRef] [PubMed]
9. Tai, Y.; Shen, J.; Luo, Y.; Qu, H.; Gong, X. Research progress on the ethanol precipitation process of traditional Chinese medicine. *Chin. Med.* **2020**, *15*, 84. [CrossRef] [PubMed]
10. Gong, X.; Li, Y.; Guo, Z.; Qu, H. Control the effects caused by noise parameter fluctuations to improve pharmaceutical process robustness: A case study of design space development for an ethanol precipitation process. *Sep. Purif. Technol.* **2014**, *132*, 126–137. [CrossRef]
11. Zhao, F.; Li, W.; Pan, J.; Qu, H. Process characterization for ethanol precipitation of Salviae miltiorrhizae Radix et Rhizoma (Danshen) using 1H NMR spectroscopy and chemometrics. *Process Biochem.* **2021**, *101*, 218–229. [CrossRef]
12. Zhou, F.; Li, J.; He, Y.; Mu, R.; Fu, C. Simultaneous determination of eight components in Siwu decoction by HPLC and analysis of transmitting of the components in water extraction and ethanol precipitation process. *Chin. J. Pharm. Anal.* **2019**, *39*, 983–991.
13. Jiang, M.; Zhang, X.; Shao, F.; Shang, Y.; Yang, M.; Liu, R.; Mei, H. Effect of ethanol to material ratio on ethanol precipitation and sediment morphology of Shuanghuanglian preparation. *Chin. Tradit. Herb. Drugs* **2020**, *51*, 4954–4959.
14. Pan, H.; Deng, H.; Chen, Z.; Zhang, Y.; Wang, L. Study on balance of process of alcohol precipitation of ganmaoling granules. *Chin. J. Chin. Mater. Med.* **2016**, *41*, 1376–1379.
15. Shao, F.; Yu, M.; Jiang, M.; Yang, M.; Shang, Y.; Liu, R.; Zhang, X. Establishment of determination of fractal dimension of ethanol-precipitated flocs of two root medicinal herbs. *Chin. J. Exp. Tradit. Med. Form.* **2019**, *25*, 103–107.
16. Zhang, Y.; Liu, L.; Chang, X.; Wu, Y.; Xiao, W.; Hu, J.; Chao, E. Optimization of alcohol precipitation technology of Biqiu granules based on index components and pharmacodynamics. *Chin. J. Chin. Mater. Med.* **2016**, *41*, 4598–4604.
17. Koh, G.; Chou, G.; Liu, Z. Purification of a water extract of Chinese sweet tea plant (*Rubus suavissimus* S. Lee) by alcohol precipitation. *J. Agric. Food Chem.* **2009**, *57*, 5000–5006. [CrossRef]
18. Gong, X.; Huang, S.; Jiao, R.; Pan, J.; Li, Y.; Qu, H. The determination of dissociation constants for active ingredients from herbal extracts using a liquid–liquid equilibrium method. *Fluid Phase Equilibr.* **2016**, *409*, 447–457. [CrossRef]
19. Xu, R.; Cong, Y.; Zheng, M.; Chen, G.; Chen, J.; Zhao, H. Solubility and modeling of hesperidin in cosolvent mixtures of ethanol, isopropanol, propylene glycol, and n-propanol + water. *J. Chem. Eng. Data* **2018**, *63*, 764–770. [CrossRef]
20. Mo, F.; Ma, J.; Zhang, P.; Zhang, D.; Fan, H.; Yang, X.; Zhi, L.Q.; Zhang, J. Solubility and thermodynamic properties of baicalein in water and ethanol mixtures from 283.15 to 328.15 K. *Chem. Eng. Commun.* **2019**, *208*, 183–196. [CrossRef]
21. Shen, Y.; Farajtabar, A.; Xu, J.; Wang, J.; Xia, Y.; Zhao, H.; Xu, R. Thermodynamic solubility modeling, solvent effect and preferential solvation of curcumin in aqueous co-solvent mixtures of ethanol, n-propanol, isopropanol and propylene glycol. *J. Chem. Thermodyn.* **2019**, *131*, 410–419. [CrossRef]
22. Pan, J.; Shao, J.; Qu, H.; Gong, X. Ethanol precipitation of Codonopsis Radix concentrate with a membrane dispersion micromixer. *J. Clean. Prod.* **2020**, *251*, 119633. [CrossRef]
23. Pan, J.; Tai, Y.; Qu, H.; Gong, X. Optimization of membrane dispersion ethanol precipitation process with a set of temperature control improved equipment. *Sci. Rep.* **2020**, *10*, 19010. [CrossRef] [PubMed]
24. Pan, J.; He, S.; Zheng, J.; Shao, J.; Li, N.; Gong, Y.; Gong, X. The development of an herbal material quality control strategy considering the effects of manufacturing processes. *Chin. Med.* **2019**, *14*, 38. [CrossRef]
25. Zhang, H.; Yan, A.; Gong, X.; Qu, H. Study on quality indicators for concentration process of supernatant obtained in first ethanol precipitation in production of Danshen injection. *Chin. J. Chin. Mater. Med.* **2011**, *36*, 1436–1440.
26. Yan, A.; Gong, X.; Qu, H. Method for discriminating key quality control indicators of concentrated solution before traditional Chinese medicine ethanol precipitation. *Chin. J. Chin. Mater. Med.* **2012**, *37*, 1558–1563.
27. Zou, Y.; Zhang, Y.; Paulsen, B.; Rise, F.; Chen, Z.; Jia, R.; Li, L.X.; Song, X.; Feng, B.; Tang, H.; et al. Structural features of pectic polysaccharides from stems of two species of Radix Codonopsis and their antioxidant activities. *Int. J. Biol. Macromol.* **2020**, *159*, 704–713. [CrossRef]

28. Bailly, C. Anticancer properties of lobetyolin, an essential component of Radix Codonopsis (Dangshen). *Nat. Prod. Bioprospect.* **2021**, *11*, 143–153. [CrossRef]
29. Tai, Y.; Qu, H.; Gong, X. Design space calculation and continuous improvement considering a noise parameter: A case study of ethanol precipitation process optimization for Carthami Flos extract. *Separations* **2021**, *8*, 74. [CrossRef]
30. Li, B.; Kang, Q.; Chen, C.; Wang, Y.; Zhao, Y.; He, B.; Wu, Q. Optimization of alcohol precipitation technology for Fufang Shuanghua oral liquid based on FAHP-entropy method. *Cent. South Pharm.* **2019**, *17*, 414–419.
31. Zhu, Y.; Yu, S.; Zhang, X.; Zhou, C.; Wei, J. Optimization of alcohol-precipitation technology for Fufang Roucongrong Mixture by AHP combined with orthogonal test. *China Pharm.* **2019**, *22*, 1257–1260.
32. Zhang, L.; Gong, X.; Qu, H. Optimizing the alcohol precipitation of Danshen by response surface methodology. *Sep. Purif. Technol.* **2013**, *48*, 977–983. [CrossRef]
33. Xu, Z.; Huang, W.; Gong, X.; Ye, T.; Qu, H.; Song, Y.; Liu, D.; Wang, G. Design space approach to optimize first ethanol precipitation process of Dangshen. *Chin. J. Chin. Mater. Med.* **2015**, *40*, 4411–4416.
34. Peng, Y.; Lei, C.; Tang, Y.; Zhou, L.; Xia, X. Effects of Chitosan Flocculation Clarification Process and Alcohol Precipitation Process on Water Extract of Codonopsis Radix. *Chin. J. Inform. Tradit. Chin. Med.* **2017**, *24*, 81–84.
35. Deineka, V.; Deineka, L.; Sidorov, A.; Kostenko, M.; Blinova, I. Estimating the solubility of anthocyanins using cartridges for solid-phase extraction. *Russ. J. Phys. Chem. A* **2016**, *90*, 861–863. [CrossRef]
36. Zhang, X.; Chen, W.; Zeng, Y.; Mi, S.; Wang, Q.; Li, K.; Zhang, L. The synthesis and identification of Codonopsis Pilosula polyferose. *Acta Chin. Med. Pharmacol.* **2011**, *39*, 77–81. [CrossRef]
37. Gong, X.; Wang, S.; Li, Y.; Qu, H. Separation characteristics of ethanol precipitation for the purification of the water extract of medicinal plants. *Sep. Purif. Technol.* **2013**, *107*, 273–280. [CrossRef]

Article

Determination of Genotoxic Azide Impurity in Cilostazol API by Ion Chromatography with Matrix Elimination

Boglárka Páll [1], Zsuzsa Gyenge [1], Róbert Kormány [1] and Krisztián Horváth [2,*]

[1] Drug Substance Analytical Development Division, Egis Pharmaceuticals Plc., Keresztúri út 30-38, H-1106 Budapest, Hungary; pall.boglarka@egis.hu (B.P.); gyenge.zsuzsa@egis.hu (Z.G.); kormany.robert@egis.hu (R.K.)
[2] Research Group of Analytical Chemistry, University of Pannonia, Egyetem utca 10, H-8200 Veszprém, Hungary
* Correspondence: raksi@almos.uni-pannon.hu

Abstract: Cilostazol is a commonly used active pharmaceutical ingredient (API) to treat and reduce the symptoms of intermittent claudication in peripheral vascular disease. Recently, it was found to be a potential medicine in the effective treatment of COVID-19. In addition to the positive effects of this API, genotoxic sodium azide is used in the synthesis of cilostazol that can appear in the API. In this work, a method was developed for the determination of sodium azide (as azide anion) in cilostazol API at 7.5 ppm limit level by using ion chromatography (IC) and liquid–liquid extraction (LLE) sample preparation. The liquid–liquid extraction allows the application of high sample concentrations. Because of the low limit concentration (7.5 ppm), 500 mg sample was dissolved in 5 mL solvent. By using LLE for sample preparation, the huge amount of cilostazol was omitted and column overload was avoided. The developed method was validated in accordance with the relevant guidelines. Specificity, accuracy, precision, limit of detection and limit of quantification parameters were evaluated. The calculated limit of detection was 0.52 ppm (S/N:3) and the limit of quantification was 1.73 ppm (S/N:10) for sodium azide. The recovery of the sodium azide was 102.4% and the prepared solutions were stable in the sample holder for 24 h.

Keywords: azide; cilostazol; COVID-19; genotoxic impurity; validation; LLE

1. Introduction

The cilostazol is a platelet-aggregation inhibitor and arterial vasodilator, its long-term use may prevent stroke [1]. Recently, a network-based ranking was used to prioritize drugs to treat COVID-19 symptoms [2]. Cilostazol was the fourth on this list. Several studies were published about cilostazol long-term treatment safety. It was found by W.R. Hiatt et al. that the mortality was not higher in the treatment group, than the placebo group during the examined 42 months [3]. In addition to the long-term treatment safety of cilostazol, attention should be drawn to the possible impurities because these can cause deteriorate side effect for patients in the long term. Figure 1 shows a possible synthesis pathway of cilostazol. It can be seen that sodium azide is used for forming the tetrazole ring [4]. Sodium azide is toxic and genotoxic. Its lowest lethal dose is 10 mg/kg. After poisoning, death can occur within an hour by hypotension [5].

According to the European Medicine Agency (EMA) and International Council for Harmonization (ICH), the maximum daily intake of potential genotoxic impurities for more than 12 months of exposure is 1.5 µg. The limit concentration of genotoxic impurity depends on the daily dose of API [6] that varies depending on whether it is used alone or in combination with other anti-platelet agents [1]. If it is taken alone the daily dose is 200 mg/day, if it is taken with CYP2C19 inhibitors it is 100 mg/day [7]. Considering the daily dose mentioned before, this means that the permitted sodium azide concentration is only 7.5 ppm.

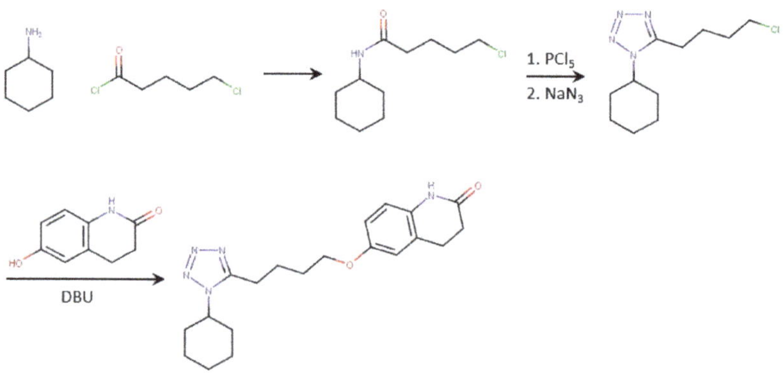

Figure 1. Possible synthesis path of cilostazol API.

Due to its low limit concentration, determination of sodium azide is a problem in APIs and pharmaceutical products. Vinkovic and Drevenkar developed an ion chromatographic method for azide determination in protein samples [8]. The USP refer to a method for determining azide impurity in ibersartan API, however, the sample preparation is not applicable for other APIs in general. In this method [9], azide anions are separated by ion chromatography with 0.1 N sodium hyroxide eluent and conductivity detector. The specified column (L31) is a strong anion exchange column with quaternary amine groups [10]. This method was applied for the determination of sodium azide in a range of "sartan" drugs. The sodium azide was quantified at 15 ppm level while the retention time of azide anion was about 42 min [11]. A reversed phase liquid chromatography method was developed for sartan APIs by Gricar and Andrensek [12]. In this RP-HPLC method, UV detection was used for determination of azide at 10 ppm limit level. During the sample preparation, the APIs were precipitated and removed by filtration. In case of cilostazol the limit is lower, and the sample preparation should be carried out in different solution. In order to control this impurity at 7.5 ppm level, a large amount of the sample must be dissolved. This high cilostazol API concentration can decrease the efficiency of the analytical method by overloading the chromatographic system. Therefore, during the sample preparation, the amount of the API needs to be decreased while the quantity of sodium azide should remain the same.

The aim of this work was to develop and validate an analytical procedure for the determination of sodium azide content in cilostazol API. In our method, 500 mg cilostazol is dissolved in 5 mL of methylene chloride. Since sodium azide is hydrophilic and cilostazol is insoluble in water, aqueous phase liquid–liquid extraction (LLE) was used for the sample preparation followed by the ion chromatographic determination of the extracted azide anions. The validation was carried out in accordance with ICH Q2(R1) guideline recommendations [13].

2. Materials and Methods

Analytical grade dimethyl sulfoxide, methylene chloride (Fisher Scientific, Loughborough, UK) and sodium azide (Sigma-Aldrich,, Darmstadt, Germany) was used for sample preparations. Cilostazol originates from the synthesis of Egis Pharmaceuticals Plc. Water was prepared freshly using ELGA Purelab system (ELGA, Lane End, UK). Mettler Toledo analytical balances were used for weighing (Greifensee, Switzerland) and Eppendorf automatic pipettes were used for liquid handling (Hamburg, Germany).

Dionex ICS 5000 HPIC system equipped with eluent generator (EGC) and suppressed conductivity detector (CD) was used for IC measurements (Thermo Scientific, Waltham, MA, USA). The anion exchange column was Dionex IonPac AS11HC (2 × 250 mm) with a guard column AG11HC (2 × 50 mm) (Thermo Scientific, Waltham, MA, USA). The chromatograms were processed with Chromeleon 7. (Thermo Scientific, Waltham, MA, USA).

3. Results

3.1. Sample Preparation with Liquid-Liquid Extraction

Because of the low concentration limit of sodium azide, large amount of cilostazol API is needed for the determination. The high concentration of API may have a negative impact on the ion chromatographic determination of azide ion. Cilostazol can precipitate in the eluent and deteriorate the column performance or overload it. A suitable sample preparation technique should be developed to avoid overloading problems. Due to the significantly different solubilities of cilostazol and sodium azide in the immiscible solutions of methylene chloride and dimethyl-sulfoxide/water, liquid–liquid extraction can be used efficiently for sample preparation. The methylene chloride solution is the lower phase and the aqueous solution is the upper phase. Then, 500 mg cilostazol is dissolved in 5 mL of methylene chloride (100 mg/mL). The sodium azide is extracted by 5 mL of dimethyl-sulfoxide/water. The extraction efficiency was tested on three temperatures (20, 25, and 30 °C). It was found that the extraction efficiency did not depend on the temperature. Even if the temperature affected slightly the solubility of cilostazol in the aqueous phase, it did not affected the extraction of the azide anion.

3.2. Ion Chromatographic Analysis of Sodium Azide

3.2.1. Effect of Flow Rate

The examined flow rates were over the optimum velocity. However, the results met the system suitability requirements, and the theoretical plates did not change significantly at different flow rates. Accordingly, so either flow rate could be applicable.

3.2.2. Effect of Eluent Concentration

The retention factor azide anion was determined at different eluent concentrations (5, 10, 15, 20, 25, and 30 mM). Figure 2 shows that as retention factors of azide anions decreased by the increasing concentration of the eluent. In ion exchange chromatography, the plot of logarithm (with base 10) of retention factor *versus* logarithm of eluent concentration should be linear for isocratic separations. The slope of equation (−0.989) that is fitted for the measured data points verifies that the retention behavior of azide ion is in line with equilibrium theory.

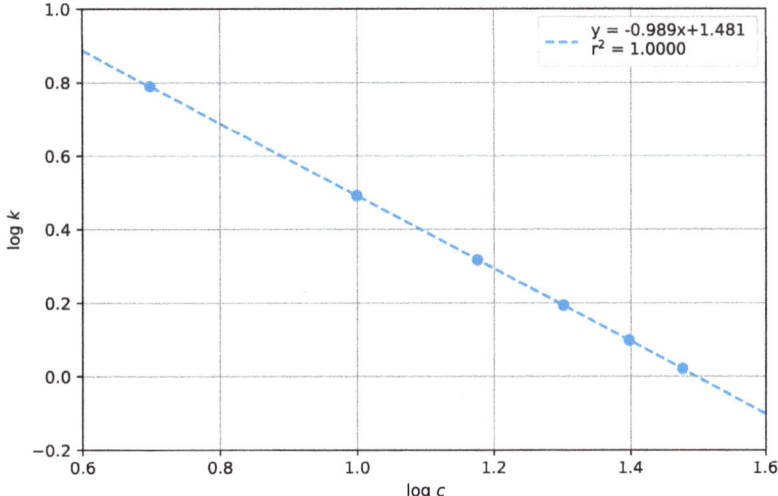

Figure 2. Logarithm (with base 10) of retention factor of azide anions as a function of logarithm of eluent concentration.

3.2.3. Effect of Column Temperature

The effect of column temperature on the separation of azide anions were studied in the range of 25 °C to 40 °C in six points. The results showed that the azide ion retention time did not depend significantly on the column temperature. The relative standard deviation of the six retention times (measured on six different column temperature) was only 0.18%.

3.3. Validation of the Chromatographic Method

A developed and optimized method can only be used for quantitative measurements of raw materials, intermediates or APIs if the applicability of the method was proved earlier so the method is validated. The validation of this method was performed according to ICH Q2(R1) guideline [13] for limit tests.

3.3.1. Solvent Preparations

The following solutions were used during chromatographic method validation:
- Blank: mixture of 5 mL purified water and 5 mL methylene chloride. It was homogenized by shaking for at least one minute;
- Sodium azide solution: 75 µg/mL of sodium azide in dimethyl sulfoxide;
- Sodium azide reference solution: 5 mL of methylene chloride and 50 µL of the sodium azide solution was added into a HS vial. After the dissolution, 5 mL of purified water was added. It was homogenized by shaking for at least one minute. For the measurement the upper phase was used;
- Test solution: 100 mg/mL of cilostazol in methylene chloride. After the dissolution 5 mL purified water was added. It was homogenized by shaking for at least one minute. For the measurement the upper phase was used;
- Spiked test solution: 500 mg of cilostazol and 50 µL of the sodium azide solution was measured into 5.0 mL methylene chloride. After the dissolution 5 mL purified water was added. It was homogenized by shaking for at least one minute. For the measurement the upper phase ws used;
- Limit of detection: 5 mL of methylene chloride and 50 µL of 22.5 µg/mL of sodium azide solution was added into a HS vial. After the dissolution 5 mL purified water was measured. It was homogenized by shaking for at least one minute. For the measurement the upper phase was used.

3.3.2. Ion Chromatographic Method for Azide Determination

As a result of the preliminary experiments the final anion chromatographic method can be seen in Table 1.

Table 1. HPIC method parameters.

Run time	40 min
Eluent	KOH solution
Eluent flow rate	0.50 mL/min
Gradient program $15\,\text{mM}\,(5\,\text{min}) \xrightarrow{3\,\text{min}} 80\,\text{mM}\,(12\,\text{min}) \xrightarrow{3\,\text{min}} 15\,\text{mM}\,(17\,\text{min})$	
Temperatures	
Autosampler temperature	15 °C
Column temperature	40 °C
CD detector cell temperature	35 °C
Compartment temperature	30 °C
Suppressor $19\,\text{mA}\,(7\,\text{min}) \longrightarrow 99\,\text{mA}\,(18\,\text{min}) \longrightarrow 19\,\text{mA}\,(15\,\text{min})$	
Injection volume	5.0 µL

3.4. Validation Measurements

The following measurements were carried out during limit test validation:

- Specificity. The specificity test has to be made for the limit validation. It can verify that the method is specific and selective for sodium azide;
- Limit of quantification (LQ) and limit of detection (LD). Limit of detection (LD) and limit of quantification (LQ) were specified as the minimum concentration, at which the signal of the investigated component was at least three times (LD) and ten times (LQ) greater than the noise level;
- System precision. System precision was demonstrated by calculating the repeatability of six replicate injections of the reference solution at limit level;
- Accuracy and stability. In the study of accuracy, the sample was spiked with sodium azide at limit level (7.5 ppm). The concentrations were determined, and the recoveries of the spiked quantities were calculated in each case.

The stability was determined by analyzing the prepared solutions over a period of 24 h in closed plastic vials in the sampler holder.

Results of validation are presented in Table 2. Figure 3. shows that no interference from blank and peak due to any impurity was observed at the retention time of sodium azide peaks. The sodium azide reference solution and sample solution stability were measured and the result was, that these were stable for 24 h in the sampler holder. The method usefulness was also proved by measuring four consecutive production batches.

Table 2. Parameters and results of the validation of method developed for the analysis of sodium azide.

Parameters	Results	Requirement
Specificity (ppm)	7.5	7.5
Retention time (t_R, min)	4.41	–
Retention factor (k; t_0 = 1.5 min)	2.94	1–10
Plate number	5097	2000
Symmetry factor (As)	1.1	1.5
Limit of Detection (S/N = 3)	0.52	2.25
Limit of Quatification (S/N = 10)	1.73	7.5
System precision (at 7.5 ppm)		
Retention time (RSD%)	0.14	5
Peak area (RSD%)	16.5	20
Recovery (%, at 7.5 ppm)	102.4	75–125

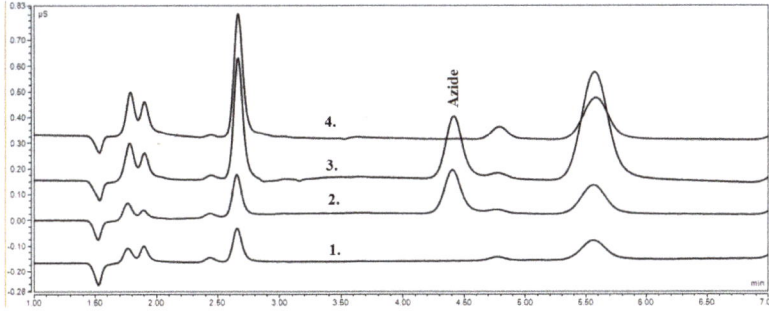

Figure 3. Representative chromatograms (1. Blank solution, 2. Limit solution, 3. Spiked sample solution, 4. Sample solution).

4. Conclusions

A fast and effective high performance ion chromatographic (HPIC) method was developed for the determination of sodium azide content of cilostazol API. For the appropriate detection, a liquid–liquid extraction (LLE) step was necessary. The proposed new HPIC method developed for quantitative determination of sodium azide in cilostazol drug substance is accurate, precise, robust and selective. The solutions, which were made with LLE sample preparation, are stable at least 24 h. The method produced satisfactory validation data for the tested parameters for the appropriate ICH guidelines. The developed method is simple, cost-effective, and provides the possibility to reduce the limit concentration by up to three quarters from the current 7.5 ppm, if necessary.

Author Contributions: Conceptualization, R.K. and B.P.; methodology, B.P.; software, K.H.; validation, B.P. and Z.G.; writing—original draft preparation, B.P. and R.K.; writing—review and editing, K.H.; visualization, B.P. and K.H.; supervision, R.K. and K.H.; project administration, K.H.; funding acquisition, K.H. All authors have read and agreed to the published version of the manuscript.

Funding: This work was supported by the Cooperative Doctoral Programme granted by the Ministry for Innovation and Technology from the source of the National Research, Development and Innovation Fund. Financial support of the Hungarian National Research, Development and Innovation Fund (NKFIH FK128350) is also greatly acknowledged.

Institutional Review Board Statement: Not applicable.

Informed Consent Statement: Not applicable.

Data Availability Statement: The data presented in this study are available on request from the corresponding author.

Conflicts of Interest: The authors declare no conflict of interest. The funders had no role in the design of the study; in the collection, analyses, or interpretation of data; in the writing of the manuscript, or in the decision to publish the results.

Abbreviations

The following abbreviations are used in this manuscript:

MDPI	Multidisciplinary Digital Publishing Institute
DOAJ	Directory of open access journals
API	Active pharmaceutical ingredient
CD	Conductivity detector
EMA	European medicine agency
EGC	Eluent generator cartridge
HPIC	High-performance ion chromatography
HPLC	High-performance liquid chromatography
IC	Ion chromatography
ICH	International Council for Harmonization
LD	Limit of detection
LQ	Limit of quantification
LLE	Liquid–liquid extraction
QbD	Quality by design
RP-HPLC	Reversed-phase high-performance liquid chromatography
UV	Ultra-violet

References

1. Cilostazol Monograph for Professionals. American Society of Health-System Pharmacists. Available online: https://www.drugs.com/monograph/cilostazol.html (accessed on 5 February 2021).
2. Gysi, D.; Do Valle, I.; Zitnik, M.; Ameli, A.; Gan, X.; Varol, O.; Ghiassian, S.; Patten, J.; Davey, R.; Loscalzo, J.; et al. Network medicine framework for identifying drug-repurposing opportunities for COVID-19. *Proc. Natl. Acad. Sci. USA* **2021**, *118*, e2025581118. [CrossRef] [PubMed]
3. Hiatt, W.; Money, S.; Brass, E. Long-term safety of cilostazol in patients with peripheral artery disease: The CASTLE study (Cilostazol: A Study in Long-term Effects). *J. Vasc. Surg.* **2008**, *47*, 330–336. [CrossRef] [PubMed]

4. Kleemann, A. Cardiovascular Drugs. In *Ullmann's Encyclopedia of Industrial Chemistry*; American Cancer Society: Atlanta, GA, USA, 2008. [CrossRef]
5. Chang, S.; Lamm, S. Human health effects of sodium azide exposure: A literature review and analysis. *Int. J. Toxicol.* **2003**, *22*, 175–186. [CrossRef] [PubMed]
6. ICH Guidance for Industry M7(R1), Assessment and Control of DNA Reactive (Mutagenic) Impurities in Pharmaceuticals to Limit Potential Carcinogenic Risk. 2017. Available online: https://database.ich.org/sites/default/files/M7_R1_Guideline.pdf (accessed on 27 July 2021).
7. European Medicines Agency, Cilostazol-Containing Medicines, Cilostazol-Containing Medicines—Article-31 referral—Annex III (Cilostazol), Published: 11 September 2013. Available online: https://www.ema.europa.eu/en/documents/referral/cilostazol-containing-medicines-article-31-referral-annex-iii-cilostazol_en.pdf (accessed on 27 July 2021).
8. Vinković, K.; Drevenkar, V. Ion chromatography of azide in pharmaceutical protein samples with high chloride concentration using suppressed conductivity detection. *J. Chromatogr. B* **2008**, *864*, 102–108. [CrossRef] [PubMed]
9. United States Pharmacopeia. *The National Formulary. Irbesartan. USP 31-NF 26*; United States Pharmacopeia: Rockville, MD, USA, 2008; p. 2446.
10. USP Monographs for Dionex Columns (2008 USP Pharmaciopeia). Available online: http://tools.thermofisher.com/content/sfs/manuals/Man-025203-USP-Monographs-Man025203-EN.pdf (accessed on 27 July 2021).
11. Kushwah, D.K.; Kohle, P.Y.; Joshi, R.D.; Rajyaguru, B.; Pandey, R.; Vishwakarma, B. Validated Method for the Quantification of Sodium Azide in a Range of 'Sartan' Drugs by Ion Chromatography. *Res. J. Pharm. Tech.* **2010**, *3*, 82–86.
12. Gričar, M.; Andrenšek, S. Determination of azide impurity in sartans using reversed-phase HPLC with UV detection. *J. Pharm. Biomed. Anal.* **2016**, *125*, 27–32. [CrossRef]
13. ICH Guidance for Industry Q2(R1), Validation of Analytical Procedures: Text and Methodology. 2005. Available online: https://database.ich.org/sites/default/files/Q2(R1)Guideline.pdf (accessed on 27 July 2021).

Review

Research Progress on Quality Control Methods for Xiaochaihu Preparations

Guangzheng Xu [1], Hui Wang [1], Yingqian Deng [1], Keyi Xie [1], Weibo Zhao [1] and Xingchu Gong [1,2,*]

1. Pharmaceutical Informatics Institute, College of Pharmaceutical Sciences, Zhejiang University, Hangzhou 310058, China; 3200105953@zju.edu.cn (G.X.); 3200105076@zju.edu.cn (H.W.); 3200102962@zju.edu.cn (Y.D.); 3200100601@zju.edu.cn (K.X.); 3200103181@zju.edu.cn (W.Z.)
2. Innovation Center, State Key Laboratory of Component-Based Chinese Medicine, Zhejiang University, Hangzhou 310058, China
* Correspondence: gongxingchu@zju.edu.cn

Abstract: Xiaochaihu (XCH) is a classic Chinese medicine formula. XCH tablet, XCH granule, XCH capsule, and XCH effervescent tablet are included in the Chinese Pharmacopoeia. In this review, the formula and quality standards of XCH preparations at home and abroad were compared. The differences in manufacturing process of XCH preparations are discussed. The progress of research on the qualitative identification, quantitative detection and fingerprint chromatogram/specific chromatogram of XCH preparations was reviewed. The characteristic components of Pinelliae Rhizoma Praeparatum Cum Zingibere Et Alumine and Jujubae Fructus was rarely analyzed for XCH preparations. It is suggested that the specificity of drug quality detection methods should be improved. Considering drug safety and drug efficacy, it is suggested to set the upper and lower limits of the content of saikosaponins. The standards for heavy metals and other limited items for XCH preparations are also suggested to be set.

Keywords: Xiaochaihu (XCH) formula; preparations; quality control; standard

1. Introduction to XCH Formula

Xiaochaihu (XCH) formula, which was created by Zhang Zhongjing in the East Han Dynasty, is capable of inducing sweat to dispel heat, channeling the liver, regulating the spleen, soothing the stomach [1], etc. Traditionally, the recipe is composed of *Bupleuri Radix*, *Scutellariae Radix*, *Ginseng Radix Et Rhizoma* (*Ginseng Radix*), *Glycyrrhizae Radix Et Rhizoma Praeparata Cum Melle* (*Glycyrrhizae Radix*), *Zingiberis Rhizoma Recens*, *Jujubae Fructus*, and *Pinelliae Rhizoma* [2]. According to the principle of JUN-CHEN-ZUO-SHI (emperor-minister-assistant-courier in English), in this formula, *Bupleuri Radix* is JUN, *Scutellariae Radix* is CHEN, *Glycyrrhizae Radix* is SHI, and the others are ZUO. Modern research has verified that XCH has anti-inflammatory [3] and antitumor [4] functions and regulates the endocrine system [5]. Clinically, the formula is applied to treat various diseases of the respiratory system [6], digestive system [7], urogenital system [8], immune system [9], circulatory system [10], etc. The mechanism of XCH acting on the human body can be preliminarily explored by means of liquid chromatography-mass spectrometry, network pharmacology, and animal experiments. For fever, the widest application of XCH, potential antipyretic mechanism includes the reduction of inflammation level, inhibition of endogenous pyrogen and COX-2 [11]. Some active ingredients of XCH including quercetin, baicalein, and hanbaicalein can significantly inhibit the growth of hepatocellular carcinoma and induce apoptosis of hepatocellular carcinoma cells [12]. In recent years, many novel applications have been reported, including the prevention and treatment of methicillin-resistant *Staphylococcus aureus* [13], syncytial virus, and adenovirus [14], as well as the inhibition of influenza A virus [15], etc. For the period from 2000 to 2020, an overall trend of a steady rise in the numbers of publications in the field of XCH could be found. In the database of www.cnki.net, the number has grown annually and ranged from about 150 to nearly 400 works [16].

2. Formula Differences of Existing XCH Preparations

Capsules, granules, pills, tablets, and other XCH preparations are all on the Chinese domestic market. Among those, XCH tablets, XCH effervescent tablets, XCH capsules, and XCH granules were included in the 2020 edition of the Chinese Pharmacopoeia [17]. The Pharmaceuticals and Medical Devices Agency in Japan [18] has published more than 10 kinds of XCH preparations, which are mainly granules or tablets. The Japanese Pharmacopoeia includes two different specifications of the Shosaikoto extract [19]. In Korea, Soshiho-Tang is widely used as a classic recommendation, which is mainly sold in granules [20].

The raw materials of the Japanese XCH preparations are Pinelliae Rhizoma, Ginseng Radix, Bupleuri Radix, Scutellariae Radix, Glycyrrhizae Radix, Zingiberis Rhizoma Recens, and Jujubae Fructus. However, the main XCH preparations on the Chinese market use Pinelliae Rhizoma Praeparatum Cum Zingibere Et Alumine (Jiangbanxia), Codonopsis Radix, Bupleuri Radix, Scutellariae Radix, Glycyrrhizae Radix, Zingiberis Rhizoma Recens, and Jujubae Fructus as raw materials. Table 1 shows four XCH preparations that are listed in the Chinese Pharmacopoeia. The JUN material Bupleuri Radix is the highest in mass ratio among the four dosage forms included in Chinese Pharmacopoeia, accounting for approximately 30%. The raw material mass ratio of XCH tablets and XCH capsules is exactly the same, and the mass ratio of Jiangbanxia is higher than those of XCH effervescent tablets and XCH granules. Regarding the materials of Japanese XCH preparations, the mass ratio of Pinelliae Rhizoma is lower than that of Bupleuri Radix but higher than that of any other herb. The mass ratio values of Glycyrrhizae Radix and Zingiberis Rhizoma Recens are both lower than 10%.

Pinelliae Rhizoma can cause adverse reactions, such as mucosal irritation [21], hepatorenal toxicity [22], and pregnancy toxicity [23,24]. It has been reported that the needle crystals of calcium oxalate and its lectin protein contained in *Pinelliae Rhizoma* are the main irritant toxic substances [25,26]. In China, there is a long history to use *Zingiberis Rhizoma Recens* to alleviate the toxicity of *Pinellia ternata*. The processing standards for preparing Jiangbanxia have been established [27]. Therefore, the use of Jiangbanxia in XCH preparations in China is conducive to improving drug safety [28].

In Table 1, we compared the amount and mass ratio of raw materials in different XCH preparations which were included in the Chinese and Japanese Pharmacopoeia [17,19]. By having materials divided by the total weight, the mass ratios are calculated and listed.

Table 1. Formula amount of raw materials, their mass ratio and preparation amount in different XCH preparations in pharmacopoeias.

Raw Materials	XCH Tablets & XCH Capsules		XCH Effervescent Tablets		XCH Granules		Shosaikoto Extract (Japanese)			
	Amount (g)	Mass Ratio (%)	Amount (g)	Mass Ratio (%)	Amount (g)	Mass Ratio (%)	Amount (g)	Mass Ratio (%)	Amount (g)	Mass Ratio (%)
Bupleuri Radix	445	29.6	1550	31.0	150	31.0	7	29.2	6	26.1
Jianghanxia	222	14.8	575	11.5	56	11.5	-	-	-	-
Pinelliae Rhizoma	-	-	-	-	-	-	5	20.8	5	21.7
Scutellariae Radix	167	11.1	575	11.5	56	11.5	3	12.5	3	13.0
Codonopsis Radix	167	11.1	575	11.5	56	11.5	-	-	-	-
Ginseng Radix	-	-	-	-	-	-	3	12.5	3	13.0
Glycyrrhizae Radix	167	11.1	575	11.5	56	11.5	2	8.33	2	8.70
Zingiberis Rhizoma Recens	167	11.1	575	11.5	56	11.5	1	4.17	1	4.35
Jujubae Fructus	167	11.1	575	11.5	56	11.5	3	12.5	3	13.0
Preparation amount	XCH tablets: 1000 tablets, 0.4 g each; XCH capsules: 1000 capsules, 0.4 g each;		XCH effervescent tablets: 1000 tablets, 2.5 g each		XCH granules: 1000 g (combined with sucrose); 400 g (combined with mannitol); 250 g (combined with lactose)		Not specified			

3. Differences in XCH Preparation Methods

There are different manufacturing processes for preparing XCH [29–32] preparations. Manufacturing processes included in the Chinese Pharmacopoeia [17] are shown in Figures 1–4. For Codonopsis Radix, Glycyrrhizae Radix, Bupleuri Radix, Scutellariae Radix, and Jujubae Fructus, the plants are extracted with water decoction. Jiangbanxia and Zingiberis Rhizoma Recens are extracted with ethanol solution with percolation. Compared with water decoction process, the percolation process is time consuming and solvent consuming. However, the volatilization or degradation of active components can be effectively decreased with the percolation process because it is operated at a low temperature. It has been reported that gingerols are easily degraded at a high temperature [33]. Therefore, it is reasonable to extract active components from Jiangbanxia and Zingiberis Rhizoma Recens with a percolation process [34]. Gingerol and other components in Zingiberis Rhizoma Recens have low solubility in water [35]. Therefore, ethanol solution is generally used as the percolation solvent [36].

In the production process for XCH capsules and XCH tablets, part of Codonopsis Radix and Glycyrrhizae Radix are directly crushed and added, which is significantly different from the process of XCH effervescent tablets and XCH granules. Codonopsis Radix and Glycyrrhizae Radix powder can play a role similar as excipients [37]. The excipients in XCH preparations vary depending on the formulation forms.

Apart from the manufacturing processes included in the Chinese Pharmacopoeia mentioned above, there are several other manufacturing processes for different XCH preparations, such as XCH sustained release tablets [38], nano XCH preparations [39], and others [40,41].

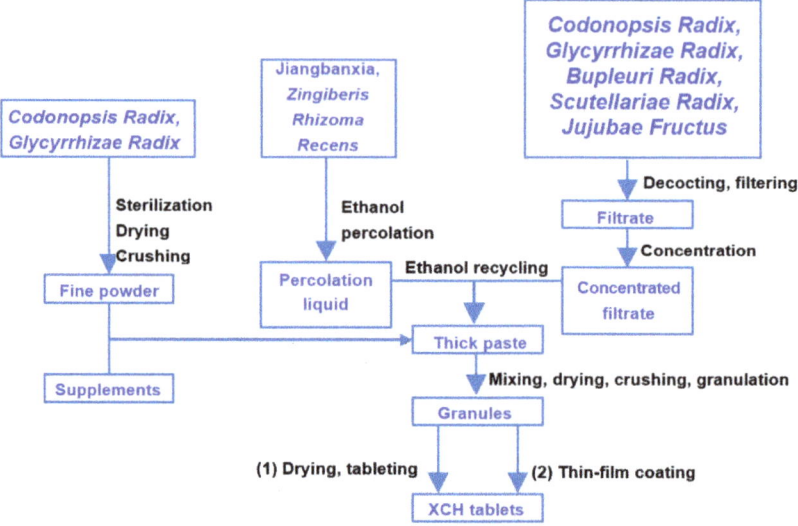

Figure 1. XCH tablets Production Process.

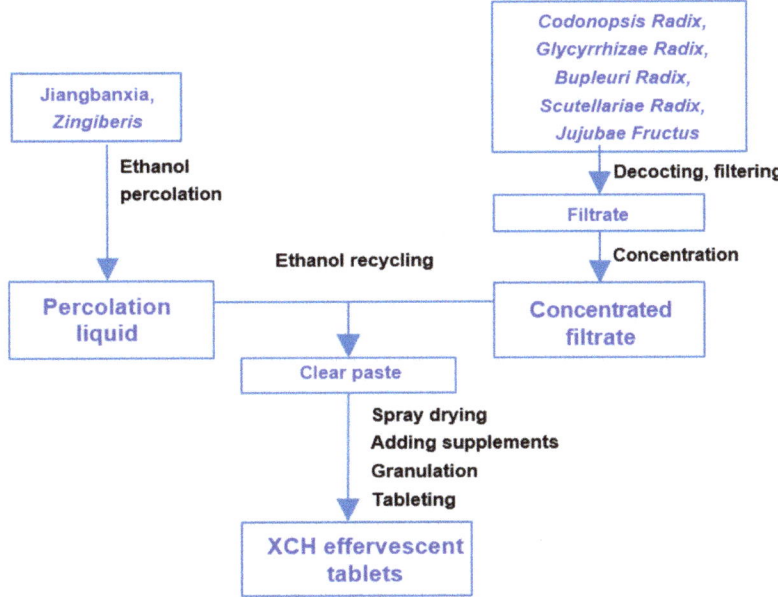

Figure 2. XCH effervescent tablets Production Process.

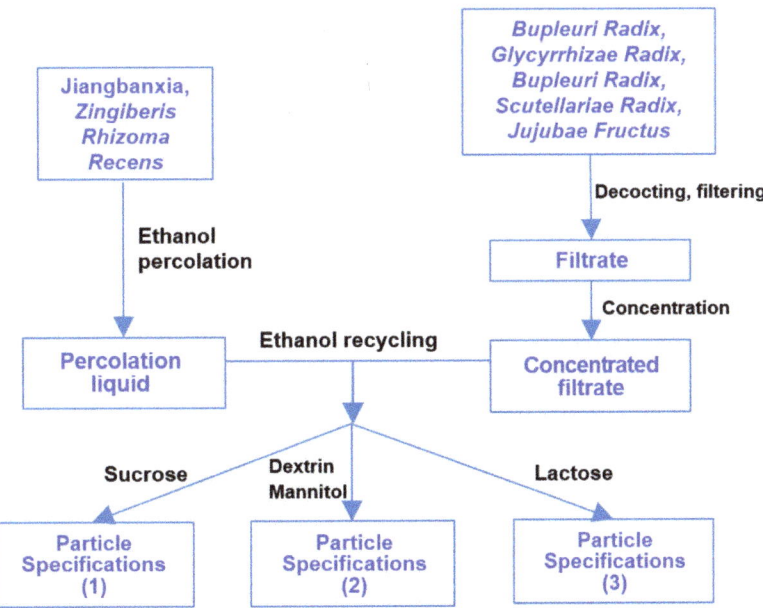

Figure 3. XCH granules Production Process.

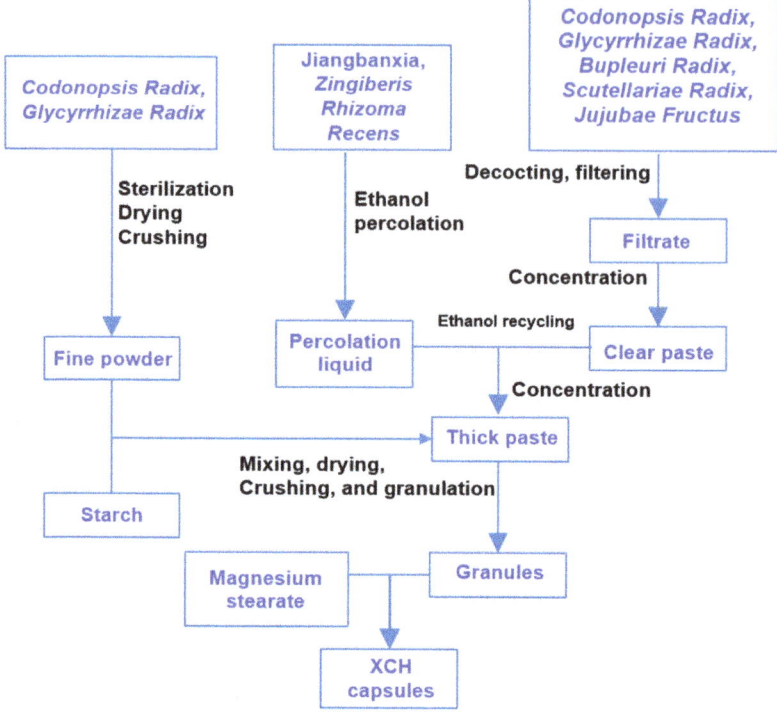

Figure 4. XCH capsules Production Process.

4. Differences in Quality Test Indices and Limits of Different XCH Preparations

4.1. Indicators for Qualitative Identification

There are two kinds of qualitative methods for the preparations of XCH in production: microscopic identification and thin-layer chromatography (TLC) identification. The identification methods and reference materials from the Chinese Pharmacopoeia and the Japanese Pharmacopoeia are summarized in Table 2.

The method involving the microscopic identification of medicinal materials is suitable for fragmentary medicinal materials or powdered medicinal materials. Raw powdered medicinal materials of *Glycyrrhizae Radix* and *Codonopsis Radix* are used in the manufacturing of XCH tablets and XCH capsules, which is suitable to be analyzed with microscopic identification. Both Chinese and Japanese XCH preparations adopt thin-layer identification, but there are obvious differences. First, in Chinese Pharmacopoeia, thin-layer identification uses reference medicinal materials, including *Glycyrrhizae Radix*, *Bupleuri Radix*, and *Codonopsis Radix*. However, thin-layer identification uses only reference substances in Japanese Pharmacopoeia. Second, Japanese thin-layer identification uses more reference substances, including the index components from *Bupleuri Radix*, *Zingiberis Rhizoma Recens*, *Scutellariae Radix*, *Glycyrrhizae Radix* and *Ginseng Radix*. In contrast, the use of reference medicinal materials in thin-layer identification can provide more information than using reference substances, which is conducive to assessing the authenticity of the medicinal materials used. Both Chinese and Japanese XCH preparations quantitatively analyze baicalin contents. Therefore, it seems unnecessary to use baicalin as the reference substance in thin-layer identification.

Silica gel G thin layer plate is mostly widely used in TLC identification. A mixed solvent of ethyl acetate-butanone-formic acid-water is usually used for baicalin identification. A mixed solvent of chloroform-methanol-water is usually used for the identification of *Glycyrrhizae Radix*.

Table 2. Qualitative identification method and comparison of XCH preparations in the Chinese and Japanese Pharmacopoeias.

Preparations	Identification Method	TLC Reference Substance	TLC Control Crude Drug
XCH tablets	Microscopic Identification, Thin-Layer Chromatography Identification	Baicalin	*Glycyrrhizae Radix*
XCH effervescent tablets	Thin-Layer Chromatography Identification	Baicalin	*Glycyrrhizae Radix, Bupleuri Radix, Codonopsis Radix*
XCH capsules	Microscopic Identification, Thin-Layer Chromatography Identification	-	*Glycyrrhizae Radix, Bupleuri Radix, Codonopsis Radix*
XCH granules	Thin-Layer Chromatography Identification	Baicalin	*Glycyrrhizae Radix, Bupleuri Radix*
Shosaikoto Extract (Japanese)	Thin-Layer Chromatography Identification	Saikosaponin B2, 6-Gingerol, Wogonin, Ginsenoside Rb1, Liquiritin	-

4.2. Quantitative Determination

Table 3 lists the quantitative detection methods for XCH preparations in Chinese Pharmacopoeia and Japan Pharmacopoeia. According to the Chinese Pharmacopoeia, the XCH tablets weight 0.4 g per tablet, the XCH capsules weight 0.4 g per capsule, and the XCH effervescent tablets weight 2.5 g per tablet. The XCH granules have three specifications (10 g/4 g/2.5 g per bag) due to the various preparation methods. The only quantitative determined index component for XCH preparations mentioned in the 2020 edition of the Chinese Pharmacopoeia is baicalin, while the determination of baicalin, saikosaponin B2, and glycyrrhizic acid are required in Japan Pharmacopoeia. Considering that there are seven medicinal materials in the formula for XCH preparations, more index components should be determined to control drug quality. The Chinese Pharmacopoeia specifies a lower limit for baicalin, while the Japanese Pharmacopoeia specifies both the upper and lower limits for the contents of saikosaponin B2, baicalin, and glycyrrhizic acid. Herbal materials of XCH preparations were often decocted before quantitative analysis. Methanol is a common solvent for sample preparation.

Table 3. Comparison of quantitative detection methods for XCH preparations in the Chinese and Japanese Pharmacopoeias.

Preparations	Detection Component	Prescribed Limit
XCH tablets	Baicalin	Not lower than 2.0 mg per tablet/capsule
XCH capsules	Baicalin	Not lower than 2.0 mg per tablet/capsule
XCH effervescent tablets	Baicalin	Not lower than 20.0 mg per tablet/package
XCH granules	Baicalin	Not lower than 20.0 mg per tablet/package
Shosaikoto Extract (Japanese)	Saikosaponin B2, baicalin and glycyrrhizic acid	Saikosaponin B2: 2–8 mg

Table 4 lists the published works on the quantitative detection of XCH preparations. In such work, the raw materials often went under decoction treatment before they were put into use. In addition, methanol is a common solvent in the procedure. HPLC technology is used to separate the components of XCH preparations. The detectors stated in the literature are mostly ultraviolet detectors, and a few are diode array detectors (DAD) and mass spectrometer detectors. Since the content of a saikosaponin is low, mass spectrometer detectors are used more often to analyze it. In some papers, the method of quantitative analysis of multiple components by a single marker (QAMS) was used, which can reduce the cost of testing. There are more reports on the detection of index components of *Bupleuri Radix* and *Scutellariae Radix*, which reflects the emphasis on JUN and CHEN drugs. Some literature has detected gingerol, liquiritin, glycyrrhizic acid, lobetyolin, and other substances, which can help control the contents of chemical components in *Zingiberis Rhizoma Recens*, *Glycyrrhizae Radix*, and *Codonopsis Radix*. However, there are still few detections of index components in *Jujubae Fructus* and Jiangbanxia.

Table 4. Quantitative detection methods for the XCH formula.

	Year	Apparatus	Quantified Components	Other Instructions	Reference
1	2020	HPLC	Baicalin, Wogonoside, Baicalein, Wogonin, Ammonium Glycyrrhizinate, Saikosaponin B2, Saikosaponin B1	QAMS	[42]
2	2018	HPLC	6-Gingerol	-	[43]
3	2018	HPLC	Ginsenoside Rg1, Ginsenoside Re, Ginsenoside Rb1, Saikosaponin A, Saikosaponin D	QAMS	[44]
4	2018	HPLC	Baicalin, Baicalein, Wogonin	-	[45]
5	2017	HPLC	Saikosaponin A, Saikosaponin D, Saikosaponin B1, Baicalin, Ginsenoside Rb1, Ginsenoside Re, 6-Gingerol, Liquiritin, Ammonium Glycyrrhizinate	-	[29]
6	2017	HPLC	Lobetyolin, Liquiritin, Baicalin, Baicalein	Detection wavelength switched	[46]
7	2016	HPLC	Baicalin	-	[47]
8	2016	UPLC	Baicalin	-	[48]
9	2015	HPLC-MSMS	Saikosaponin A, Saikosaponin D	-	[49]
10	2015	HPLC	Liquiritin, Baicalin, Wogonoside, Baicalein, Wogonin	-	[50]
11	2015	HPLC	Baicalin	-	[51]
12	2015	HPLC	Saikosaponin, Baicalin, Ginsenoside Rg1, Liquiritin, Ephedrine, 6-Gingerol	-	[52]
13	2014	HPLC	Saikosaponin A, Baicalin	-	[53]
14	2014	HPLC	Liquiritin, Baicalin, Wogonoside, Baicalein, Ammonium Glycyrrhizinate, Saikosaponin A, Wogonin	-	[54]
15	2014	HPLC	Saikosaponin B2	-	[55]
16	2013	HPLC	Saikosaponin, Baicalin, Glycyrrhizic acid	-	[56]
17	2012	Capillary electrophoresis	Saikosaponin A, Saikosaponin D	-	[57]
18	2012	HPLC	Baicalin	-	[58]
19	2012	HPLC	Baicalin	-	[59]
20	2010	HPLC	Baicalin, Wogonoside	-	[60]
21	2010	HPLC	Baicalin, Baicalein, Wogonoside, Wogonin, Glycyrrhizic acid	-	[61]
22	2010	HPLC	Saikosaponin A	-	[62]
23	2007	HPLC-DAD-MS	Saikosaponin A, Baicalin, Glycyrrhizic acid	-	[63]
24	2007	HPLC	Baicalin	-	[64]
25	2006	HPLC/DAD	Baicalin, Glycyrrhizic acid	-	[65]
26	2006	HPLC-MSMS	Cytidine, Tyrosine, Uridine, Adenine, Guanosine, Phenylalanine, Adenosine, Tryptophan	-	[66]
27	2004	HPLC-MSMS	Saikosaponin A, B1, B2, C, D, G, H, I	-	[67]

"-" means there is no special instruction that is necessary to be presented.

4.3. Fingerprint and Specific Chromatogram

Fingerprint and specific chromatogram detection methods can reflect the overall characteristics of Chinese medicines and are widely used in drug quality analysis. At present, the application of these two methods represents a significant research progress. Both qualitative identification and quantitative detection can be carried out on the basis of fingerprint and specific chromatograms. A summary of the fingerprint and specific chromatogram detection methods for XCH preparations is shown in Table 5. Compared with the quantitative methods listed in Table 4, fingerprint and specific chromatogram detection can identify more components in chromatographic peaks, therefore providing more information. At present, the most identified components are the saponins in *Bupleuri Radix*, flavonoids of *Scutellariae Radix*, gingerol in *Zingiberis Rhizoma Recens*, liquiritin, and glycyrrhizic acid in *Glycyrrhizae Radix*, etc. However, the characteristic components of *Jujubae Fructus* and Jiangbanxia have not been identified.

In some of the research works, a quantitative fingerprint or specific chromatogram of the XCH preparation was obtained. The quantitatively determined components are mainly from *Scutellariae Radix*, *Glycyrrhizae Radix*, *Bupleuri Radix*, and *Codonopsis Radix*. Wang et al. [68] compared the HPLC spectra of XCH granules at different wavelengths and concluded that spectral analysis at a single ultraviolet absorption wavelength is not suitable for quality detection. Liu et al. used charged aerosol detector (CAD) to analyze saikosaponins [69]. Compared with using evaporative light scattering detector (ELSD), lower detection limit and wider detection range can be realized with CAD.

In some studies, the active substances in *Bupleuri Radix* such as saikosaponin A and saikosaponin D were not detected in XCH granules, which may be due to the hydrolysis of saikosaponin during decoction [70]. There are also reports that the existing detection methods often add an acid to the mobile phase, and saikosaponin A and saikosaponin D are prone to degrade under acidic conditions, which makes them difficult to detect [71].

Table 5. Fingerprint/specific chromatogram detection methods for XCH preparation.

	Year	Apparatus	Quantitative Determined Components	Qualitatively Identified Components	Detection Method	Reference
1	2021	HPLC-CAD	Saikosaponin A, B1, B2, C, G, H, I	Saikosaponin A, B1, B2, C, G, H, I	Fingerprint chromatogram	[69]
2	2021	UHPLC	Liquiritin, Baicalin, Wogonin, Baicalein, Glycyrrhizin G2, Glycyrrhizic acid, Saikosaponin B2, Saikosaponin B1	-	Fingerprint chromatogram	[71]
3	2018	UPLC	Liquiritin, Baicalin, Berberine, Wogonoside, Baicalein, Ammonium Glycyrrhizinate	Liquiritin, Baicalin, Berberine, Wogonoside, Baicalein, Ammonium Glycyrrhizinate	Specific chromatogram	[72]
4	2017	HPLC	-	-	Fingerprint chromatogram	[73]
5	2017	HPLC	-	-	Fingerprint chromatogram	[74]
6	2016	HPLC	Lobetyolin, Saikosaponin A	Lobetyolin, Saikosaponin A	Specific chromatogram	[75]
7	2016	HPLC	Baicalin, Ammonium Glycyrrhizinate	Baicalin, Ammonium Glycyrrhizinate	Specific chromatogram	[68]
8	2015	HPLC-ELSD	-	Liquiritin, Ginsenoside Re, Baicalin, Wogonoside, Baicalein, Ginsenoside Rb1	Fingerprint chromatogram	[76]

Table 5. Cont.

	Year	Apparatus	Quantitative Determined Components	Qualitatively Identified Components	Detection Method	Reference
9	2014	HPLC	Liquiritin, Baicalin, Wogonoside, Baicalein, Ammonium Glycyrrhizinate, Saikosaponin A, Wogonin	Liquiritin, Baicalin, Wogonoside, Baicalein, Ammonium Glycyrrhizinate, Saikosaponin A, Wogonin	Specific chromatogram	[54]
10	2013	HPLC	-	Liquiritin, Baicalin, Wogonoside, Baicalein, Ammonium Glycyrrhizinate, Saikosaponin A, Wogonin	Fingerprint chromatogram	[77]
11	2013	HPLC	Baicalin, Glycyrrhizic acid	Baicalin, Glycyrrhizic acid	Fingerprint chromatogram	[78]
12	2013	HPLC	-	Liquiritin, Baicalin, Ononin, Wogonoside, Saikosaponin A, Skullcapflavone II, etc.	Fingerprint chromatogram	[79]
13	2012	HPLC-DAD-ESI-MS	Homogentisic acid, Baicalin, Glycyrrhizic acid, Saikosaponin A, 6-Gingerol, Ginsenoside Rg3	-	Fingerprint chromatogram	[80]
14	2012	HPLC	Baicalin, Wogonoside, Baicalein, Wogonin, Glycyrrhetic acid	Baicalin, Wogonoside, Baicalein, Wogonin, Glycyrrhetic acid	Specific chromatogram	[81]
15	2012	UPLC	-	Glycyrrhizin, Ginsenoside Rg1, Baicalin, Isowogonin, Baicalein, Saikosaponin A	Fingerprint chromatogram	[82]
16	2012	HPLC	Baicalin, Wogonoside, Baicalein, Wogonin	Baicalin, Wogonoside, Baicalein, Wogonin	Specific chromatogram	[83]
17	2011	HPLC-DAD	-	-	Fingerprint chromatogram	[84]
18	2009	HPLC-TOF/MS	-	Liquiritin, Baicalin, Wogonoside, Ginsenoside Rg1, Glycyrrhizic acid	Fingerprint chromatogram	[85]
19	2009	HPLC	-	-	Fingerprint chromatogram	[86]

"-" means there is no quantitative determined or qualitatively identified component.

In conclusion, a fingerprint/specific chromatogram can be used to characterize multiple chemical component information of XCH preparations. However, fingerprint/specific chromatogram is not included in Chinese Pharmacopoeia. Further development of quality control technology with use of the fingerprint/specific chromatogram is required for XCH preparations.

5. Prospect on the Development Direction of Quality Control of XCH Preparations

5.1. Improvement in the Specificity of Quality Testing

According to Chinese Pharmacopoeia, *Glycyrrhizae Radix*, *Bupleuri Radix* and *Codonopsis Radix* are used as TLC reference materials, and baicalin is used as a TLC reference substance in qualitative identification. However, less attention has been given to *Zingiberis Rhizoma Recens*, *Jujubae Fructus*, and Jiangbanxia. The specific components of Jiangbanxia and *Jujubae Fructus* are not quantitatively analyzed in literature. Recently, guanosine, uridine, hypoxanthine and several other components were analyzed [87], which do not especially belong to Jiangbanxia, but it still suggests a way to improve the specificity of HPLC detection by detecting these compositions with strong polarity.

It is essential to distinguish the authenticity of *Bupleuri Radix*. There are 36 species, 17 varieties, and seven forms distributed all over China [88]. Among them, *Bupleurum marginatum var. stenophyllum* and even poisonous *Bupleurum longiradiatum* are common varieties that are all easy to mix up [89]. To confirm whether or not *Bupleurum marginatum var. stenophyllum* had been added, Liu et al. using the retention time and peak area of the specific ion detected in the mass spectrum as standards [90]. Liang et al. tried to establish near-infrared spectrum models to distinguish products of different factories, which provided a practical technology for low-cost and rapid detection [91]. Lai et al. used a polymerase chain reaction (PCR) method based on the site specificity of the Internal Transcribed Spacer (ITS) sequence to identify *Bupleurum marginatum var. stenophyllum* from *Bupleurum chinense* DC [92]. These new technologies provide ideas for improving the specificity of analytical methods. *Bupleurum scorzonerifolium Willd* and *Bupleurum chinense* DC are both included in the Chinese Pharmacopoeia, but National Institutes for Food and Drug Control can provide only the reference material of *Bupleurum chinense* DC. Hence, the lack of reference material of *Bupleurum scorzonerifolium Willd* is a problem for quality control of *Bupleuri Radix*.

5.2. Setting Reasonable Content Range of Index Components from Bupleuri Radix

Bupleuri Radix is the JUN of XCH formula. Thus far, qualitative identification using the reference material of *Bupleuri Radix* was adopted in Chinese Pharmacopoeia. However, considering drug safety and efficacy, the contents of saikosaponins should be controlled in specific ranges. Studies have indicated that saikosaponins are important active ingredients of *Bupleuri Radix*, which has antipyretic, anti-inflammatory and antitumor activities. Therefore, it is necessary to set up lower limits for their contents [93,94]. Moreover, some reference materials have reported that *Bupleuri Radix* has a certain degree of toxicity when taken in a large dose for a long period of time, and its toxic side effects are often caused by its saponins and volatile substances, which mainly affect the liver [95]. Therefore, from the perspective of drug safety, it is necessary to set up upper limits for saikosaponins. At present, the upper and lower limits of the saikosaponin B2 content are set up in the Japanese Pharmacopoeia, which is worth referencing. When setting up the lower limit, companies can consider collecting big data from clinical practice. Accordingly, the needs of drug quality control indicators can be taken into consideration, such as drug interactions and medications for special populations.

5.3. Strengthening the Standard of Limited Detected Items

In recent years, great progress in the control of heavy metals, pesticides, and biological toxins in Chinese medicines and extracts was achieved. The Chinese Pharmacopoeia has specially listed items General Principle for Inspection of Crude Drugs and Decoction Pieces and Guidelines for Establishment of Limit for Harmful Residue of Traditional Chinese Medicine, which have provided guidance for controlling heavy metals, pesticides and biological toxins for medicinal materials. The Chinese Pharmacopoeia stipulates that *Jujubae Fructus* needs to be tested for aflatoxin, *Glycyrrhizae Radix* needs to be tested for heavy metals, harmful elements and pesticide residues, and *Codonopsis Radix* needs to be tested for sulfur dioxide residues, all of which help to guarantee the safety of XCH preparations. However, the current guidelines for XCH preparations still require more relevant limiting items for heavy metals, pesticides and biological toxins, and other toxic ingredients. The Japanese Pharmacopoeia stipulates the limits of heavy metals and arsenic in XCH preparations. It takes the increase in heavy metals during the production process into account, which is more rigorous and improves the level of quality control.

Therefore, from the perspective of drug safety, XCH preparations require an upper limit for the amounts of certain active ingredients, heavy metals, pesticides, biotoxins, and other toxic components. Similar quality control problems exist for many other Chinese medicines. Therefore, the development direction of quality control presented in this work can also be referenced for that of other Chinese medicines.

Author Contributions: Analysed the data: G.X., Y.D. and H.W.; collected the materials: G.X., Y.D., H.W., K.X., W.Z.; wrote the paper: G.X., H.W., Y.D.; editing: X.G. and G.X. All authors have read and agreed to the published version of the manuscript.

Funding: This research was funded by the National S&T Major Project of China, grant number 2018ZX09201011, and the National Project for Standardization of Chinese Materia Medica, grant number ZYBZH-C-GD-04.

Acknowledgments: The authors would like to thank the support from Innovation Group of Component-based Chinese Medicine and Intelligent Manufacturing with multi-crossed disciplines.

Conflicts of Interest: The authors declare no conflict of interest.

References

1. Liu, T.T.; Yao, K.W.; Duan, J.L.; Yu, Z.; Zhang, L.D. Efficacy of Xiaochaihu Decoction on Treating Coronary Heart Disease. *Lishizhen Med. Mater. Med. Res.* **2019**, *30*, 2216–2218.
2. Zhang, W.Z.; Nie, H.; Wang, Z.T.; Mao, D.X. Discussion of ginseng used in Xiaochaihu Decoction. *Lishizhen Med. Mater. Med. Res.* **2013**, *24*, 2480–2481.
3. Zhang, Y.; Zhou, X.L.; Shao, Q.; Fang, Z.E.; Wang, S.S. Anti-inflammatory effect of Xiaochaihutang in rats with collagen-induced arthritis and the mechanism. *Immunol. J.* **2015**, *31*, 781–785.
4. Zhai, Y.X.; Zhao, Y.; Xiang, R.W.; Zhai, F.; Zeng, Y.; Liang, J.K. Study on anti-heatoma mechanism of Xiaochaihu decoction based on weighted network pharmacology. *Chin. J. Med. Chem.* **2020**, *30*, 658–668.
5. Lu, X.W.; Zhu, L.H.; Huang, H.; He, K. Clinical Observations on Modified Xiao Chaihutang in Treatment of Subacute Thyroiditis. *Chin. J. Exp. Tradit. Med. Formulae* **2018**, *22*, 153–158.
6. Zhang, W.Z.; Tian, H.Z.; Zhang, L.J.; Yang, S.J. Clinical Study and Analysis of Modified Xiaochaihu Decoction in Treating Cough Caused by Variant Asthma. *World J. Integr. Tradit. Western Med.* **2020**, *15*, 790–793.
7. Liu, L.; Wang, C.X.; Cui, S.Y.; Zhang, H.X.; Wu, X.X.; Zhang, W. Observation of curative effect of lamivudine combined with compound glycyrrhizin and Xiaochaihu Decoction in the treatment of chronic hepatitis B. *Modern J. Integr. Tradit. Chin. Western Med.* **2020**, *29*, 4042–4045.
8. Luo, X.M. Clinical observation on treating chronic renal failure with the Xiaochaihu decoction. *Clin. J. Chin. Med.* **2012**, *4*, 99.
9. Wang, D.X.; Liu, X.Z.; Lv, C.Y.; Fan, H.; Sun, Y.F.; Yang, S.L. Changes of Platelet- Associated Parameters in Patients with Refractory Immune Thrombocytopenia Treated with Xiaochaihu Decoction and Erzhi Pill. *J. Hebei North Univ. Nat. Sci. Ed.* **2020**, *36*, 5–9.
10. Jiang, M. Effect observation of Minor Decoction of Bupleurum treating acute viral myocarditis. *World Latest Med. Inf.* **2015**, *15*, 1–14.
11. Ma, R.; Liu, H.; Zheng, R.W.; Xu, Z.T.; Jiang, Z.Q.; Zhang, J.H.; Jiang, Y.F.Y.; Bi, C. Pharmacological analysis and confirmation of antipyretic mechanism of Xiaochaihu Granules. *J. Pharm. Res.* **2021**, *8*, 497–503.
12. Li, X.; Li, Y.Y.; Wen, C.X.; Liang, Y.Y.; Zhou, Z.L. Study on the mechanism of Xiaochaihu Decoction in treatment of hepatocellular carcinoma based on network pharmacology and bioinformatics. *Anti-Tumor Pharm.* **2021**, *4*, 456–462.
13. Ma, R.; Liu, H.; Jiang, Y.F.Y.; Zheng, R.; Jiang, Z.Q.; Bi, C.; Zhang, J.H.; He, Y.X.; Du, H.Y. A New Use of Xiaochaihu Granules Combined with Antibiotics. Guangdong Province Patent CN111529671A, 14 August 2020.
14. Liu, H.; Huang, W.; Cheng, X.Y.; Bi, C.; Zheng, R.W.; Jiang, Z.Q.; Zhang, J.H.; He, Y.X.; Du, H.Y. New Application of Xiaochaihu Granules in the Control of Syncytial Virus and Adenovirus. Patent CN111467466A, 31 July 2020.
15. Zheng, R.W.; Liu, H.; Situ, W.H.; Jiang, Z.Q.; Bi, C.; Zhang, J.; He, Y.X.; Du, H.Y. New Application of Xiaochaihu Granules in Inhibiting Influenza a Virus. Patent CN111358929A, 3 July 2020.
16. China National Knowledge Infrastructure. Available online: https://kns.cnki.net/kns8/defaultresult/index (accessed on 9 October 2021).
17. Chinese Pharmacopoeia Commission. *The Chinese Pharmacopoeia*; China Medical Science Press: Beijing, China, 2020; pp. 602–606.
18. Pharmaceuticals and Medical Devices Agency. Available online: https://www.pmda.go.jp/PmdaSearch/iyakuSearch/ (accessed on 22 September 2021).
19. Japanese Pharmacopoeia Committee. *The Japanese Pharmacopoeia*, 17th ed.; The Ministry of Health, Labour and Welfare: Tokyo, Japan, 2016; pp. 1988–1990.
20. Kim, H.M.; Kim, Y.Y.; Jang, H.Y.; Moon, S.J.; An, N.H. Action of Sosiho-Tang on systemic and local anaphylaxis by anal administration. *Immunopharm. Immunot.* **1999**, *21*, 635–643. [CrossRef]
21. Zhong, L.Y.; Wu, H. Current researching situation of mucosal irritant components in Acacae family plants. *China J. Tradit. Chin. Med. Pharm.* **2006**, *18*, 1561–1563.
22. Zhang, L.M.; Bao, Z.Y.; Huang, Y.Y.; Huang, W.; Zhang, Y.N.; Sun, R. Experimental Study on the "Dosage-Time-Toxicity" Relationship of Acute Hepatotoxicity induced by Water Extraction from *Rhizoma Pinelliae* in Mice. *Chin. J. Pharmacovigil.* **2011**, *8*, 11–15.
23. Jin, X.Q.; Huang, C.Q.; Zhang, G. Toxic Components and Processing Mechanism of *Rhizoma Pinelliae*. *Lishizhen Med. Mater. Med. Res.* **2019**, *30*, 1717–1720.

24. Zhang, K.; Shan, J.J.; Xu, J.Y.; Wang, M.M. Research progress and prospects of pregnancy toxicity of pinellia. *China J. Tradit. Chin. Med. Pharm.* **2016**, *31*, 938–941.
25. Zhu, F.G.; Yu, H.L.; Wu, H.; Shi, R.J.; Tao, W.T.; Qiu, Y.Y. Correlation of *Pinellia ternata agglutinin* and *Pinellia ternata raphides'* toxicity. *China J. Chin. Mater. Med.* **2012**, *37*, 1007–1011.
26. Zhong, L.Y.; Wu, H.; Zhang, K.W.; Wang, Q.R. Study on irritation of calcium oxalate crystal in raw Pinellia ternate. *China J. Chin. Mater. Med.* **2006**, *20*, 1706–1710.
27. Su, T.; Zhang, W.W.; Zhang, Y.M.; Cheng, C.Y.B.; Fu, X.Q.; Li, T.; Guo, H.; Li, Y.X.; Zhu, P.L.; Cao, H.; et al. Standardization of the manufacturing procedure for *Pinelliae Rhizoma Praeparatum cum Zingibere et Alumine*. *J. Ethnopharmacol.* **2016**, *193*, 663–669. [CrossRef]
28. Shi, R.J.; Wu, H.; Yu, H.L.; Chen, L. The advance of the research that zingiber officinale rosc detoxify the tuber of pinellia. *Chin. J. Inf. Tradit. Chin. Med.* **2010**, *17*, 108–110.
29. Jiang, X.; Yan, F.J.; Jiang, L.; Si, G.M. Different Decocting Methods Influence 9 Kinds of Ingredients of Xiao Chaihutang by HPLC. *Chin. J. Exp. Tradit. Med. Formulae* **2017**, *23*, 98–103.
30. Megumi, S.; Yuko, S.; Fumio, I.; Yoshiro, H.; Takao, N. Extraction Efficiency of Shosaikoto (Xiaochaihu Tang) and Investigation of the Major Constituents in the Residual Crude Drugs. *Evid.-Based Compl. Altern. Med.* **2012**, *2012*, 7.
31. Sakata, K.; Kim, S.J.; Yamada, H. Effects of commercially available mineral waters on decoction of Kampo Medicines. *Kampo Med.* **2000**, *51*, 225–232. [CrossRef]
32. Honma, S.; Ogawa, A.; Kobayashi, D.; Ueda, H.; Fang, L.; Numajiri, S.; Kimura, S.; Teresawa, S.; Morimoto, Y. Effect of hardness on decoction of Chinese medicine. *J. Tradit. Med.* **2003**, *20*, 208–215.
33. Jiang, S.Z.; Fu, S.Q.; Wang, N.S. Overview of chemical constituents of gingerol. *Tradit. Chin. Drug Res. Clin. Pharmacol.* **2006**, *05*, 386–389.
34. Yu, N.; Zeng, H.Y.; Deng, X.; Zeng, M.X.; Feng, B. Study on Extraction, Identification and Antioxidation of Gingerol. *Food Sci.* **2007**, *8*, 201–204.
35. Zhao, W.Z.; Zhang, R.X.; Yu, Z.P.; Wang, X.K.; Li, J.R.; Liu, J.B. Research process in ginger chemical composition and biological activity. *Sci. Technol. Food Ind.* **2016**, *37*, 383–389.
36. Liu, D.; Zhang, C.H.; An, R.H.; Feng, X.Q.; Zhou, F.; Deng, Y.J.; Zhang, J.L.; Liu, H. Advances on extraction and application of ginger bioactive ingredient. *Sci. Technol. Food Ind.* **2016**, *37*, 391–400.
37. Zhang, D.K.; Fu, C.M.; Lin, J.Z.; Ke, X.M.; Zou, W.Q.; Xu, R.C.; Han, L.; Yang, M. Study on theory and application value of "unification of medicines andexcipients" in Chinese materia medica preparations. *Chin. Tradit. Herb. Drugs* **2017**, *40*, 1921–1929.
38. Du, Y.; Xue, J.L.; Zhang, L.P.; Liang, G.X.; Ding, H. A Xiaochaihu Sustained Release Tablet along with Its Preparation Methods. Patent CN105106914B, 12 March 2019.
39. Yang, M.J. A Nanoscale Xiaochaihu Pharmaceutical Preparation along with Its Preparation Method. Patent CN1362246, 7 August 2002.
40. Li, C.H. A Fast Classification and Screening Device for Xiaochaihu Granules. Patent CN208960363U, 11 June 2019.
41. Xu, D.J.; Chi, H.C.; Feng, Z.W.; Li, W.Z. A Preparation Method of Xiaochaihu Granules. Patent CN108853027A, 23 November 2018.
42. Liu, A.X.; Xu, T.T.; Yan, Y.; Wu, Q.Y.; Zhou, D.D.; Sha, Y.W. Study on determination of seven components in Xiaochaihu Granules by QAMS. *Drug Eval. Res.* **2020**, *43*, 2217–2221.
43. Qu, N.N. Determination of 6-Gingerol Content in Minor Bupleurum Tablets by HPLC Method. *Heilongjiang Med. J.* **2018**, *31*, 718–720.
44. Shao, J.X.; Lin, L.F.; Liu, Y.L.; Wang, C.M.; Li, H.; Yang, Y.J. Determination of five saponins in *Ginseng Radix et Rhizoma-Bupleuri Radix* herb pair and its preparations by QAMS method. *Chin. Tradit. Herb. Drugs* **2018**, *49*, 2873–2877.
45. Chen, L.L.; Lin, R.X.; Li, R.M. Effects of Different Decoction Methods of Xiaochaihu Decoction on Active Chemical Constituents of Scutellaria Baicalensis Georgi. *World Chin. Med.* **2018**, *13*, 743–745+750.
46. Li, C.D.; Wang, X.; Pan, X.B.; Zhang, W. Detection of Four Kinds of Ingredients in Modified Xiao ChaiHu Capsules by HPLC Wavelength Switching Method Simultaneously. *Western J. Tradit. Chin. Med.* **2017**, *30*, 26–28.
47. Tang, F.S.; Zhang, X.H.; Lan, X.; Meng, C.; Wang, Y.H. Study on the Dissolution of Xiaocaihu Pill from Different Manufacturers. *China Pharm.* **2016**, *27*, 4272–4274.
48. Li, J.; Yang, P.P.; Tang, W.X.; Li, H.J.; Xiang, W.; Jiang, H.; Yang, S.H.; Li, W.; Hou, Y. NIR Calibration Model to Detect Baicalin Content of Xiaochaihu Granules by Using Characteristic Spectrum Selection. *Chin. J. Exp. Tradit. Med. Formulae* **2016**, *22*, 72–77.
49. Wang, J.L.; Wang, S.T. Simultaneous determination the saikosaponin a and saikosaponin d in Sho-Saiko Granule with HPLC-MSMS. *Anhui Med. Pharm. J.* **2015**, *19*, 453–456.
50. Yuan, H.X.; Wang, T.T.; Yan, Y. Simultaneous Determination of Five Components in Ethyl Acetate Extract of Xiao Chaihu Tang by HPLC. *Chin. J. Exp. Tradit. Med. Formulae* **2015**, *21*, 64–66.
51. Lan, X.; Tang, F.H.; Chen, Z.N.; Zhang, Y.; Wang, W.; Zhou, X.M. Determination of baicalin in Xiaochaihu decoction by HPLC. *J. Zunyi Med. Univ.* **2015**, *38*, 435–438.
52. Xu, C.C. Determination of active components in Xiaochaihu Decoction by HPLC. *Asia-Pac. Tradit. Med.* **2015**, *11*, 39–40.
53. Jiang, S.Z.; Huang, X.B.; Chen, C.L.; Wu, H.X.; Xiang, Y.Y.; Qiu, Z.J.; Qiu, J.T. Effect of Ginger on Dissolution of Saikosaponin a and Baicalin in Xiaochaihu Decoction Determined by HPLC. *Chin. Arch. Tradit. Chin. Med.* **2014**, *32*, 1652–1654.

54. Zhang, H.; Yan, L.; Lin, L.F.; Dong, X.X.; Shen, M.R.; Qu, C.H.; Ni, J. Determination of seven index components in Xiaochaihu Granules by HPLC. *Drugs Clin.* **2014**, *29*, 162–165.
55. Li, P.T.; Lin, L.R.; Lin, L.H.; Wang, S.L.; Liu, S.Q.; Gao, X.L. Determination of Saikosaponin b2 in Xiaochaihu Granules by HPLC. *Tradit. Chin. Drug Res. Clin. Pharm.* **2014**, *25*, 356–359.
56. Zhang, S.Y. Determination of Saikosaponin, Baicalin and Glycyrrhizic Acid Content in Xiaochaihu Granule by HPLC. *J. Shanxi Univ. Chin. Med.* **2013**, *14*, 20–22.
57. Yi, R.Q.; Song, F.Y. Determination of Saikosaponin a and Saikosaponin d in Xiaochaihu Granules by Capillary Electrophoresis. *Chin. J. Pharm.* **2012**, *43*, 47–50.
58. Zhang, Y.J.; Mai, Y.M.; Mo, Z.J. Investigation of Determination and Evaluation for in vitro Dissolution of Chinese Medicine Preparations by Co-Chromatography and UV. *Chin. J. Exp. Tradit. Med. Formulae* **2013**, *19*, 17–21.
59. Liu, L.N.; Qi, Z.H.; Guo, Z.W.; Zhang, Y.; Zhang, H.F.; Zhang, W.J. Study on quality standard of xiaochaihutang dripping pill. *Pract. Pharm. Clin. Remedies* **2012**, *15*, 742–744.
60. Ji, S.G.; Liu, X.F.; Zhu, Z.Y.; Liang, S.S.; Zhao, L.; Zhang, H.; Chai, Y.F. High performance liquid chromatography in determination of baicalin and wogonoside contents in Xiaochaihu decoction. *Acad. J. Sec. Mil. Med. Univ.* **2010**, *31*, 1010–1013. [CrossRef]
61. Liu, Q.C.; Zhao, J.N.; Yan, L.C.; Yi, J.H.; Song, J. Simultaneous determination in Xiaochaihu Tang by HPLC. *China J. Chin. Mater. Med.* **2010**, *35*, 708–710.
62. Zhang, J.Y.; Xie, H.; Pan, S.L. Determination of saikosaponin a in four common bupleurum preparations by HPLC. *Chin. Tradit. Pat. Med.* **2010**, *32*, 81–83.
63. Zhou, B.Y.; Zhu, Z.Y.; Wang, B.; Zhao, L.; Zhang, H.; Xu, Q.; Chai, Y.F. HPLC-DAD combined with TOF/MS technique in determination of saikosaponin a, baicalin and glycyrrhizic acid in Xiaochaihu Decoction. *Acad. J. Sec. Mil. Med. Univ.* **2007**, *5*, 527–530.
64. Shi, Y. Determination of baicalin in Xiaochaihu Decoction pill by HPLC. *Chin. Tradit. Pat. Med.* **2007**, *7*, 1115–1117.
65. Zhu, Z.Y.; Liu, Y.; Zhao, B.Y.; Zhang, H.; Li, X.; Chai, Y.F. Simultaneous determination of two effective components in Xiaochaihu decoction by HPLC/DAD. *J. Pharm. Pract.* **2006**, *6*, 331–334.
66. Chen, P.; Li, C.; Liang, S.P.; Song, G.Q.; Sun, Y.; Shi, Y.H.; Xu, S.L.; Zhang, J.W.; Sheng, S.Q.; Yang, Y.M. Characterization and quantification of eight water-soluble constituents in tubers of *Pinellia ternata* and in tea granules from the Chinese multiherb remedy Xiaochaihu-tang. *J. Chrom. Banalytical Technol. Biomed. Life Sci.* **2006**, *843*, 83–193. [CrossRef] [PubMed]
67. Bao, Y.W.; Li, C.; Shen, H.W.; Nan, F.J. Determination of Saikosaponin Derivatives in *Radix bupleuri* and in Pharmaceuticals of the Chinese Multiherb Remedy Xiaochaihu-tang Using Liquid Chromatographic Tandem Mass Spectrometry. *Anal. Chem.* **2004**, *76*, 4208–4216. [CrossRef]
68. Wang, D.Q.; Chang, F.M.; Wang, Y.H.; Zhao, P.; Wang, Y.G. Multi-wavelength HPLC analysis of Xiaochaihu Granules and medicinal materials-preparation peak pattern matching analysis. *J. Chin. Med. Mater.* **2016**, *39*, 810–812.
69. Liu, A.X.; Xu, T.T.; Yang, W.N.; Zhou, D.D.; Sha, Y.W. Quantitative Determination of 7 Saikosaponins in Xiaochaihu Granules Using High-Performance Liquid Chromatography with Charged Aerosol Detection. *J. Anal. Methods Chem.* **2021**, *2021*, 6616854. [CrossRef] [PubMed]
70. Guo, Z.; Peng, B.; Li, Z.Y.; Pan, R.L. Study on Transformation Rule of Saikosaponin d Under Different Extraction Conditions by UPLC-QTof-MS. *Nat. Prod. Res. Dev.* **2014**, *26*, 716–720.
71. Zhang, X.; Wu, H.W.; Lin, L.N.; Tang, S.H.; Liu, H.H.; Yang, H.J. The development and application of a quality evaluation method for Xiaochaihu Granules granules which is based on "benchmark sample". *China J. Chin. Mater. Med.* **2021**, *7*, 1–11. [CrossRef]
72. Yang, M.L.; Wu, H.H.; Zhou, A.J.; Chen, S.H.; Liu, Y.W. Study on the Quality Standard for Modified Xiaochaihu Granules. *China Pharmacist.* **2018**, *21*, 1299–1303.
73. Zheng, R.W.; Jiang, Z.Q.; Cheng, G.H.; Hu, W.L.; Zhang, J.H. Study on HPLC Fingerprint for Xiaochaihu Granules. *Eval. Anal. Drug-Use Hosp. China.* **2017**, *17*, 1089–1093.
74. Li, X.D.; Gong, W. Multiwavelength fingerprints of Xiaochaihu Granules preparation and analysis of medicinal materials-preparation peak pattern matching analysis. *China Health Ind.* **2017**, *14*, 42–43.
75. Wang, X.; Li, C.D.; Pan, X.B.; Zhang, W. Quality Standard for Flavored Xiaochaihu Yigan Capsules by HPLC with Wavelength Switch. *China Pharmacist.* **2016**, *19*, 1597–1599.
76. Yuan, H.X.; Liu, Y.Q.; Li, H.F.; Pei, M.R. Fingerprint of n-butanol effective fraction of Xiao Chaihu Tang by HPLC-ELSD. *J. Shenyang Pharm. Univ.* **2015**, *32*, 623–632.
77. Yan, L.; Lin, L.F.; Zhang, H.; Dang, X.F.; Ni, J. Study on fingerprint of Xiaochaihu granules sold in the market. *China J. Chin. Mater. Med.* **2013**, *38*, 3498–3501.
78. Chen, W.X.; Wu, J.L.; Ma, D.; Zhang, P.; Gao, X.; Hu, F.D. Fingerprint study on Xiaochaihu Granules-medicinal materials based on peak pattern matching. *Chin. Tradit. Herb. Drugs* **2013**, *44*, 3154–3161.
79. Jin, Y.; Sun, L.; Qiao, S.Y. Establishment of HPLC fingerprint of Xiaochaihu Decoction. *J. Inter. Pharm. Res.* **2013**, *40*, 224–232.
80. Yang, H.J.; Ma, J.Y.; Weon, J.B.; Lee, B.Y.; Ma, C.J. Qualitative and quantitative simultaneous determination of six marker compounds in soshiho-tang by HPLC-DAD-ESI-MS. *Arch. Pharm. Res.* **2012**, *35*, 1785–1791. [CrossRef]
81. Zhuang, Y.H.; Cai, H.; Liu, X.; Cai, B.C. Simultaneous determination of five main index components and specific chromatograms analysis in Xiaochaihu granules. *Acta Pharm. Sin.* **2012**, *47*, 84–87.

82. Yang, J.; Huang, D.X.; Lu, X.M.; Wang, F.; Li, F.M. Analysis on chemical constituents in Xiaochaihu Decoction and their in vivo metabolites in depressed rats. *Chin. Tradit. Herb. Drugs* **2012**, *43*, 1691–1698.
83. Li, C.H.; Mei, Z.Q.; He, B.; Tian, J. Simultaneous determination of 4 components in Xiaochaihu granules by HPLC. *J. Southwest Med. Univ.* **2012**, *35*, 399–401.
84. Wu, Y.Q.; Gou, Y.Q.; Han, J.; Bi, Y.Y.; Feng, S.L.; Hu, F.D.; Wang, C.M. Evaluation preparation technology of Xiaochaihu granules using fingerprint-peak pattern matching. *J. Pharm. Anal.* **2011**, *1*, 119–124. [CrossRef]
85. Liu, X.F.; Lou, Z.Y.; Zhu, Z.Y.; Zhang, H.; Zhao, L.; Chai, Y.F. HPLC-TOF/MS in rapid identification of chemical compositions in Xiaochaihu decoction. *Acad. J. Sec. Mil. Med. Univ.* **2009**, *30*, 941–946. [CrossRef]
86. Li, J.; Jiang, H.; Yin, W.P. Study on the HPLC Fingerprint of Saikosaponins in *Radix Bupleuri* and Its Water Extract. *Lishizhen Med. Mater. Med. Res.* **2009**, *20*, 2205–2206.
87. Chen, W.; Zhang, C.; Sun, L.; Xue, M. Combination of fingerprint and chemometrics for identification of four processed products of *Pinelliae Rhizoma* based on nucleosides and nucleobases. *Chin. J. Pharm. Anal.* **2021**, *41*, 919–928.
88. Chinese Academy of Sciences. Flora of China. Available online: http://www.iplant.cn/info/Bupleurum?t=z (accessed on 28 July 2021).
89. Zhao, J.F.; Guo, Y.Z.; Meng, X.S. The toxic principles of *Bupleurum longiradiatum*. *Acta Pharm. Sin.* **1987**, *7*, 507–511.
90. Liu, X.X.; Luo, Z.Y.; Ji, Z.Z.; Li, H.; Zhang, W.Q.; Chen, F. A Detection Method of Adulteration of B. Marginatum var.Stenophyllum in Xiaochaihu Granules Preparation and Its Chinese Herbal Medicine Raw Materials. Patent CN110779993A, 11 February 2020.
91. Liang, H.L.; Tan, C.C.; Jiang, X.J.; Xiao, X.Y.; Xu, W.B. Study on Rapid Identification on Xiaochaihu Granules from Different Manufacturers by Near Infrared Spectroscopy. *Guiding J. Tradit. Chin. Med. Pharm.* **2021**, *27*, 62–64.
92. Lai, J.; Shi, Z.F.; Song, P.S.; Wang, Z.X.; Ni, L.; Teng, B.X. Identification of *Bupleurum marginatum var.stenophyllum* Adulterated *Bupleurum chinese* by PCR Amplification of Based on ITS Sequences. *Res. Pract. Chin. Med.* **2021**, *35*, 18–21.
93. Xie, D.H.; Cai, B.C.; An, Y.Q.; Li, X.; Jia, X.B. The research progerss in saikosaponin and its pharmacology action. *J. Nanjing Univ. Tradit. Chin. Med.* **2007**, *1*, 63–65.
94. Lv, X.H.; Sun, Z.X.; Su, R.Q.; Fan, J.W.; Zhao, Z.Q. The pharmacology research progress in bupleurum and its active constituents. *Chin. J. Inf. Tradit. Chin. Med.* **2012**, *29*, 105–107.
95. Liu, Y.M.; Liu, X.M.; Pan, R.L. The research progress in toxic effect of bupleurum. *Chin. Tradit. Pat. Med.* **2012**, *34*, 1148–1151.

MDPI
St. Alban-Anlage 66
4052 Basel
Switzerland
www.mdpi.com

Separations Editorial Office
E-mail: separations@mdpi.com
www.mdpi.com/journal/separations

Disclaimer/Publisher's Note: The statements, opinions and data contained in all publications are solely those of the individual author(s) and contributor(s) and not of MDPI and/or the editor(s). MDPI and/or the editor(s) disclaim responsibility for any injury to people or property resulting from any ideas, methods, instructions or products referred to in the content.

www.ingramcontent.com/pod-product-compliance
Lightning Source LLC
LaVergne TN
LVHW070047120526
838202LV00101B/1138